T0186001

Understanding Electromagnetic Waves

Ming-Seng Kao · Chieh-Fu Chang

Understanding Electromagnetic Waves

 Springer

Ming-Seng Kao
National Chiao-Tung University
Hsinchu, Taiwan

Chieh-Fu Chang
National Applied Research Laboratory
National Space Organization
Hsinchu, Taiwan

ISBN 978-3-030-45710-5   ISBN 978-3-030-45708-2   (eBook)
https://doi.org/10.1007/978-3-030-45708-2

This Springer imprint is published by the registered company Springer Nature Switzerland AG
The registered company address is: Gewerbestrasse 11, 6330 Cham, Switzerland

Preface

This book adopts a different approach to help students learn electromagnetic waves.

We believe that the most effective way for undergraduate students to learn EM waves should begin directly with Maxwell's equations, especially when the course takes only ONE semester, so that students can grasp the core ideas of EM waves instantly. Therefore, we begin with Maxwell's Equations in Chap. 1 to lay down the solid background of EM waves.

Unlike traditional EM textbooks, which try to cover all the materials including static Electromagnetism and inevitably make students learn Maxwell's Equations at the latter half of the course, our approach avoid this significant drawback since undergraduate students usually take General Physics which already introduces static Electromagnetism. Besides, most EM textbooks make college students difficult to learn EM waves because they usually emphasize mathematics and formulate the principles in a complete and rigorous way. It makes students spend a lot of time and effort learning mathematics instead of physics. In contrast, we try to introduce EM waves in a more heuristic way. For example, in the beginning of each topic, we introduce each fundamental principle in a simple and understandable way. Usually it is an example in our daily life or something one can "imagine" or "see". Then we develop the concept step by step with acceptable mathematics for undergraduate students. Moreover, we provide just-enough mathematics so that students can concentrate on physical insights of critical concepts and parameters.

This book is organized as the following flowchart:

In order to achieve our goal described above, in the beginning of Chap. 1, we introduce two principle operators of Maxwell's equations: divergence and curl. After students get the ideas of divergence and curl, the physical meaning and mathematical principles of Maxwell's Equations are provided to help them obtain the relevant concepts and insight.

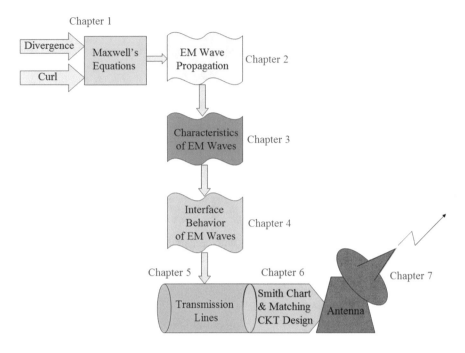

After learning Maxwell's Equations, we proceed to study EM waves based on these equations in Chap. 2. The wave equation and the propagation of uniform plane wave will be learned. Next, we explore important parameters and characteristics of EM waves in Chap. 3, and introduce behaviors of EM waves across an interface between two different media in Chap. 4. These chapters provide background knowledge for students to access advanced topics and various applications of EM waves.

In Chap. 5, we introduce the principles of an important application—transmission line, which can guide and convey EM waves to a long distance. Moreover, Smith chart is introduced in Chap. 6 so that readers can get the insight and use it to solve many practical problems. The impedance matching circuits and the design of high frequency amplifiers are also included. Finally in Chap. 7, we introduce antennas which are the key elements in a wireless communication system. We start from the simplest dipole antenna and then develop useful principles step by step. It ends with reciprocity theorem depicting the relationship between transmitting antennas and receiving antennas.

This book contains seven chapters given below:
Chapter 1 Maxwell's Equations
Chapter 2 EM Wave Propagation
Chapter 3 Characteristics of EM Waves
Chapter 4 Interface Behavior of EM Waves
Chapter 5 Transmission Lines
Chapter 6 Smith Chart and Matching Circuit Design
Chapter 7 Antenna

These chapters include the most important knowledge which undergraduate students need in EM waves. Through the study of these contents, we believe students will have a solid background in EM waves and be ready to get into high frequency circuit design as well as wireless communication systems.

Hsinchu, Taiwan   Ming-Seng Kao
                                                                                              Chieh-Fu Chang

Contents

List of Figures

Chapter 1
Maxwell's Equations

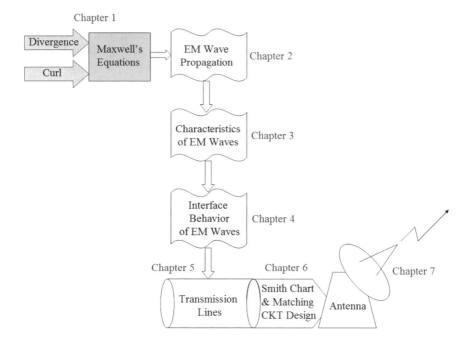

© Springer Nature Switzerland AG 2020
M.-S. Kao and C.-F. Chang, *Understanding Electromagnetic Waves*,
https://doi.org/10.1007/978-3-030-45708-2_1

Abstract **Maxwell's equations** are the key to understand electromagnetic (EM) waves because these equations summarize the most important and useful knowledge of EM fields. In order to help readers get the key easily and quickly, two mathematical operators: **divergence** and **curl**, are introduced by ignoring lengthy mathematics but focusing on their physical meanings. Unlike traditional approach, we provide a lot of examples and illustrations to help readers catch the core ideas.

Furthermore, we introduce Maxwell's equations in two steps: **physical perspective** first, and then **mathematical perspective**, so that readers can catch the core ideas effectively. Once readers have a good understanding of Maxwell's equations, a solid background for learning EM waves is established!

Keywords Maxwell's equations · Electromagnetic wave · Vector field · Divergence · Divergence theorem · Curl · Stokes' theorem · Electric field intensity · Magnetic field intensity · Electric flux density · Magnetic flux density · Faraday's law · Ampere's law · Gauss's law · Gauss's law of magnetic field

Maxwell's equations are the key to understand electromagnetic (EM) waves because these equations summarize the most important and useful knowledge so far we have for electromagnetic fields. In the beginning two sections, we introduce the principle operators of Maxwell's equations: divergence and curl. Then the physical meaning and mathematical principles of Maxwell's equations are provided to get the useful concepts and insight of these equations.

Background

In order to understand the key idea of Maxwell's equations, we need to have proper mathematical and physical background. Here we suppose you have basic knowledge of static electricity and magnetism learned in fundamental physics. We further assume that you have required background knowledge in vector calculus such as vector differentiation and integration. If you are not familiar with these, please refer to Appendix A.

1.1 Divergence

EM waves can be fully interpreted by the time-varying behavior of electric field (E-field) and magnetic field (M-field) in three-dimensional (3-D) space. Since E-field and M-field are vector fields in 3-D space, we need to realize two important operators of a vector field: divergence and curl. Once we get the idea of these two operators, we can measure the important characteristics of vector fields and then open the door of EM waves. Let us start from divergence.

A Usage of Vector Field

In real world, we can observe many physical phenomena having magnitude and direction, and they may change with position too. For example, when you stand along seashore, the wind breezes you can feel on your face with varying pressure and direction. Suppose we have a vector $\vec{A}$ representing the wind. Then the direction of $\vec{A}$ represents the direction of the wind and the magnitude of $\left|\vec{A}\right|$ represents the strength of the wind. Since $\vec{A}$ may change with its position, we let $\vec{A}$ be a function of x, y, z and denoted by $\vec{A} = \vec{A}(x, y, z)$. Similarly, we can use another vector field $\vec{B} = \vec{B}(x, y, z)$ representing the water flow in a river. From these two examples, we can use vector fields to represent various physical effects in the nature.

From mathematical viewpoint, a vector field is an efficient and effective tool for us to describe the behavior of EM wave in 3-D space. For example, if we put a charge in 3-D space, it will create a vector field $\vec{E}(x, y, z)$ representing the resultant electric field. Based on this, we can analyze and explore the world of EM waves.

After understanding the applications of a vector field, we are ready to introduce the first key operator of a vector field—divergence.

B Meaning of Divergence

For simplicity, we first consider a vector field $\vec{A}(x, y)$ in two-dimensional (2-D) plane. An example is shown in Fig. 1.1, where $\vec{A}$ is a uniform vector field. In this figure, $\vec{A}$ has identical magnitude and direction throughout the plane. On the other hand, a non-uniform vector field is shown in Fig. 1.2, where $\vec{A}$ has different magnitudes and directions at different points. Other examples of non-uniform vector fields are shown in Figs. 1.3 and 1.4.

In Fig. 1.3, it is not difficult to see the vectors "diverge" from the center point P. Hence, P can be regarded as a divergent point of the vector field $\vec{A}$. On the other hand, in Fig. 1.4, the point P can be regarded as a convergent point of $\vec{A}$ since the vectors converge to P. For a point in a vector field, it may be a divergent point or a convergent point, depending on its surroundings. For example, in Fig. 1.5, P is

Fig. 1.1 Example of a
uniform vector field

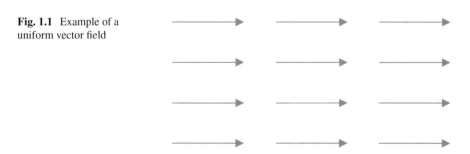

Fig. 1.2 Example of a
non-uniform vector field

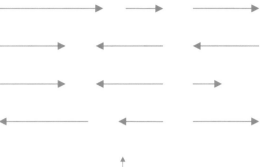

Fig. 1.3 Illustrating a
divergent point

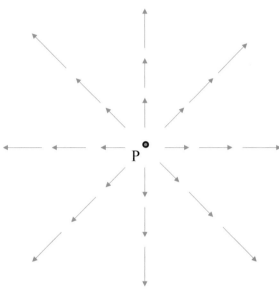

obviously a divergent point of $\vec{A}$ while Q is a convergent point of $\vec{A}$. From those plots shown in Figs. 1.1, 1.2, 1.3, 1.4 and 1.5, we can easily determine if a point is a divergent point or a convergent point in a vector field. However, in most cases, we do not have such a plot of the observed vector field. Then how can we determine whether a point P in 3-D space is a convergent point or a divergent point in a vector field $\vec{A}$?

To answer this, mathematicians have a very interesting and creative idea: imagine that we have a scarf enclosing the point P. For a point (x_1, y_1, z_1) on the surface of the scarf, if the direction of $\vec{A}$ is outgoing, we give the point a positive value that is proportional to $\left|\vec{A}\right|$. On the other hand, if the direction of $\vec{A}$ is incoming, we give the point a negative value that is proportional to $\left|\vec{A}\right|$. Finally, we sum up all the corresponding values of points on the surface of the scarf. If the sum is positive, then P is a divergent point. If the sum is negative, it is a convergent point. If the sum is zero, it is neither a divergent point nor a convergent point. Note that we need to make the scarf as small as possible so that only the point P is enclosed in the scarf.

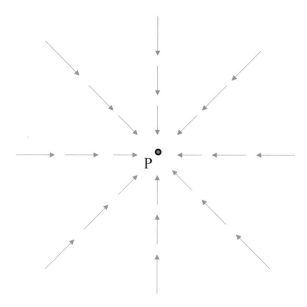

Fig. 1.4 Illustrating a convergent point

Fig. 1.5 Illustrating a divergent point P and a convergent point Q

Next, we transfer the above idea into mathematical formula. As illustrated in Fig. 1.6, suppose S is the surface of the scarf and we define a scalar η given by

$$\eta = \lim_{S \to 0} \oint_S \vec{A} \cdot d\vec{s}, \tag{1.1}$$

where the area of S approaches zero and is denoted by $S \to 0$, which implies only the point P is enclosed in the scarf. In Eq. (1.1), η is the surface integral of $\vec{A}$ over S, and $d\vec{s}$ is an outward area vector perpendicular to S. Obviously, if a vector $\vec{A}$ goes outward, $\vec{A} \cdot d\vec{s}$ would be positive. Otherwise, $\vec{A} \cdot d\vec{s}$ would be negative or zero. Evidently, Eq. (1.1) is a perfect formulation of the idea mentioned above.

From above, the sign and magnitude of η would indicate the divergence property of $\vec{A}$ at the point P. The physical meaning of η can be summarized as follows:

1. From the sign of η, we can determine if P is a convergent point or a divergent point:

Fig. 1.6 Schematic plot explaining the definition of divergence

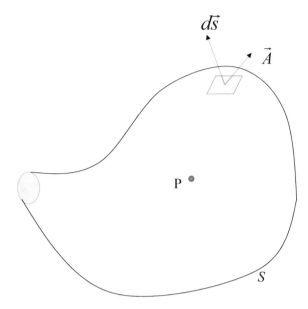

(i)   If $\eta > 0$, P is a divergent point.
(ii) If $\eta < 0$, P is a convergent point.
(iii) If $\eta = 0$, P is neither a divergent point nor a convergent point.

2. From the magnitude of η, we can determine the "strength" of the divergence at the point P.

Hence, we can quantize the divergence ("outgoingness" or "incomingness") of $\vec{A}$ at a point P without plotting the vector field of concern.

In addition, when we think about a convergent point, it is naturally regarded as a "sink" of $\vec{A}$ because the flux comes inward. On the other hand, when we consider a divergent point, it is naturally regarded as a "source" of $\vec{A}$ because the flux goes outward.

From the previous explanations, the physical meaning of Eq. (1.1) is simple and clear. But a significant flaw exists in the mathematical expression of Eq. (1.1): since $\vec{A}$ is finite and when S approaches zero, the surface integral will approach zero as well. Thus, the result of Eq. (1.1) is always zero! In order to correct this flaw, we modify Eq. (1.1) and re-define η as follows:

$$\eta = \lim_{S \to 0} \frac{\oint_S \vec{A} \cdot d\vec{s}}{V}, \tag{1.2}$$

where V denotes the volume enclosed by S. In the new definition of Eq. (1.2), when $S \to 0$, V approaches zero too. Hence, η is not necessarily zero. This new definition preserves the physical meaning of divergence and meanwhile it is "rigorous" in mathematics.

In vector calculus, the divergence in a vector field $\vec{A}$ is generally denoted by $\nabla \cdot \vec{A}$ and is defined as

$$\nabla \cdot \vec{A} = \lim_{S \to 0} \frac{\oint_S \vec{A} \cdot d\vec{s}}{V}, \tag{1.3}$$

where S is the surface enclosing the point of interest and V is the volume of S. Notice that $\nabla \cdot \vec{A}$ is a scalar, instead of a vector.

C   Calculation of Divergence

Let $\vec{A}$ be a 3-D vector field denoted by

$$\vec{A} = A_x \cdot \hat{x} + A_y \cdot \hat{y} + A_z \cdot \hat{z}, \tag{1.4}$$

where (A_x, A_y, A_z) are axial components of $\vec{A}$, and they are functions of x, y, and z, i.e., $A_x = A_x(x, y, z)$, $A_y = A_y(x, y, z)$, $A_z = A_z(x, y, z)$. Based on Eq. (1.3) and the physical meaning of divergence, it can be shown that[1]

$$\begin{aligned} \nabla \cdot \vec{A} &= \lim_{S \to 0} \frac{\oint_S \vec{A} \cdot d\vec{s}}{V} \\ &= \frac{\partial A_x}{\partial x} + \frac{\partial A_y}{\partial y} + \frac{\partial A_z}{\partial z}. \end{aligned} \tag{1.5}$$

Compared with Eq. (1.3), Eq. (1.5) provides a simple and efficient way to calculate the divergence. If we know axial components A_x, A_y, and A_z of a vector field $\vec{A}$, we can simply use partial differentiation to derive $\nabla \cdot \vec{A}$, instead of calculating complicated surface integral in Eq. (1.3). Therefore, Eq. (1.5) can help us measure the divergence of a point in space easily. For example, suppose P is a point at (x_0, y_0, z_0) in 3-D space. The divergence of a vector field $\vec{A}$ at P is given by

$$\nabla \cdot \vec{A}|_{(x_0, y_0, z_0)} = \frac{\partial A_x}{\partial x} + \frac{\partial A_y}{\partial y} + \frac{\partial A_z}{\partial z}\bigg|_{(x_0, y_0, z_0)}. \tag{1.6}$$

From above,

(i)   if $\nabla \cdot \vec{A}|_{(x_0, y_0, z_0)} > 0$, P is a divergent point.
(ii) if $\nabla \cdot \vec{A}|_{(x_0, y_0, z_0)} < 0$, P is a convergent point.
(iii) if $\nabla \cdot \vec{A}|_{(x_0, y_0, z_0)} = 0$, P is neither a divergent point nor a convergent point.

---

[1] George Thomas, *Calculus*, 13th edition, Pearson, 2012.

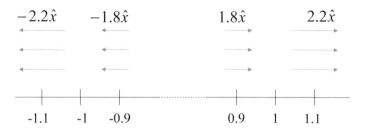

Fig. 1.7 Illustrating Example 1.1

Besides, from the magnitude of $\nabla \cdot \vec{A}|_{(x_0, y_0, z_0)}$, we know the "strength" of divergence characteristic at the point P. Hence, divergence is a useful measure to characterize a vector field.

Example 1.1

Suppose we have a vector field $\vec{A}(x, y, z) = 2x \cdot \hat{x}$. Please calculate the divergence when (i) $x = 1$ (ii) $x = -1$.

Solution

First, we have three axial components as below:

$$A_x = 2x$$
$$A_y = 0$$
$$A_z = 0.$$

From Eq. (1.6), we have

$$\nabla \cdot \vec{A} = \frac{\partial A_x}{\partial x} + \frac{\partial A_y}{\partial y} + \frac{\partial A_z}{\partial z} = 2.$$

Because $\nabla \cdot \vec{A} = 2$, the divergence of vector field $\vec{A}$ is independent of the location and the divergence at all points equals 2. Hence, both cases of $x = 1$ and $x = -1$ have identical divergence. This value is positive and implies both cases are "divergent" points. ∎

Figure 1.7 shows the vector field in Example 1.1 near $x = 1$ and $x = -1$. It can be seen when $x = 1$, it is a divergent point because the inward flux is less than the outward flux. For example, the flux $\vec{A} = 1.8 \cdot \hat{x}$ when $x = 0.9$ and $\vec{A} = 2.2 \cdot \hat{x}$ when $x = 1.1$. Hence, the outward flux is larger than the inward one. Similarly, the flux $\vec{A} = -1.8 \cdot \hat{x}$ when $x = -0.9$ and $\vec{A} = -2.2 \cdot \hat{x}$ when $x = -1.1$. Hence, for $x = -1$, the outward flux is also greater than the inward flux.

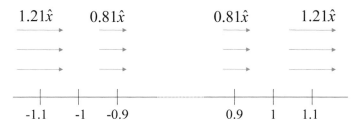

Fig. 1.8 Illustrating Example 1.2

Example 1.2
Suppose we have a vector field $\vec{A}(x, y, z) = x^2 \cdot \hat{x}$. Please calculate the divergence when (i) $x = 1$ and (ii) $x = -1$.

Solution
First, we have three axial components as below:

$$A_x = x^2$$
$$A_y = 0$$
$$A_z = 0.$$

From Eq. (1.6), we have

$$\nabla \cdot \vec{A} = \frac{\partial A_x}{\partial x} + \frac{\partial A_y}{\partial y} + \frac{\partial A_z}{\partial z} = 2x.$$

Hence, when $x = 1$,
$\nabla \cdot \vec{A}|_{x=1} = 2 \Rightarrow x = 1$ is a divergent point.
On the other hand, when $x = -1$,
$\nabla \cdot \vec{A}|_{x=-1} = -2 \Rightarrow x = -1$ is a convergent point.   ■

Figure 1.8 shows the vector field around $x = 1$ and $x = -1$. It can be seen that $x = 1$ is a divergent point because the outward flux is greater than the inward flux. For example, the flux $\vec{A} = 0.81 \cdot \hat{x}$ when $x = 0.9$ and $\vec{A} = 1.21 \cdot \hat{x}$ when $x = 1.1$. Hence, for $x = 1$, the outward flux is greater than the inward flux. On the other hand, $x = -1$ is a convergent point because the inward flux is greater than the outward flux. For example, the flux $\vec{A} = 0.81 \cdot \hat{x}$ when $x = -0.9$ and $\vec{A} = 1.21 \cdot \hat{x}$ when $x = -1.1$. Hence, for $x = -1$, the inward flux is greater than the outward flux. It is worth to compare the difference between Examples 1.1 and 1.2, which will help readers understand the meaning of divergence.

Example 1.3

Suppose we have a vector field $\vec{A}(x, y, z) = (xy^2) \cdot \hat{x} + (yz + x) \cdot \hat{y} + (x + y^3) \cdot \hat{z}$. Please calculate the divergence when (i) $(x, y, z) = (3, 1, -2)$ and (ii) $(x, y, z) = (2, 3, -4)$.

Solution

First, we have three axial components as below:

$$A_x = xy^2$$
$$A_y = yz + x$$
$$A_z = x + y^3.$$

From Eq. (1.6), we have

$$\nabla \cdot \vec{A} = \frac{\partial A_x}{\partial x} + \frac{\partial A_y}{\partial y} + \frac{\partial A_z}{\partial z} = y^2 + z + 0 = y^2 + z.$$

Hence when $(x, y, z) = (3, 1, -2)$, we have
$\nabla \cdot \vec{A}|_{(3,1,-2)} = -1 \Rightarrow (3, 1, -2)$ is a convergent point.
On the other hand, when $(x, y, z) = (2, 3, -4)$, we have
$\nabla \cdot \vec{A}|_{(2,3,-4)} = 5 \Rightarrow (2, 3, -4)$ is a divergent point.
Besides, because $5 > |-1|$, the strength of divergence at $(x, y, z) = (2, 3, -4)$ is greater than the strength of convergence at $(x, y, z) = (3, 1, -2)$.   ∎

From the above examples, once we know the axial components of a vector field $\vec{A}(x, y, z)$, we can utilize Eq. (1.6) to calculate the divergence immediately.

D   Divergence Theorem

Now we introduce an important theorem related to divergence, that is very useful in EM field. It is called the divergence theorem and can be formulated as follows. Let V be an arbitrary volume with any shape and S be the surface of V. The divergence theorem states that

$$\int_V (\nabla \cdot \vec{A}) dv = \oint_S \vec{A} \cdot d\vec{s}. \tag{1.7}$$

In the left side of Eq. (1.7), we have volume integral of the scalar $\nabla \cdot \vec{A}$ over V; in the right side of Eq. (1.7), we have surface integral of the vector field $\vec{A}$ over S. Equation (1.7) holds for any vector field regarding any volume V and the associated surface S.

Fig. 1.9 Subdivided cubes for proof of divergence theorem

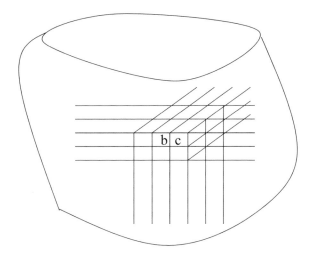

The divergence theorem can be interpreted as that the volume integral of the divergence $\nabla \cdot \vec{A}$ inside a closed surface is equal to the outward flux of $\vec{A}$ through the surface. The proof of divergence theorem is quite complicated and readers who are interested in it may refer to the reference.[2] Here we would like to provide a heuristic way to understand the physical meaning of this theorem in order to help readers get the insight.

Suppose V is an arbitrary volume in space as shown in Fig. 1.9. First, we divide it into infinite number of small cubes and each cube has a tiny volume dv so that $dv \to 0$. From the definition of divergence in Eq. (1.3), we have

$$\left. \left(\nabla \cdot \vec{A} \right) dv \right|_{v_i} = \lim_{s_i \to 0} \oint_{s_i} \vec{A} \cdot d\vec{s}, \tag{1.8}$$

where v_i denotes the i-th small cube and s_i is its surface. As we sum $\left. (\nabla \cdot \vec{A}) dv \right|_{v_i}$ of all the small cubes inside V, we have

$$\sum_{i=1}^{\infty} \left. (\nabla \cdot A) dv \right|_{v_i} = \sum_{i=1}^{\infty} \lim_{s_i \to 0} \oint_{s_i} \vec{A} \cdot d\vec{s}$$

$$= Sum \ of \ surface \ integral \ of \ all \ the \ cubes. \tag{1.9}$$

In Fig. 1.10, suppose b and c are two neighboring cubes. When we focus on the inter-surface s_{bc} between b and c, and assume $\vec{A}$ goes from cube b to cube c at the small interface, then we have the sub-surface integral $\vec{A} \cdot \vec{s}_{bc}$ for cube b and the sub-surface integral $-\vec{A} \cdot \vec{s}_{bc}$ for cube c. Hence, the inter-surface integrals of two

---

[2]James Stewart, *Calculus*, 6th edition, Thomson, 2008 (Sect. 16.9).

Fig. 1.10 Two neighboring
cubes for proof of divergence
theorem

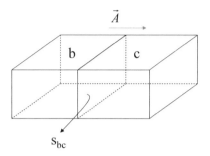

neighboring cubes will be mutually cancelled. Thus, for the sum of surface integral
of all the cubes, only those at the surface of V will not be cancelled. Therefore, we
obtain the following equality:

$$Sum\ of\ surface\ integral\ of\ all\ the\ cubes = \oint_S \vec{A} \cdot d\vec{s}. \qquad (1.10)$$

On the other hand, as $dv \to 0$, we have the following equality:

$$\sum_{i=1}^{\infty} (\nabla \cdot A)dv|_{v_i} = \int_V \left(\nabla \cdot \vec{A}\right)dv. \qquad (1.11)$$

Finally, from Eqs. (1.9)–(1.11), we have

$$\int_V (\nabla \cdot \vec{A})dv = \oint_S \vec{A} \cdot d\vec{s}. \qquad (1.12)$$

It is worth to mention that the divergence theorem is very useful because from
Eq. (1.12), we can reduce our computation from 3-D volume integral to 2-D surface
integral. In addition, it provides different perspectives to interpret several important characteristics of EM field. The applications of the divergence theorem in
electromagnetics will become clear later.

1.2 Curl

In the previous section, we have learned the first key operator of a vector field—
divergence, which measures the divergent property of a vector field. Now, we are
going to learn the second key operator—curl, which measures the "rotation" property
of a vector field. These two operators will help us open the door of EM wave.

A   Physical Meaning of Curl

Figure 1.11 shows a 2-D vector field and obviously there is no rotation in this field. Figure 1.12 shows another 2-D vector field, and it can be easily seen that there is rotation in this field. Then we may ask ourselves a question: how do we describe this "rotation" property for a vector field? And a more challenging question is: how do we quantize this property in 3-D space?

First, let us imagine a rotational vector field in 3-D space as shown in Fig. 1.13. From Fig. 1.13, we find that a rotation field occurs on a "plane". For example, the rotation of the earth around the sun occurs on the ecliptic plane. Apparently, the plane of rotation is the first characteristic of a rotational vector field.

Next, from Fig. 1.14, we learn that a rotational vector field on the plane of rotation may have "clockwise" rotation or "counter-clockwise" rotation. Hence, the direction of rotation forms the second characteristic of a rotational vector field.

Fig. 1.11 Illustrating a
vector field without rotation

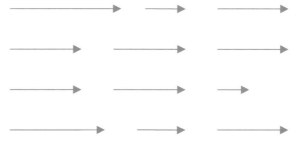

Fig. 1.12 Illustrating a
rotational vector field

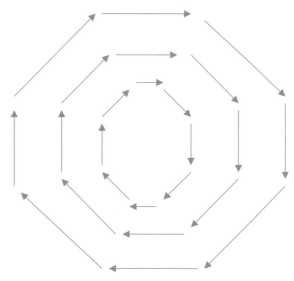

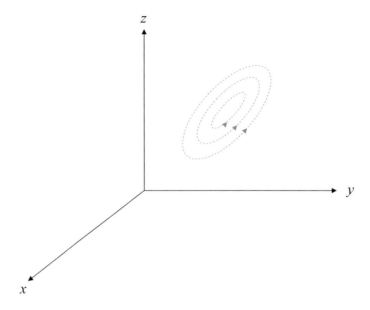

Fig. 1.13 Illustrating a rotational vector field in 3-D space

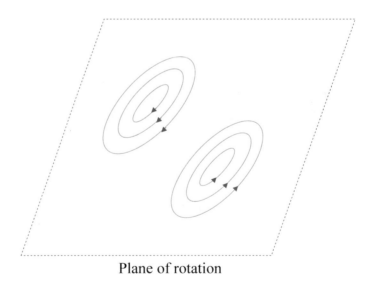

Plane of rotation

Fig. 1.14 Illustrating clockwise and counter-clockwise rotational vector fields

Finally, the strength of rotation at different points in a rotational vector field may be different as illustrated in Fig. 1.15. In this figure, it is easy to see that the inner

Fig. 1.15 Illustrating
strength of a rotational vector
field at different points

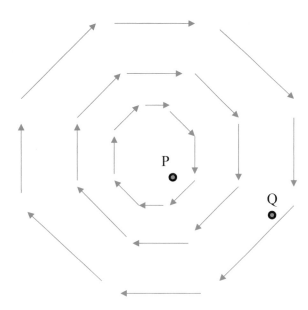

point P has "stronger" rotation than the point Q. Hence, the strength of rotation forms
the third characteristic of a rotational vector field.

From above, if we want to define a measure which can describe and quantize the
property of a rotational vector field, this measure must include the following three
characteristics:

1. Plane of rotation,
2. Direction of rotation,
3. Strength of rotation.

Now, we start to define this measure. First, we focus on the plane of rotation and
the direction of rotation. From mathematical viewpoint, it is hard to formulate these
two factors. Fortunately, we find a clever and convenient way to resolve the problem.
That is, the "right-hand rule".

The very idea of right-hand rule is to **define the plane of rotation and the
associated direction of rotation simultaneously with a specific vector**. Suppose
we show our right hand as in Fig. 1.16, where the thumb is perpendicular to the
other four fingers. Then, imagine that the plane of rotation is attached on an axis
such that it is perpendicular to this axis. An example is shown in Fig. 1.17, where the
plane of rotation is the xy-plane and it is attached to the z-axis. Assume the direction
of rotation goes from x-axis to y-axis. We can define the plane of rotation and the
associated direction of rotation with our right thumb and the other four fingers as
follows.

Let $\vec{K}$ be a vector pointing to the direction of thumb of the right-hand rule. We
let $\vec{K} = \hat{z}$ in Fig. 1.17, then the plane of rotation is a plane perpendicular to $\vec{K}$ (the
xy-plane) while the curl of the four fingers (except the thumb) is the direction of

Fig. 1.16 Illustrating
right-hand rule (I)

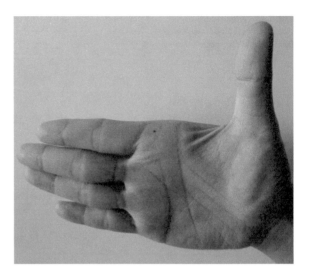

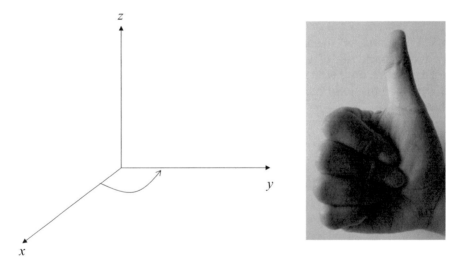

Fig. 1.17 Illustrating right-hand rule (II)

rotation goes from x-axis to y-axis. Consequently, we can use the vector $\vec{K}$ to define
the plane of rotation and the associated direction of rotation simultaneously. This is
the intelligent idea behind the right-hand rule.

Another example is shown in Fig. 1.18, where the arrow denotes the direction of
the thumb in right-hand rule. It is easy to see if the direction of the thumb is given,
then the plane of rotation and the direction of rotation are well defined.

Fig. 1.18 Illustrating
right-hand rule (III)

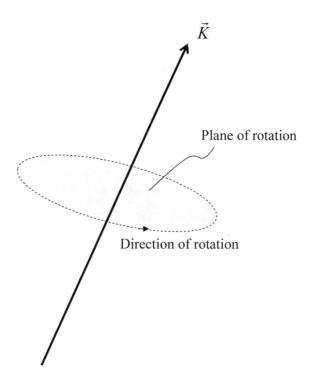

Next, we handle the third characteristic of a rotational vector field—the strength
of rotation. Suppose $\vec{A}$ is a rotational vector field in Fig. 1.19, where P is a point on the
plane of rotation and C is a very small contour surrounding P. When we perform line
integral over C along the direction of rotation of $\vec{A}$, the result would meaningfully
represent the strength of rotation. As we think about it deeply, it is actually consistent
with our intuition: if the integral is large, the vector field has "strong" rotation around
P; if the integral is small, the vector field has "weak" rotation around P; and if the
integral is zero, there is no rotation around P.

Finally, in order to let the contour C stand for the point P well, C should be as
small as possible so that only the point P is enclosed in C.

Let β denote the strength of rotation at the point P in a vector field $\vec{A}$. We formulate
the above concept as below:

$$\beta = \lim_{C \to 0} \oint_C \vec{A} \cdot d\vec{l}, \tag{1.13}$$

where $\beta \geq 0$ since the line integral is performed along the direction of rotation and
$C \to 0$ implies only the point P is enclosed in C. If β is large, the strength of rotation
at P is strong; if β is small, the strength of rotation at P is weak; and if $\beta = 0$, there
is no rotation at P. Thus, we can anticipate the strength of rotation at P via β.

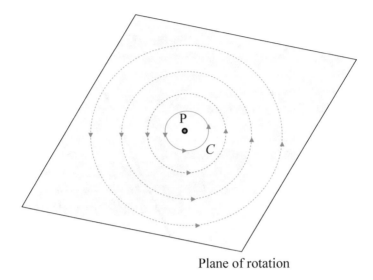

Plane of rotation

Fig. 1.19 Illustrating line integration of a rotational vector field

For Eq. (1.13), the physical meaning is well defined and clear. However, it is not rigorous in mathematics. The reason is that the vector field $\vec{A}$ is finite and when $C \to 0$, the line integral of the right-hand side in Eq. (1.13) approaches zero. Hence, the result of Eq. (1.13) is always zero, i.e., $\beta = 0$. In order to resolve this issue, we re-define β as

$$\beta = \lim_{C \to 0} \frac{\oint_c \vec{A} \cdot d\vec{l}}{U}, \tag{1.14}$$

where U is the area surrounded by C. In Eq. (1.14), when $C \to 0$, the area U also approaches zero. Thus, β may not be zero. This new definition preserves the physical meaning of β and is rigorous in mathematics.

Based on the above results, we can define an important measure of rotation, called "curl", of a vector field as follows. The curl of a point in a vector field $\vec{A}$ is denoted by $\nabla \times \vec{A}$, which is given by

$$\nabla \times \vec{A} = \beta \cdot \hat{n} = \left(\lim_{C \to 0} \frac{\oint_c \vec{A} \cdot d\vec{l}}{U} \right) \cdot \hat{n}, \tag{1.15}$$

where $\hat{n}$ is a unit vector $\left(\left| \hat{n} \right| = 1 \right)$ pointing to the direction of the thumb in the right-hand rule and β is the strength of rotation. For a point in $\vec{A}$, $\nabla \times \vec{A}$ characterizes the rotation at that point. The direction of $\nabla \times \vec{A}$, i.e., the direction of $\hat{n}$, determines both the plane of rotation and the direction of rotation. Meanwhile, the magnitude $|\nabla \times \vec{A}| = \beta$ represents the strength of rotation at that point. For example, suppose we have $\nabla \times \vec{A} = 3 \cdot \hat{z}$ at a point P. It tells us the rotational property at P as follows:

1. The plane of rotation is a plane containing P and perpendicular to z-axis.
2. As we use right-hand rule and let the thumb points to $\hat{z}$, the curl of the remaining four fingers indicates the direction of rotation.
3. The strength of rotation at P is $|\nabla \times \vec{A}| = 3$.

From the above example, we find that $\nabla \times \vec{A}$ clearly describes the three salient characteristics of a rotational vector field. Note that the curl $\nabla \times \vec{A}$ is a vector while the divergence $\nabla \cdot \vec{A}$ is a scalar, which signifies the difference between these two operators in vector fields.

B Calculation of Curl

Suppose $\vec{A}$ is a vector field in 3-D space given by

$$\vec{A} = A_x \cdot \hat{x} + A_y \cdot \hat{y} + A_z \cdot \hat{z}, \tag{1.16}$$

where (A_x, A_y, A_z) represent the axial components of $\vec{A}$. They are functions of x, y, and z, i.e., $A_x = A_x(x, y, z)$, $A_y = A_y(x, y, z)$, and $A_z = A_z(x, y, z)$. Based on Eq. (1.15) and the physical meaning of curl, it can be shown that[3]

$$\nabla \times \vec{A} = \left(\lim_{C \to 0} \frac{\oint_c \vec{A} \cdot \vec{dl}}{U} \right) \cdot \hat{n}$$

$$= \left(\frac{\partial A_z}{\partial y} - \frac{\partial A_y}{\partial z} \right) \cdot \hat{x} + \left(\frac{\partial A_x}{\partial z} - \frac{\partial A_z}{\partial x} \right) \cdot \hat{y} + \left(\frac{\partial A_y}{\partial x} - \frac{\partial A_x}{\partial y} \right) \cdot \hat{z}. \tag{1.17}$$

Hence, we can simply derive the curl by Eq. (1.17), instead of performing complicated line integral as defined in Eq. (1.15). For example, let P be a point at (x_0, y_0, z_0) in 3-D space. The curl of a vector field $\vec{A}$ at P can be calculated by

$$\nabla \times \vec{A}|_{(x_0, y_0, z_0)} = \left(\frac{\partial A_z}{\partial y} - \frac{\partial A_y}{\partial z} \right) \cdot \hat{x} + \left(\frac{\partial A_x}{\partial z} - \frac{\partial A_z}{\partial x} \right) \cdot \hat{y} + \left(\frac{\partial A_y}{\partial x} - \frac{\partial A_x}{\partial y} \right) \cdot \hat{z} \Big|_{(x_0, y_0, z_0)}. \tag{1.18}$$

From the result of Eq. (1.18), we can easily understand the rotational properties of a vector field at a given point.

---

[3] George Thomas, *Calculus*, 13th edition, Pearson, 2012.

Example 1.4

Suppose we have a vector field $\vec{A} = 2x \cdot \hat{x} + 2y \cdot \hat{y}$.

(a) Please derive $\nabla \times \vec{A}$ at the origin $(0, 0, 0)$.
(b) On xy-plane, plot the vector field $\vec{A}$ at the four points $(a, 0)$, $(0, a)$, $(-a, 0)$, and $(0, -a)$ surrounding the origin O, where a $= 0.1$. Observe if there is rotation around the origin.

Solution

(a) First, we have the axial components of $\vec{A}$ as below:

$$A_x = 2x$$
$$A_y = 2y$$
$$A_z = 0.$$

Thus

$$\frac{\partial A_z}{\partial y} - \frac{\partial A_y}{\partial z} = 0$$
$$\frac{\partial A_x}{\partial z} - \frac{\partial A_z}{\partial x} = 0$$
$$\frac{\partial A_y}{\partial x} - \frac{\partial A_x}{\partial y} = 0.$$

From Eq. (1.18), we obtain

$$\nabla \times \vec{A}|_{(0,0,0)} = \left(\frac{\partial A_z}{\partial y} - \frac{\partial A_y}{\partial z}\right) \cdot \hat{x} + \left(\frac{\partial A_x}{\partial z} - \frac{\partial A_z}{\partial x}\right) \cdot \hat{y} + \left(\frac{\partial A_y}{\partial x} - \frac{\partial A_z}{\partial y}\right) \cdot \hat{z}\Big|_{(0,0,0)}$$
$$= 0.$$

Because $\nabla \times \vec{A} = 0$, there is "no rotation" around the origin.

(b) From above, we plot the associated vectors in Fig. 1.20, where we have four vectors with identical magnitude 0.2. Obviously, there is no rotation around the origin O. It is consistent with the result in (a).

■

Fig. 1.20 Illustrating
Example 1.4

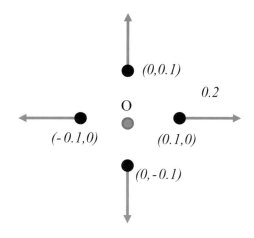

Example 1.5

Suppose we have a vector field $\vec{A} = -2y \cdot \hat{x} + 2x \cdot \hat{y}$.

(a) Calculate $\nabla \times \vec{A}$ at the origin.
(b) On xy-plane, plot the vector field $\vec{A}$ at the four points (a, 0), (0, a), (−a, 0), and
      (0, −a) surrounding the origin O, where a = 0.1. Observe if there is rotation
      around the origin.

Solution

(a) First, we have the axial components of $\vec{A}$ as below:

$$A_x = -2y,$$
$$A_y = 2x,$$
$$A_z = 0.$$

Thus

$$\frac{\partial A_z}{\partial y} - \frac{\partial A_y}{\partial z} = 0,$$
$$\frac{\partial A_x}{\partial z} - \frac{\partial A_z}{\partial x} = 0,$$
$$\frac{\partial A_y}{\partial x} - \frac{\partial A_x}{\partial y} = 2 - (-2) = 4.$$

From Eq. (1.18), we have

Fig. 1.21 Illustrating
Example 1.5

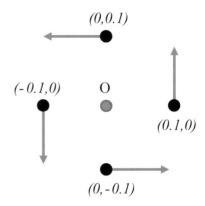

$$\nabla \times \vec{A}|_{(0,0,0)} = \left(\frac{\partial A_z}{\partial y} - \frac{\partial A_y}{\partial z}\right) \cdot \hat{x} + \left(\frac{\partial A_x}{\partial z} - \frac{\partial A_z}{\partial x}\right) \cdot \hat{y} + \left(\frac{\partial A_y}{\partial x} - \frac{\partial A_z}{\partial y}\right) \cdot \hat{z}\Bigg|_{(0,0,0)}$$

$$= 4 \cdot \hat{z}.$$

Because $\nabla \times \vec{A} = 4 \cdot \hat{z}$, there is rotation around the origin. Besides, from the direction of $\nabla \times \vec{A}$, which is $\hat{z}$, the plane of rotation is xy-plane according to the right-hand rule.

(b) From above, we plot the associated vectors in Fig. 1.21. Obviously, there is rotation around the origin O. It is consistent with the result in (a).

■

Example 1.6
Suppose we have a vector field $\vec{A} = (xy^2) \cdot \hat{x} + (yz + x) \cdot \hat{y} + (x + y^3) \cdot \hat{z}$. Calculate the curl at point $(3, 1, -2)$.

Solution
First, we have

$$A_x = xy^2,$$
$$A_y = yz + x,$$
$$A_z = x + y^3.$$

Then

$$\frac{\partial A_z}{\partial y} - \frac{\partial A_y}{\partial z} = 3y^2 - y,$$

$$\frac{\partial A_x}{\partial z} - \frac{\partial A_z}{\partial x} = 0 - 1 = -1,$$

$$\frac{\partial A_y}{\partial x} - \frac{\partial A_x}{\partial y} = 1 - 2xy.$$

From Eq. (1.18), we derive the curl at the point $(3, 1, -2)$ by

$$\nabla \times \vec{A}|_{(3,1,-2)} = \left(\frac{\partial A_z}{\partial y} - \frac{\partial A_y}{\partial z}\right) \cdot \hat{x} + \left(\frac{\partial A_x}{\partial z} - \frac{\partial A_z}{\partial x}\right) \cdot \hat{y} + \left(\frac{\partial A_y}{\partial x} - \frac{\partial A_z}{\partial y}\right) \cdot \hat{z}\Big|_{(3,1,-2)}$$

$$= (3y^2 - y) \cdot \hat{x} + (-1) \cdot \hat{y} + (1 - 2xy) \cdot \hat{z}|_{(3,1,-2)}$$

$$= 2 \cdot \hat{x} - \hat{y} - 5 \cdot \hat{z}.$$

∎

Example 1.7
Suppose we have a vector field $\vec{A} = \cos(x + y) \cdot \hat{x} + \sin(x + y) \cdot \hat{y}$. Compare the strength of rotation at the point P_1 (2, 1, 3) and the point P_2 (1, 0, 5).

Solution
First, we have

$$A_x = \cos(x + y),$$
$$A_y = \sin(x + y),$$
$$A_z = 0.$$

Thus

$$\frac{\partial A_z}{\partial y} - \frac{\partial A_y}{\partial z} = 0,$$

$$\frac{\partial A_x}{\partial z} - \frac{\partial A_z}{\partial x} = 0,$$

$$\frac{\partial A_y}{\partial x} - \frac{\partial A_x}{\partial y} = \cos(x + y) + \sin(x + y).$$

Since we can derive the strength of rotation by curl, from Eq. (1.18), for the point P_1, we have

$$\nabla \times \vec{A}|_{(2,1,3)} = \left(\frac{\partial A_z}{\partial y} - \frac{\partial A_y}{\partial z}\right) \cdot \hat{x} + \left(\frac{\partial A_x}{\partial z} - \frac{\partial A_z}{\partial x}\right) \cdot \hat{y} + \left(\frac{\partial A_y}{\partial x} - \frac{\partial A_x}{\partial y}\right) \cdot \hat{z}\Big|_{(2,1,3)}$$

$$= [\cos(x + y) + \sin(x + y)] \cdot \hat{z}|_{(2,1,3)}$$

$$= (\cos 3 + \sin 3) \cdot \hat{z}.$$

And for the point P$_2$, we have

$$\nabla \times \vec{A}|_{(1,0,5)} = \left(\frac{\partial A_z}{\partial y} - \frac{\partial A_y}{\partial z}\right) \cdot \hat{x} + \left(\frac{\partial A_x}{\partial z} - \frac{\partial A_z}{\partial x}\right) \cdot \hat{y} + \left(\frac{\partial A_y}{\partial x} - \frac{\partial A_x}{\partial y}\right) \cdot \hat{z}\bigg|_{(1,0,5)}$$
$$= [\cos(x + y) + \sin(x + y)] \cdot \hat{z}|_{(1,0,5)}$$
$$= (\cos 1 + \sin 1) \cdot \hat{z}.$$

The strength of rotations are given by

$$|\nabla \times \vec{A}|_{(2,1,3)} = |\cos 3 + \sin 3| = 0.85,$$

$$|\nabla \times \vec{A}|_{(1,0,5)} = |\cos 1 + \sin 1| = 1.38,$$

respectively. Hence, the strength of rotation at the point P$_2$ is greater than that of P$_1$.
■

C Stokes' Theorem

A useful theorem regarding curl in EM field is called Stokes' theorem and formulated as follows. Let Ś be an arbitrary surface and C be the contour surrounding S as shown in Fig. 1.22. Then, we have

$$\int_S (\nabla \times \vec{A}) \cdot d\vec{s} = \oint_C \vec{A} \cdot d\vec{l}. \tag{1.19}$$

Fig. 1.22 Illustrating Stokes' theorem

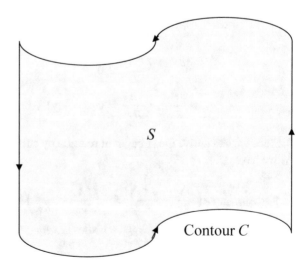

S

Contour C

Fig. 1.23 Subdivided grids
for proof of Stokes' theorem

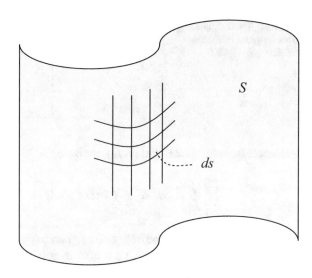

In Eq. (1.19), the left side is the surface integral of $\nabla \times \vec{A}$ over S and the right side is the line integral of $\vec{A}$ over C. The equality holds for any $\vec{A}$ and S. The proof of Stokes' theorem is quite complicated. For readers who are interested in it, they may refer to the reference.[4] Here we simply explain the concept geometrically in a heuristic way so that readers may get the insight of this theorem.

First, suppose S is an arbitrary surface, and we can divide it into infinite number of small grids as shown in Fig. 1.23. Let each grid has an area ds and $ds \to 0$. From Eq. (1.15), we have the curl of the i-th grid given by

$$(\nabla \times \vec{A})_i = \frac{\oint_{c_i} \vec{A} \cdot d\vec{l}}{U} \cdot \hat{n} = \frac{\oint_{c_i} \vec{A} \cdot d\vec{l}}{ds} \cdot \hat{n}, \tag{1.20}$$

where $\hat{n}$ is the unit vector perpendicular to the i-th grid and c_i denotes the contour surrounding the i-th grid. Let $d\vec{s}$ be the area vector perpendicular to the grid given by $d\vec{s} = ds \cdot \hat{n}$. Then from Eq. (1.20), we have

$$(\nabla \times \vec{A})_i \cdot d\vec{s} = \left(\frac{\oint_{c_i} \vec{A} \cdot d\vec{l}}{ds} \cdot \hat{n} \right) \cdot d\vec{s} = \left(\frac{\oint_{c_i} \vec{A} \cdot d\vec{l}}{ds} \cdot \hat{n} \right) \cdot ds \cdot \hat{n} = \oint_{c_i} \vec{A} \cdot d\vec{l}. \tag{1.21}$$

Hence, the $(\nabla \times \vec{A})_i \cdot d\vec{s}$ of the i-th grid equals the line integral along the boundary of the grid. As we sum $(\nabla \times \vec{A})_i \cdot d\vec{s}$ of all the grids in S, we have the surface integral of $\nabla \times \vec{A}$ over S. That is,

$$\sum_{i=1}^{\infty} (\nabla \times \vec{A})_i \cdot d\vec{s} = \int_S (\nabla \times \vec{A}) \cdot d\vec{s}. \tag{1.22}$$

---

[4]James Stewart, *Calculus*, 6th edition, Thomson, 2008 (Sect. 16.8).

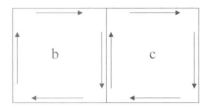

Hence, from Eq. (1.21) and (1.22), we have

$$\sum_{i=1}^{\infty} \oint_{c_i} \vec{A} \cdot d\vec{l} = \sum_{i=1}^{\infty} (\nabla \times \vec{A})_i \cdot d\vec{s} = \int_{S} (\nabla \times \vec{A}) \cdot d\vec{s} \qquad (1.23)$$

In Fig. 1.24, suppose b and c are two neighboring grids. At the boundary between b and c, we find that the line integral of $\vec{A}$ over b is equal to that of $\vec{A}$ over c with opposite direction. Hence, two integrals at boundary will be cancelled! Therefore, as we sum up the line integrals of $\vec{A}$ over each grid in S, only those line integrals along the contour C in Fig. 1.22 will not be cancelled, which is exactly the line integral of $\vec{A}$ over C. Hence, we have

$$\sum_{i=1}^{\infty} \oint_{c_i} \vec{A} \cdot d\vec{l} = \oint_{C} \vec{A} \cdot d\vec{l}. \qquad (1.24)$$

A schematic explanation of Eq. (1.24) is shown in Fig. 1.25, where the sum of line integrals of the nine grids is equal to the line integral over the contour surrounding these grids.

Finally, from Eqs. (1.23) and (1.24), we obtain Stokes' theorem as follows:

$$\int_{S} (\nabla \times \vec{A}) \cdot d\vec{s} = \oint_{C} \vec{A} \cdot d\vec{l}. \qquad (1.25)$$

In mathematical point of view, Stokes' theorem simplifies the 2-D surface integral to 1-D line integral. As will be clear later, Stokes' theorem is a useful tool providing a different perspective to interpret EM field.

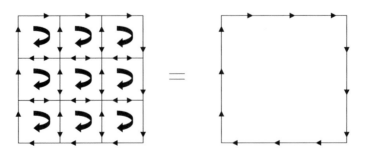

Fig. 1.25 Schematic plot for illustrating Eq. (1.24)

1.3 Maxwell's Equations in Physical Perspective

In 1861, a young Scottish scientist J. C. Maxwell published an important paper. He collected all the known EM theories and experimental results and tried to formulate them in a unified way. Considering displacement current, he corrected the original Ampere's law and formulated a set of equations describing electromagnetic phenomena precisely. These equations are finally summarized as four classical equations called "Maxwell's equations".

With Maxwell's equations, we can use a mathematical approach to describe the behavior of EM waves. Not only interpreting the phenomena of EM waves, we can also quantize the characteristics of EM waves and explore the potential of EM waves. A promising example is the modern wireless communication using EM waves. This is why a smartphone can be accessed almost everywhere. Therefore, learning Maxwell's equations not only grants us a key to understand EM phenomena, but also helps us realize many appliances of EM wave in daily life. In the following, we start from the physical meaning of Maxwell's equations in order to get the principal ideas.

A Maxwell's Equations

For electricity, we have two sources: charge and current. Generally speaking, a charge generates the electric field and a current generates the magnetic field. In an open space, we use the charge density ρ to represent the amount of charge at a point. The unit of ρ is $Coulomb/m^3$, which means the charge per unit volume. Besides, we use the current density $\vec{J}$ to represent the current at a point. The unit of $\vec{J}$ is $Ampere/m^2$, which means the current per unit area. Note that because the current is a quantity having both magnitude and direction, the current density $\vec{J}$ is a vector.

In an EM field, when sources ρ and $\vec{J}$ are given, we can use Maxwell's equations to formulate the behavior of the four key components of EM field:

1. $\vec{E}$: **electric field intensity**,
2. $\vec{H}$: **magnetic field intensity**,
3. $\vec{D}$: **electric flux density**,
4. $\vec{B}$: **magnetic flux density**.

Note that because electric field and magnetic field are quantities having both magnitude and direction, the above four components $\vec{E}$, $\vec{H}$, $\vec{D}$, and $\vec{B}$ are vectors. The background knowledge of $\vec{E}$, $\vec{H}$, $\vec{D}$, and $\vec{B}$ can be found in Appendix B.

To summarize in the beginning, the Maxwell's equations are given by

$$\nabla \times \vec{E} = -\frac{\partial \vec{B}}{\partial t}, \tag{1.26}$$

$$\nabla \times \vec{H} = \vec{J} + \frac{\partial \vec{D}}{\partial t},\tag{1.27}$$

$$\nabla \cdot \vec{D} = \rho,\tag{1.28}$$

$$\nabla \cdot \vec{B} = 0,\tag{1.29}$$

Equations (1.26)–(1.29) formulate the behavior of ρ, $\vec{J}$, $\vec{E}$, $\vec{H}$, $\vec{D}$, and $\vec{B}$ in time and in space. When we take a look at these four equations, they seem quite complicated. Indeed, from the mathematical viewpoint, it is not easy to access the insight of Maxwell's equations. But if we start from the physical viewpoint, it becomes much easier to get the most critical idea behind Maxwell's equations as given in the following.

Equations (1.26)–(1.29) intuitively describe the relationship between EM sources and their resultant effects.

In Eqs. (1.26)–(1.29), the left side is the resultant EM effect (phenomenon) and the right side is the source which generates this effect. In order to help readers get the idea visually, we "formulate" this idea in the following:

EM effect = EM source.

For example, in Eq. (1.26), $\nabla \times \vec{E}$ is the resultant EM effect and $-\partial \vec{B}/\partial t$ is the source which generates this effect. Another example is in Eq. (1.27), where $\nabla \times \vec{H}$ is the resultant EM effect as $\vec{J}$ and $\partial \vec{D}/\partial t$ are the sources which generate this effect. With this point in mind, we get the most critical idea to understand Maxwell's equations.

B   Faraday's Law

The first Maxwell's equation is originally discovered by the British scientist M. Faraday and given by

$$\nabla \times \vec{E} = -\frac{\partial \vec{B}}{\partial t},\tag{1.30}$$

where $\nabla \times \vec{E}$ denotes the curl of electric field $\vec{E}$ and $\partial \vec{B}/\partial t$ denotes the time derivative of the magnetic flux density $\vec{B}$. Equation (1.30) is well known as **Faraday's law**.

Before interpreting the physical meaning of Eq. (1.30), we review the meaning of curl learned in Sect. 1.2. First, in Fig. 1.26, suppose $\vec{A}$ is a vector field and the curl at a point P in this filed is given by

Fig. 1.26 Illustrating the
meaning of curl in a vector
field

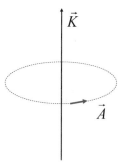

Fig. 1.27 Illustrating
Faraday's law

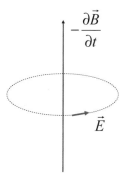

$$\nabla \times \vec{A} = \vec{K}, \tag{1.31}$$

where $\vec{K}$ is a vector. Equation (1.31) tells us that at the point P, the vector field $\vec{A}$ has the following characteristics:

(1) $\vec{A}$ is a rotational vector field if $\vec{K} \neq 0$.
(2) The rotation plane of $\vec{A}$ is perpendicular to $\vec{K}$.
(3) The rotational direction of $\vec{A}$ is determined by the right-hand rule—when the thumb points to the direction of $\vec{K}$, then the curl of the other four fingers naturally indicates the rotational direction of $\vec{A}$.
(4) The larger the magnitude of $\vec{K}$ (i.e., $|\vec{K}|$), the stronger the rotational strength.

After reviewing the meaning of curl, we are ready to get the insight of Faraday's law. Recall the left side of Eq. (1.30) is the resultant EM effect and the right side represents the source generating this effect. It means the negative of time derivative of magnetic flux density $-\partial \vec{B}/\partial t$ generates the electric field $\vec{E}$. When $\vec{B}$ changes with time, the electric field $\vec{E}$ is generated. This field is a rotational field whose rotational plane is perpendicular to $-\partial \vec{B}/\partial t$. The clockwise or counter-clockwise rotation can be determined by the right-hand rule. As shown in Fig. 1.27, if $-\partial \vec{B}/\partial t$ points to + z direction, it will generate a rotational electric field on the xy-plane.

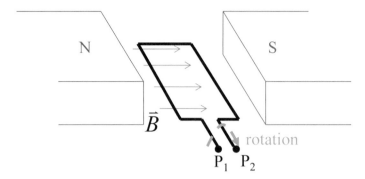

Fig. 1.28 Illustrating generation of electric voltage

Faraday's law is very important and useful because it tells us **how to generate an electric field**. As shown in Fig. 1.28, first we use magnets to generate a magnetic flux density $\vec{B}$. Then we try to rotate the coil to make the magnetic flux inside the coil vary with time. Thus, we have $-\partial \vec{B}/\partial t$. Finally, according to Faraday's law, an electric field $\vec{E}$ will be generated on the coil in Fig. 1.28. This electric field will drive free electrons in the coil to move and thus create an electric voltage between two end points P_1 and P_2. Finally, the voltage is converted to 110 V or 220 V as we usually use in daily life!

What we described in the above example is actually the principle of all modern electrical generators. Essentially, Faraday's law tells us that "a time-varying magnetic field can generate an electric field". In the example of Fig. 1.28, we simply make a time-varying magnetic flux density and then generate the electric voltage! This breakthrough leads human beings from the darkness into the bright Electrical Age. It is a milestone in human history.

C Ampere's Law

The second Maxwell's equation is given by

$$\nabla \times \vec{H} = \vec{J} + \frac{\partial \vec{D}}{\partial t}. \tag{1.32}$$

Note that the original form of Eq. (1.32) is discovered by the French scientist A. M. Ampere, which is given by

$$\nabla \times \vec{H} = \vec{J}. \tag{1.33}$$

Fig. 1.29 Illustrating original Ampere's law with current density as source

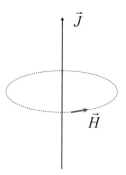

Hence, Eq. (1.32) is called **Ampere's law**. The second term $\partial \vec{D}/\partial t$ at the right side of Eq. (1.32) is called *displacement current* that is added later by Maxwell. This added term is very critical because it reveals the possibility of EM wave propagation. This point will become clear later.

According to electric conductivity, the materials are generally categorized as conductor, semiconductor, and insulator. A conductor contains a lot of free electrons and hence the conductivity is very good. On the other hand, an insulator contains almost none free electron and its conductivity is very poor. A semiconductor contains few free electrons and hence the conductivity is between that of a conductor and an insulator.

First, for a conductor, because it has a lot of free electrons, its current density $\vec{J}$ is generally much greater than $\partial \vec{D}/\partial t$. In this case, $\vec{J}$ dominates and Eq. (1.32) can be simplified as Eq. (1.33). In Eq. (1.33), the current density $\vec{J}$ is the source generating the magnetic field $\vec{H}$. As shown in Fig. 1.29, $\vec{H}$ is a rotational vector field whose rotational plane is perpendicular to $\vec{J}$.

Equation (1.33) means that the current is a source of a magnetic field. It also tells us **how to generate a magnetic field**! Suppose we have a DC voltage source V_{DC} and a resistor R in a circuit as shown in Fig. 1.30. With this simple circuit we can generate a current I and according to Ampere's law, a magnetic field $\vec{H}$ is generated. As the current I increases, the resultant magnetic field $\vec{H}$ becomes stronger. Because we can change I by adjusting R or V_{DC}, we can attain the magnetic field having the required strength.

On the other hand, there is almost none free electron in an insulator. Hence, the current density approaches zero, i.e., $\vec{J} \approx 0$, and Eq. (1.32) can be simplified as

$$\nabla \times \vec{H} = \frac{\partial \vec{D}}{\partial t}. \tag{1.34}$$

From Eq. (1.34), the time-varying electric flux density, like the current, can be a source to generate a magnetic field. This is shown in Fig. 1.31. The generated magnetic field $\vec{H}$ is on the plane perpendicular to $\frac{\partial \vec{D}}{\partial t}$. In addition, when the electric flux density varies faster with time, the resultant magnetic field is greater. For

Fig. 1.30 A simple circuit to
generate magnetic field

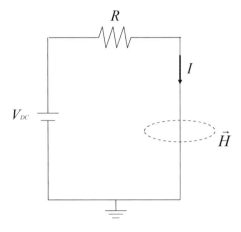

Fig. 1.31 Illustrating
Ampere's law with
time-varying electric flux
density as source

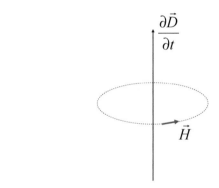

Fig. 1.32 A simple circuit to
generate magnetic field using
time-varying electric flux
density

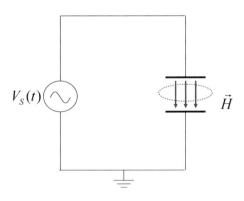

example, in Fig. 1.32, we put an AC voltage source $V_S(t)$ and two parallel metal
plates in a circuit. Then we can use $V_S(t)$ to generate a time-varying electric flux
density $\vec{D}$ between these two plates. From Eq. (1.34), a magnetic field $\vec{H}$ is generated
as shown in Fig. 1.32. In Fig. 1.32, the two parallel metal plates with an insulator in

between can be regarded as a capacitor. Hence, we can use a simple circuit consisting of merely an AC voltage source and a capacitor to generate a magnetic field.

From above, Ampere's law actually tells us that a magnetic field can be generated in two ways:

1. For an conductor, we utilize the current to generate a magnetic field, i.e., $\nabla \times \vec{H} = \vec{J}$.
2. For an insulator, we utilize the time-varying electric flux density to generate a magnetic field, i.e., $\nabla \times \vec{H} = \frac{\partial \vec{D}}{\partial t}$.

D Gauss's Law

The third Maxwell's equation essentially describes the relationship between an electric charge and its induced electric field. This is called **Gauss's law** and given by

$$\nabla \cdot \vec{D} = \rho, \tag{1.35}$$

where ρ is the charge density and $\nabla \cdot \vec{D}$ denotes the divergence of an electric flux density $\vec{D}$. In Eq. (1.35), the charge density ρ is the source generating the electric flux density $\vec{D}$. A positive charge density ($\rho > 0$) will generate a "divergent" $\vec{D}$ as shown in Fig. 1.33. On the other hand, a negative charge density ($\rho < 0$) will generate a

Fig. 1.33 Illustrating positive charge as source in Gauss's law

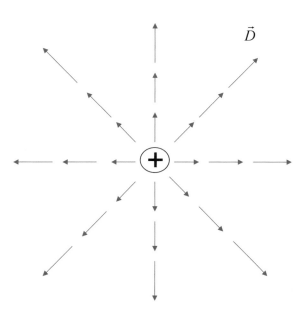

Fig. 1.34 Illustrating
negative charge as source in
Gauss's law

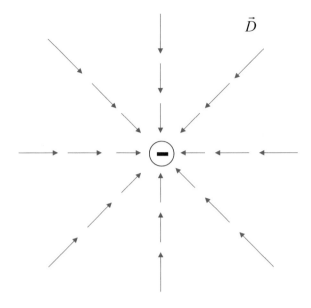

"convergent" $\vec{D}$ as shown in Fig. 1.34. Hence, Gauss's law describes the relationship between the charge density and the induced electric flux density precisely with the divergence operator.

E   *Gauss's Law of Magnetic Field*

In contrast to the third Maxwell's equation describing phenomenon of electric field, the fourth Maxwell's equation is called **Gauss's law of magnetic field** and given by

$$\nabla \cdot \vec{B} = 0. \tag{1.36}$$

Notice that in Eq. (1.36), the right side is zero. It means the divergence of any magnetic flux density is always zero! Furthermore, comparing Eq. (1.36) with Eq. (1.35), it implies that the "magnetic charge density" is always equal to zero. Besides, from Eq. (1.36), for an arbitrary point in space, the outgoing magnetic flux density $\vec{B}_{out}$ must be equal to the incoming magnetic flux density $\vec{B}_{in}$. It is illustrated in Fig. 1.35. Hence, the divergence of $\vec{B}$ is always zero. This property dictates the fundamental difference between electric flux density $\vec{D}$ and magnetic flux density $\vec{B}$. We can utilize a positive charge or a negative charge to generate a divergent $\vec{D}$ or a convergent $\vec{D}$, respectively. However, we cannot do the same thing to generate a divergent $\vec{B}$ or a convergent $\vec{B}$.

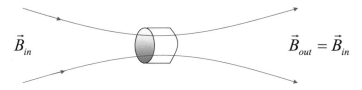

$\vec{B}_{in}$ $\vec{B}_{out} = \vec{B}_{in}$

Fig. 1.35 Illustrating Gauss's law of magnetic field

F   *Summary of Maxwell's Equations in Differential Form and Integral Form*

Finally, we summarize the physical meaning of Maxwell's equations as follows:

$\nabla \times \vec{E} = -\frac{\partial \vec{B}}{\partial t}$:   A time-varying magnetic flux density $\vec{B}$ will generate an electric field $\vec{E}$, and $\nabla \times \vec{E}$ is equal to $-\frac{\partial \vec{B}}{\partial t}$.

$\nabla \times \vec{H} = \vec{J} + \frac{\partial \vec{D}}{\partial t}$:   A current density $\vec{J}$ or a time-varying electric flux density $\frac{\partial \vec{D}}{\partial t}$ will generate a magnetic field $\vec{H}$, and $\nabla \times \vec{H}$ is equal to $\vec{J} + \frac{\partial \vec{D}}{\partial t}$.

$\nabla \cdot \vec{D} = \rho$:   A charge density ρ is the source to generate an electric flux density $\vec{D}$.

$\nabla \cdot \vec{B} = 0$:   For an arbitrary point in space, the incoming magnetic flux density must be equal to the outgoing magnetic flux density.

Example 1.8
In Fig. 1.36, suppose S is an arbitrary surface and C is the contour enclosing S. Please utilize Faraday's law to prove the following formula:

$$\oint_C \vec{E} \cdot d\vec{l} = -\frac{\partial}{\partial t} \int_S \vec{B} \cdot d\vec{s}.$$

Fig. 1.36 Plot of Example 1.8

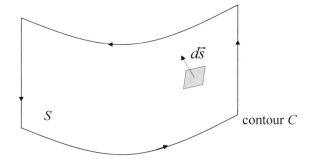

$d\vec{s}$

S contour C

Solution

First, from Faraday's law, we have

$$\nabla \times \vec{E} = -\frac{\partial \vec{B}}{\partial t}.$$

Then we do surface integral over S on both sides of the above formula to get

$$\int_S (\nabla \times \vec{E}) \cdot d\vec{s} = \int_S \left(-\frac{\partial \vec{B}}{\partial t}\right) \cdot d\vec{s} = -\frac{\partial}{\partial t} \int_S \vec{B} \cdot d\vec{s}.$$

Next, utilizing Stokes' theorem, we have

$$\int_S (\nabla \times \vec{E}) \cdot d\vec{s} = \oint_C \vec{E} \cdot d\vec{l},$$

where C is the contour enclosing S. Hence

$$\oint_C \vec{E} \cdot d\vec{l} = -\frac{\partial}{\partial t} \int_S \vec{B} \cdot d\vec{s}.$$

∎

The above formula is known as the integral form of Faraday's law. It shows the relationship between the magnetic flux density $\vec{B}$ and the electric field $\vec{E}$ over an arbitrary surface S.

Example 1.9

Suppose V is an arbitrary volume enclosed by a closed surface S. Please utilize Gauss's law to prove the following formula:

$$\oint_S \vec{D} \cdot d\vec{s} = Q,$$

where Q is the total charge inside V.

Solution

First, from Gauss's, law we have

$$\nabla \cdot \vec{D} = \rho,$$

where ρ is the charge density with the unit $Coulomb/m^3$. Then we do volume integral over V on both sides of the above equation to get

$$\int_V (\nabla \cdot \vec{D}) dv = \int_V \rho \cdot dv.$$

Since the volume integral of ρ over V is equal to the total charge inside V, we have

$$\int_V \rho \cdot dv = Q.$$

Hence

$$\int_V (\nabla \cdot \vec{D}) dv = Q.$$

On the other hand, from divergence theorem, we have

$$\int_V (\nabla \cdot \vec{D}) dv = \oint_S \vec{D} \cdot d\vec{s}.$$

Thus, we obtain

$$\oint_S \vec{D} \cdot d\vec{s} = Q.$$

∎

This formula is known as the integral form of Gauss's law. It shows that for an arbitrary volume V, the relationship between the charge Q inside V and the resultant electric flux density $\vec{D}$.

In the previous two examples, we utilize Stokes' theorem and divergence theorem to attain the integral form of Faraday's law and Gauss's law, respectively. Similarly, we can derive the integral form of all the Maxwell's equations. The results are summarized as follows:

$$\oint_C \vec{E} \cdot d\vec{l} = -\frac{\partial}{\partial t} \int_S \vec{B} \cdot d\vec{s}, \tag{1.38}$$

$$\oint_C \vec{H} \cdot d\vec{l} = \int_S \vec{J} \cdot d\vec{s} + \frac{\partial}{\partial t} \int_S \vec{D} \cdot d\vec{s}, \tag{1.39}$$

$$\oint_S \vec{D} \cdot d\vec{s} = Q, \tag{1.40}$$

$$\oint_S \vec{B} \cdot d\vec{s} = 0. \tag{1.41}$$

Comparing with the differential form of Maxwell's equations in Eqs. (1.26)–(1.29), the corresponding integral form in Eqs. (1.38)–(1.41) can be more easily understood and applied at a macroscopic level. For example, they are generally presented in static electromagnetics. On the other hand, the differential form of Maxwell's equations is more effective to analyze dynamic EM phenomena at a microscopic level. Hence, in the following context, we will mainly use Eqs. (1.26)–(1.29) to build up our knowledge of EM waves.

1.4   Maxwell's Equations in Mathematical Perspective

Physics and mathematics are two foundations of the modern science. Physics is the knowledge of nature and explains its fundamental mechanism and behavior. In contrast, mathematics originates from logics and reasoning of human beings. From the basic operation like $1 + 1 = 2$, we can construct a rigorous mathematical system step by step. Amazingly, according to uncountable findings and verifications, the way how our physical world behaves is always consistent with the induced mathematical rules! These rules are called physical laws and may be used to formulate new conjectures and analyze the physical phenomenon such as the motion of an electron inside an atom or the movement of a planet. Maxwell's equations are one of the physical laws. They are actually a promising example showing that the development of mathematical knowledge contributes to advances in science.

In the previous section, we already learned Maxwell's equations from physical perspective. In this section, we will revisit Maxwell's equations from mathematical perspective in order to help readers get the insights of them. First, for convenience, we rewrite Maxwell's equations in the following:

$$\nabla \times \vec{E} = -\frac{\partial \vec{B}}{\partial t}, \tag{1.42}$$

$$\nabla \times \vec{H} = \vec{J} + \frac{\partial \vec{D}}{\partial t}, \tag{1.43}$$

$$\nabla \cdot \vec{D} = \rho, \tag{1.44}$$

$$\nabla \cdot \vec{B} = 0, \tag{1.45}$$

where ρ denotes the charge density, $\vec{J}$ is the current density, and the four vector fields are given by

$\vec{E}$:   electric field intensity,

$\vec{H}$:   magnetic field intensity,

$\vec{D}$:   electric flux density,

$\vec{B}$:   magnetic flux density.

For a given ρ and $\vec{J}$, Eqs. (1.42)–(1.45) formulate the interaction between the four vector fields $\left(\vec{E}, \vec{H}, \vec{D}, \vec{B} \right)$. In the following, we adopt the Q&A approach step by step to build up our mathematical knowledge and get the insights of Maxwell's equations.

Q1: How many unknown variables are contained in Eqs. (1.42)–(1.45)?

Answer: Given ρ and $\vec{J}$, it can be seen that Eqs. (1.42)–(1.45) contain four unknown vectors $\left(\vec{E}, \vec{H}, \vec{D}, \vec{B} \right)$. Since each one is a 3-D vector, there are three unknown variables in each vector. For example, $\vec{E} = E_x \cdot \hat{x} + E_y \cdot \hat{y} + E_z \cdot \hat{z}$, where E_x, E_y, E_z are unknown variables. Hence, Eqs. (1.42)–(1.45) totally contain $4 \times 3 = 12$ unknown variables. ∎

Q2: How many equations are included in Eqs. (1.42)–(1.45)?

Answer: Let A_x, A_y, A_z denote the three axial components of a vector field $\vec{A}$ and then we have

$$\vec{A} = A_x \cdot \hat{x} + A_y \cdot \hat{y} + A_z \cdot \hat{z}. \tag{1.46}$$

From the previous results, the curl of $\vec{A}$ is given by

$$\nabla \times \vec{A} = \left(\frac{\partial A_z}{\partial y} - \frac{\partial A_y}{\partial z} \right)\hat{x} + \left(\frac{\partial A_x}{\partial z} - \frac{\partial A_z}{\partial x} \right)\hat{y} + \left(\frac{\partial A_y}{\partial x} - \frac{\partial A_x}{\partial y} \right)\hat{z}. \tag{1.47}$$

In Eq. (1.42), $\vec{E}$ and $\vec{B}$ are vector fields in 3-D space and can be denoted by

$$\vec{E} = E_x \cdot \hat{x} + E_y \cdot \hat{y} + E_z \cdot \hat{z}, \tag{1.48}$$

$$\vec{B} = B_x \cdot \hat{x} + B_y \cdot \hat{y} + B_z \cdot \hat{z}. \tag{1.49}$$

Hence, Eq. (1.42) can be rewritten as

$$\left(\frac{\partial E_z}{\partial y} - \frac{\partial E_y}{\partial z} \right)\hat{x} + \left(\frac{\partial E_x}{\partial z} - \frac{\partial E_z}{\partial x} \right)\hat{y} + \left(\frac{\partial E_y}{\partial x} - \frac{\partial E_x}{\partial y} \right)\hat{z} = -\left(\frac{\partial B_x}{\partial t}\hat{x} + \frac{\partial B_y}{\partial t}\hat{y} + \frac{\partial B_z}{\partial t}\hat{z} \right). \tag{1.50}$$

Since each component of x, y, z directions in both sides of Eq. (1.50) must hold, we have

$$\frac{\partial E_z}{\partial y} - \frac{\partial E_y}{\partial z} = -\frac{\partial B_x}{\partial t}, \tag{1.51}$$

$$\frac{\partial E_x}{\partial z} - \frac{\partial E_z}{\partial x} = -\frac{\partial B_y}{\partial t}, \tag{1.52}$$

$$\frac{\partial E_y}{\partial x} - \frac{\partial E_x}{\partial y} = -\frac{\partial B_z}{\partial t}. \tag{1.53}$$

Therefore, Eq. (1.42) actually consists of three equations. Similarly, we can express $\vec{H}$, $\vec{J}$, and $\vec{D}$ by

$$\vec{H} = H_x \cdot \hat{x} + H_y \cdot \hat{y} + H_z \cdot \hat{z}, \tag{1.54}$$

$$\vec{J} = J_x \cdot \hat{x} + J_y \cdot \hat{y} + J_z \cdot \hat{z}, \tag{1.55}$$

$$\vec{D} = D_x \cdot \hat{x} + D_y \cdot \hat{y} + D_z \cdot \hat{z}. \tag{1.56}$$

Therefore, Eq. (1.43) can be rewritten as

$$\left(\frac{\partial H_z}{\partial y} - \frac{\partial H_y}{\partial z}\right)\hat{x} + \left(\frac{\partial H_x}{\partial z} - \frac{\partial H_z}{\partial x}\right)\hat{y} + \left(\frac{\partial H_y}{\partial x} - \frac{\partial H_x}{\partial y}\right)\hat{z}$$
$$= \left(J_x + \frac{\partial D_x}{\partial t}\right)\hat{x} + \left(J_y + \frac{\partial D_y}{\partial t}\right)\hat{y} + \left(J_z + \frac{\partial D_z}{\partial t}\right)\hat{z}. \tag{1.57}$$

Hence, Eq. (1.43) actually consists of three equations:

$$\frac{\partial H_z}{\partial y} - \frac{\partial H_y}{\partial z} = J_x + \frac{\partial D_x}{\partial t}, \tag{1.58}$$

$$\frac{\partial H_x}{\partial z} - \frac{\partial H_z}{\partial x} = J_y + \frac{\partial D_y}{\partial t}, \tag{1.59}$$

$$\frac{\partial H_y}{\partial x} - \frac{\partial H_x}{\partial y} = J_z + \frac{\partial D_z}{\partial t}. \tag{1.60}$$

Next, from the previous results, if a vector field $\vec{A}$ is expressed by Eq. (1.46), then $\nabla \cdot \vec{A}$ is given by

$$\nabla \cdot \vec{A} = \frac{\partial A_x}{\partial x} + \frac{\partial A_y}{\partial y} + \frac{\partial A_z}{\partial z}. \tag{1.61}$$

Hence, Eqs. (1.44) and (1.45) can be rewritten as follows:

$$\nabla \cdot \vec{D} = \rho \Rightarrow \frac{\partial D_x}{\partial x} + \frac{\partial D_y}{\partial y} + \frac{\partial D_z}{\partial z} = \rho, \tag{1.62}$$

$$\nabla \cdot \vec{B} = 0 \Rightarrow \frac{\partial B_x}{\partial x} + \frac{\partial B_y}{\partial y} + \frac{\partial B_z}{\partial z} = 0. \tag{1.63}$$

Thus, each one of Eqs. (1.44) and (1.45) consists of one equation.

From above, Eqs. (1.42) and (1.43) contain three equations individually, while each one of Eqs. (1.44) and (1.45) contains one equation. Hence, Maxwell's equations totally have $3 + 3 + 1 + 1 = 8$ equations.

■

Q3: How many independent equations are contained in Eqs. (1.42)–(1.45)?

Answer: In the following, we will prove that Eq. (1.45) can be derived from Eq. (1.42), which means Eq. (1.45) is not an independent equation. Similarly, Eq. (1.44) can be derived from Eq. (1.43) so that Eq. (1.44) is not an independent equation, either.

First, for a vector field $\vec{A}$, the following equality always holds[5]:

$$\nabla \cdot \left(\nabla \times \vec{A} \right) = 0. \tag{1.64}$$

It means that the divergence of $\nabla \times \vec{A}$ must be zero.

We apply Eq. (1.64) and take divergence on both sides of Eq. (1.42). Then, we have

$$\nabla \cdot \left(\nabla \times \vec{E} \right) = 0 \Rightarrow \nabla \cdot \left(-\frac{\partial \vec{B}}{\partial t} \right) = 0. \tag{1.65}$$

Since the differentiation in time and in space can be interchanged, Eq. (1.65) can be reformulated by

$$\frac{\partial}{\partial t} \left(\nabla \cdot \vec{B} \right) = 0 \Rightarrow \nabla \cdot \vec{B} = \text{constant}. \tag{1.66}$$

From the experimental results, the constant in Eq. (1.66) is always zero. Hence, $\nabla \cdot \vec{B} = 0$. Since Eq. (1.45) can be derived from Eq. (1.42), it is not an independent equation.

Next, we prove that Eq. (1.44) can be derived from Eq. (1.43). First, we apply Eq. (1.64) again and take divergence on both sides of Eq. (1.43). Then, we have

$$\nabla \cdot \left(\nabla \times \vec{H} \right) = 0 \Rightarrow \nabla \cdot \vec{J} + \frac{\partial}{\partial t} \left(\nabla \cdot \vec{D} \right) = 0. \tag{1.67}$$

According to the conservation law of charge, the following formula always holds: (the proof is provided in Example 1.10).

$$\nabla \cdot \vec{J} = -\frac{\partial \rho}{\partial t}. \tag{1.68}$$

From Eqs. (1.67) and (1.68), we have

$$-\frac{\partial \rho}{\partial t} + \frac{\partial}{\partial t} \left(\nabla \cdot \vec{D} \right) = 0 \Rightarrow \nabla \cdot \vec{D} - \rho = \text{constant}. \tag{1.69}$$

---

[5] James Stewart, *Calculus*, 6th edition, Thomson, 2008 (Sect. 16.5).

From the experimental results, the constant in Eq. (1.69) is always zero. Hence, $\nabla \cdot \vec{D} = \rho$. Because Eq. (1.44) can be derived from Eq. (1.43), it is not an independent equation.

In above, both Eqs. (1.44) and (1.45) are dependent equations. Since Eqs. (1.42) and (1.43) consist of three independent equations individually, Maxwell's equations totally have six independent equations.   ∎

Q4: In mathematical perspective, we need N independent equations in order to solve N unknown variables. However, from above, we have 12 unknown variables in Maxwell's equations, but we have only six independent equations. How do we resolve this problem?

Answer:

Obviously, from the mathematical viewpoint, when we try to derive the unknown variables in Maxwell's equations, we still need another six independent equations. The problem is: "how do we get these six equations?".

The key lies on the medium where the EM field is in. In a homogeneous and isotropic medium, from the experimental results, the electric flux density $\vec{D}$ is proportional to the corresponding electric field $\vec{E}$. The relationship can be formulated as

$$\vec{D} = \epsilon \, \vec{E}, \tag{1.70}$$

where ϵ is the **permittivity** of the medium. The permittivity depends on the property of the medium and may vary for different media. For example, the permittivity of vacuum is

$$\epsilon = \epsilon_0 = \frac{1}{36\pi} \times 10^{-9} (\text{Farad}/\text{m}). \tag{1.71}$$

For the permittivity of other mediums, we often use ϵ_0 as a reference and represent ϵ by

$$\epsilon = \epsilon_r \cdot \epsilon_0, \tag{1.72}$$

where ϵ_r is called the **relative permittivity**. Note that the permittivity of common medium is greater than vacuum. Hence, ϵ_r is greater than 1.

Similarly, in a homogeneous and isotropic medium, the magnetic flux density $\vec{B}$ is proportional to the magnetic field $\vec{H}$. The relationship can be expressed by

$$\vec{B} = \mu \vec{H}, \tag{1.73}$$

where μ is the **permeability** of the medium. The permeability of vacuum is given by

$$\mu = \mu_0 = 4\pi \times 10^{-7} (\text{Henry}/\text{m}). \tag{1.74}$$

For a non-magnetic medium, $\mu \approx \mu_0$. Because most mediums are non-magnetic materials, we usually assume $\mu = \mu_0$ unless otherwise noted.

In above, Eqs. (1.70) and (1.73) are vector equations, where each one consists of three independent equations. A total of six independent equations are included. These six independent equations are actually what we need additionally to solve the Maxwell's equations. Since Eqs. (1.70) and (1.73) depend on the property of the medium, they are called the **constitutive equations**.   ∎

Q5: In mathematical perspective, how do we effectively express Maxwell's equations?

Answer: In mathematical perspective, the last two equations of Maxwell's equations shall be replaced by the constitutive equations given in Eqs. (1.70) and (1.73). Therefore, we have

$$\nabla \times \vec{E} = -\frac{\partial \vec{B}}{\partial t}, \tag{1.75}$$

$$\nabla \times \vec{H} = \vec{J} + \frac{\partial \vec{D}}{\partial t}, \tag{1.76}$$

$$\vec{D} = \in \vec{E}, \tag{1.77}$$

$$\vec{B} = \mu \vec{H}. \tag{1.78}$$

Equations (1.75)–(1.78) consist of 12 unknown variables and 12 independent equations. The first two equations are Faraday's law and Ampere's law. The last two equations are constitutive equations which depend on the medium of concern.

Finally, when we investigate EM field, usually μ and $\in$ are known parameters. Hence, (D_x, D_y, D_z) can be simply derived from (E_x, E_y, E_z) using Eq. (1.77), and (B_x, B_y, B_z) can be simply derived from (H_x, H_y, H_z) by using Eq. (1.78). Therefore, Maxwell's equations can be further simplified as two equations given by

$$\nabla \times \vec{E} = -\mu \frac{\partial \vec{H}}{\partial t}, \tag{1.79}$$

$$\nabla \times \vec{H} = \vec{J} + \in \frac{\partial \vec{E}}{\partial t}. \tag{1.80}$$

Hence, Eqs. (1.79) and (1.80), namely, Faraday's law and Ampere's law, are the most important equations in EM field. When ρ, $\vec{J}$, and $(\mu, \in)$ are given, the six unknown variables $(E_x, E_y, E_z, H_x, H_y, H_z)$ can be derived form these two equations. Equations (1.79) and (1.80) are the key for us to explore EM waves.   ∎

Example 1.10
Prove that $\nabla \cdot \vec{J} = -\frac{\partial \rho}{\partial t}$, where $\vec{J}$ is the current density and ρ is the charge density.

Solution
Suppose Q is the total charge enclosed by a volume V and thus we have

$$Q = \int_V \rho \cdot dv,$$

where ρ is the charge density with the unit Coulomb/m³. According to the conservation law of charge, the outgoing current I from V is equal to the decreasing rate of charge in V. This can be expressed by the following formula:

$$I = -\frac{dQ}{dt} = -\int_V \frac{d\rho}{dt} \cdot dv.$$

Let S be the surface of V and $\vec{J}$ be the surface current density over S. Since I is the outgoing current of V and the current flows through the surface S, we have

$$I = \oint_S \vec{J} \cdot d\vec{s}.$$

Thus, we get

$$\oint_S \vec{J} \cdot d\vec{s} = -\int_V \frac{d\rho}{dt} \cdot dv.$$

On the other hand, the following formula holds by the divergence theorem:

$$\oint_S \vec{J} \cdot d\vec{s} = \int_V (\nabla \cdot \vec{J}) dv.$$

From the above two formulas, we get

$$\int_V (\nabla \cdot \vec{J}) dv = -\int_V \left(\frac{d\rho}{dt}\right) dv.$$

Since this formula holds for an arbitrary volume V, the following equation always holds and it completes the proof:

$$\nabla \cdot \vec{J} = -\frac{\partial \rho}{\partial t}.$$

∎

Example 1.11

Suppose we have the magnetic flux density $\vec{B} = 3\sin(\omega t + \theta) \cdot \hat{z}$ and the induced electric field $\vec{E} = E_x \cdot \hat{x} + E_y \cdot \hat{y} + E_z \cdot \hat{z}$. Please give the associated three independent equations of $\vec{B}$ and $\vec{E}$ by using Maxwell's equations.

Solution

Let

$$\vec{B} = B_x \cdot \hat{x} + B_y \cdot \hat{y} + B_z \cdot \hat{z}.$$

Because $\vec{B} = 3\sin(\omega t + \theta) \cdot \hat{z}$, we have

$$B_x = B_y = 0$$
$$B_z = 3\sin(\omega t + \theta).$$

Since the electromagnetic source is $\vec{B}$ in this case, we apply Faraday's law, the first equation of Maxwell's equations. Then, we have

$$\nabla \times \vec{E} = -\frac{\partial \vec{B}}{\partial t}$$

$$\Rightarrow \left(\frac{\partial E_z}{\partial y} - \frac{\partial E_y}{\partial z} \right) \cdot \hat{x} + \left(\frac{\partial E_x}{\partial z} - \frac{\partial E_z}{\partial x} \right) \cdot \hat{y} + \left(\frac{\partial E_y}{\partial x} - \frac{\partial E_x}{\partial y} \right) \cdot \hat{z} = - \left(\frac{\partial B_x}{\partial t} \cdot \hat{x} + \frac{\partial B_y}{\partial t} \cdot \hat{y} + \frac{\partial B_z}{\partial t} \cdot \hat{z} \right).$$

Finally, we get the three independent equations as follows:

$$\frac{\partial E_z}{\partial y} - \frac{\partial E_y}{\partial z} = -\frac{\partial B_x}{\partial t} = 0$$

$$\frac{\partial E_x}{\partial z} - \frac{\partial E_z}{\partial x} = -\frac{\partial B_y}{\partial t} = 0$$

$$\frac{\partial E_y}{\partial x} - \frac{\partial E_x}{\partial y} = -\frac{\partial B_z}{\partial t} = -3\omega \cos(\omega t + \theta).$$

∎

Example 1.12

For a conductor, suppose $\vec{J} >> \frac{\partial \vec{D}}{\partial t}$ and $\vec{J} = 5\sin(\omega t + \phi) \cdot \hat{x}$. If the magnetic field induced by $\vec{J}$ is given by $\vec{H} = H_x \cdot \hat{x} + H_y \cdot \hat{y} + H_z \cdot \hat{z}$, please give the associated three independent equations of $\vec{H}$ and $\vec{J}$ by using Maxwell's equations.

Solution

From Ampere's law, the second equation of Maxwell's equations, we have

$$\nabla \times \vec{H} = \vec{J}.$$

Hence

$$\left(\frac{\partial H_z}{\partial y} - \frac{\partial H_y}{\partial z} \right) \cdot \hat{x} + \left(\frac{\partial H_x}{\partial z} - \frac{\partial H_z}{\partial x} \right) \cdot \hat{y} + \left(\frac{\partial H_y}{\partial x} - \frac{\partial H_x}{\partial y} \right) \cdot \hat{z} = 5 \sin(\omega t + \phi) \cdot \hat{x}.$$

Thus, we get the following three independent equations:

$$\frac{\partial H_z}{\partial y} - \frac{\partial H_y}{\partial z} = 5 \sin(\omega t + \phi)$$

$$\frac{\partial H_x}{\partial z} - \frac{\partial H_z}{\partial x} = 0$$

$$\frac{\partial H_y}{\partial x} - \frac{\partial H_x}{\partial y} = 0.$$

∎

Example 1.13

In an insulator, suppose $\vec{J} = 0$ and $\vec{D} = 2 \cos(\omega t + \theta) \cdot \hat{x} + 3 \sin \omega t \cdot \hat{z}$. If the magnetic field induced by $\vec{D}$ is given by $\vec{H} = H_x \cdot \hat{x} + H_y \cdot \hat{y} + H_z \cdot \hat{z}$, please give the associated three independent equations of $\vec{D}$ and $\vec{H}$ by using Maxwell's equations.

Solution

First, consider Ampere's law with $\vec{J} = 0$. We have

$$\nabla \times \vec{H} = \frac{\partial \vec{D}}{\partial t}.$$

Hence

$$\left(\frac{\partial H_z}{\partial y} - \frac{\partial H_y}{\partial z} \right) \cdot \hat{x} + \left(\frac{\partial H_x}{\partial z} - \frac{\partial H_z}{\partial x} \right) \cdot \hat{y} + \left(\frac{\partial H_y}{\partial x} - \frac{\partial H_x}{\partial y} \right) \cdot \hat{z}$$

$$= \frac{\partial \vec{D}}{\partial t}$$

$$= -2\omega \sin(\omega t + \theta) \cdot \hat{x} + 3\omega \cos \omega t \cdot \hat{z}.$$

Finally, we get the following three independent equations:

$$\frac{\partial H_z}{\partial y} - \frac{\partial H_y}{\partial z} = -2\omega \sin(\omega t + \theta),$$

$$\frac{\partial H_x}{\partial z} - \frac{\partial H_z}{\partial x} = 0,$$

$$\frac{\partial H_y}{\partial x} - \frac{\partial H_x}{\partial y} = 3\omega \cos \omega t.$$

■

Example 1.14

Suppose we have the electric flux density $\vec{D} = 5\cos(\omega t - kz) \cdot \hat{x} + 3\sin(\omega t - ky) \cdot \hat{y}$ at a specific point. Please derive the charge density ρ at this point.

Solution

Let

$$\vec{D} = D_x \cdot \hat{x} + D_y \cdot \hat{y} + D_z \cdot \hat{z}.$$

Then, we have

$$D_x = 5\cos(\omega t - kz),$$

$$D_y = 3\sin(\omega t - ky),$$

$$D_z = 0.$$

From Gauss's law, we have

$$\nabla \cdot \vec{D} = \rho \Rightarrow \frac{\partial D_x}{\partial x} + \frac{\partial D_y}{\partial y} + \frac{\partial D_z}{\partial z} = \rho.$$

Since

$$\frac{\partial D_x}{\partial x} + \frac{\partial D_y}{\partial y} + \frac{\partial D_z}{\partial z} = 0 - 3k\cos(\omega t - ky) + 0 = -3k\cos(\omega t - ky),$$

we have

$$\rho = -3k\cos(\omega t - ky).$$

■

Summary

In order to get an overall understanding of this chapter, we summarize the four sections in the following:

1.1: Divergence

We learn the definition and physical meaning of divergence. A useful theorem called divergence theorem is introduced.

1.2: Curl

We learn the definition and physical meaning of curl. A useful theorem called Stokes' theorem is introduced.

1.3: Maxwell's equations in physical perspective

We learn the physical meaning of Maxwell's equations. These four equations intuitively describe the relationship between EM source and its resultant effect. The left side of these equations is the resultant EM effect and the right side is the source which generates this effect. The idea can be formulated as follows:

$$\textbf{EM effect} = \textbf{EM source}.$$

1.4: Maxwell's equations in mathematical perspective

We learn Maxwell's equations from mathematical perspective and realize that Faraday's law and Ampere's law play the most important roles among Maxwell's equations. The constitutive equations depending on the medium are introduced. Using Faraday's law, Ampere's law, and constitutive equations, the EM field can be fully determined.

Exercises

1. What is "divergence" of a vector field? Explain its physical meaning in your own way. (Hint: Refer to Sect. 1.1)
2. Refer to Example 1.1 and calculate divergences of the following 1-D vector fields at the point $x = 2$. Determine if $x = 2$ is a divergent point or a convergent point.

    (a) $\vec{A} = (2x + 1) \cdot \hat{x}$,

    (b) $\vec{A} = (x^2 - 5x + 3) \cdot \hat{x}$.

3. In Exercise 2, please plot the fields at $x = 1.9$ and $x = 2.1$. Check if the result of divergence at $x = 2$ is correct or not.
4. Please calculate divergences of the following 2-D vector fields at the point (1, 2):

(a) $\vec{A} = (2x + 1) \cdot \hat{x} + (3x + y^2) \cdot \hat{y}$,

(b) $\vec{A} = (2y^2) \cdot \hat{x} + (x^2 + 3xy - 2) \cdot \hat{y}$.

5. Refer to Example 1.3 and calculate divergences of the following 3-D vector fields at the point $(2, 1, -4)$. Show whether the point is a divergent point or a convergent point.

(a) $\vec{A} = (2x^3z + y) \cdot \hat{x} + (xy + z^2) \cdot \hat{y} + (x^2) \cdot \hat{z}$,

(b) $\vec{A} = (z + y^2) \cdot \hat{x} + (xy) \cdot \hat{y} + (x^2 + 2yz) \cdot \hat{z}$,

(c) $\vec{A} = (\cos 2x) \cdot \hat{x} + [\sin 2(x + y)] \cdot \hat{y} + [\cos(x^2 + y)] \cdot \hat{z}$.

6. What is "divergence theorem"? State its properties. (Hint: Refer to Sect. 1.1)
7. What is "curl" of a vector field? Explain its physical meaning in your own way. (Hint: Refer to Sect. 1.2)
8. Refer to Example 1.4 and calculate curls of the following vector fields at the origin $(0, 0, 0)$. Note that the z-component of $\vec{A}$ is zero in the following cases:

(a) $\vec{A} = (2x) \cdot \hat{x} + (3y) \cdot \hat{y}$,

(b) $\vec{A} = (-2y) \cdot \hat{x} + (4x) \cdot \hat{y}$,

(c) $\vec{A} = (2y) \cdot \hat{x} + (4x) \cdot \hat{y}$,

(d) $\vec{A} = (3y) \cdot \hat{x} - (2x) \cdot \hat{y}$.

9. In Exercise 8, there are four points around the origin in the xy-plane: $(0.1, 0)$, $(-0.1, 0)$, $(0, 0.1)$, $(0, -0.1)$ when $z = 0$. Please plot $\vec{A}$ at these points and see if the curl can show the rotational field at the origin.
10. Refer to Example 1.6 and calculate curls of the following vector fields at the point $(1, 0, 2)$:

(a) $\vec{A} = (2x + y) \cdot \hat{x} + (y^2 + z^2) \cdot \hat{y} + (x + y) \cdot \hat{z}$,

(b) $\vec{A} = (z + y^2) \cdot \hat{x} + (xy^2) \cdot \hat{y} - (x + yz) \cdot \hat{z}$,

(c) $\vec{A} = (\cos 3x) \cdot \hat{x} - [\sin(x + y)] \cdot \hat{y} + [\cos(x + y)] \cdot \hat{z}$.

11. What is "Stokes' theorem"? State its properties. (Hint: Refer to Sect. 1.2)
12. What is Faraday's law? Explain its physical meaning in your own way.
13. What is Ampere's law? Explain its physical meaning in your own way.
14. What is Gauss's law? State its properties.
      (Hint: Refer to Sect. 1.3)
15. Write down Maxwell's equations and briefly give the mathematical perspective according to what you have learned in Sect. 1.4.
16. For an arbitrary vector field $\vec{A} = A_x \cdot \hat{x} + A_y \cdot \hat{y} + A_z \cdot \hat{z}$, prove that $\nabla \cdot \left(\nabla \times \vec{A} \right) = 0$.
17. If the electric field induced by the magnetic flux density $\vec{B}$ is $\vec{E} = \cos(\omega t - kz) \cdot \hat{x} + \sin(\omega t - kz) \cdot \hat{y}$, please derive (B_x, B_y, B_z). (Hint: Faraday's law)
18. If $\vec{E} = E_x \cdot \hat{x} + E_y \cdot \hat{y} + E_z \cdot \hat{z}$ is the electric field induced by the magnetic flux density $\vec{B} = \sin(\omega t - ky) \cdot \hat{x} + \cos(\omega t - kx) \cdot \hat{y} + 5 \cdot \hat{z}$, please provide the equations that (E_x, E_y, E_z) must satisfy.
19. If the magnetic field induced by the electric flux density $\vec{D}$ is $\vec{H} = \sin(\omega t - kz) \cdot \hat{x} + \cos(\omega t - kx) \cdot \hat{y} + 3 \cdot \hat{z}$, please derive (D_x, D_y, D_z). (Hint: Ampere's law)
20. If $\vec{H} = H_x \cdot \hat{x} + H_y \cdot \hat{y} + H_z \cdot \hat{z}$ is the magnetic field induced by the current density $\vec{J} = (2x + y) \cdot \hat{x} + (3y^2) \cdot \hat{y} + (4x - z) \cdot \hat{z}$, please provide the equations that (H_x, H_y, H_z) must satisfy.
21. If $\vec{H} = \cos(\omega t - kz + \theta) \cdot \hat{x} + \sin(\omega t - kz + \theta) \cdot \hat{y}$ is the magnetic field induced by the electric field $\vec{E}$ in free space, please derive (E_x, E_y, E_z) (Hint: $\in = \in_0$ in free space)
22. If $\vec{E} = 2xy \cdot \hat{x} + (z + y) \cdot \hat{y} + x^2 \cdot \hat{z}$ in a medium having permittivity $\in = 3 \in_0$, please derive the charge density at the points $(1, 0, 1)$ and $(2, 3, -4)$, respectively. (Hint: Example 1.14)
23. Assume S is a surface and C is the contour surrounding S. Use Ampere's law and Stokes' theorem to prove the following equality:

$$\oint_C \vec{H} \cdot d\vec{l} = I + \int_S \frac{\partial \vec{D}}{\partial t} \cdot d\vec{s},$$

where $I = \int_S \vec{J} \cdot d\vec{s}$ is the current flowing through S.

24. Suppose a 10 V DC power supply is provided. Please design a simple circuit that can produce a magnetic field intensity $H = 1$ (A/m) at a distance of 0.2 m from the circuit (Hint: Use the result of Exercise 23 and assume that $\vec{J} >> \frac{\partial \vec{D}}{\partial t}$).

Chapter 2
EM Wave Propagation

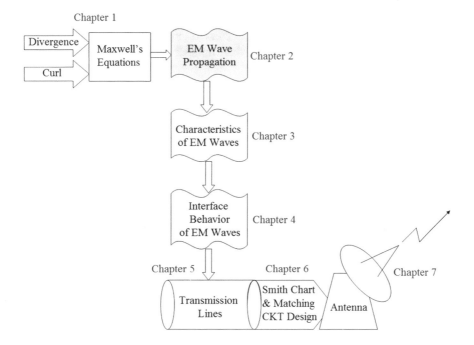

© Springer Nature Switzerland AG 2020
M.-S. Kao and C.-F. Chang, *Understanding Electromagnetic Waves*,
https://doi.org/10.1007/978-3-030-45708-2_2

Abstract Many readers may have the puzzle: why and how do EM waves propagate? Here we resolve the puzzle step by step in order to make readers catch the essential ideas quickly and effectively:

Step 1. Using a simple circuit to show that "it is reasonable and imaginable for EM waves to propagate".
Step 2. Deriving wave equation which guides EM wave propagation.
Step 3. Solving wave equation to reveal useful principles and properties of EM waves.

We use many examples and illustrations to make readers easily catch core ideas of EM waves. In fact, readers may even "see EM waves" using imagination as well as physical intuition of Maxwell's equations.

After learning this chapter, readers will understand why EM waves propagate and how they go in the space. Important parameters describing behaviors of EM waves such as phase velocity, wavelength, and wave number are also introduced.

EM wave is like a beautiful lake hidden in mystery mountains. You need to go across the forest and the fog, and then the amazing scenery will be revealed. In the following, each section will present an important property or principle of EM waves. As you go section by section, you will get more understandings of EM waves step by step. Finally, you will leave the forest and the fog behind, and enjoy the rare beauty of this lake.

In Chap. 1, we learned how an electric field and the associated magnetic field interact mutually, and Maxwell's equations are provided. In this chapter, we will develop the background knowledge of EM waves based on Maxwell's equations. This chapter consists of four sections. We start from wave equations and end with the propagation behavior of EM waves. These sections will build up our background knowledge of EM waves.

2.1 Wave Equations

The major contribution Maxwell made to electromagnetism was to amend Ampere's law. Ampere was the first person discovering that an electric current could induce a magnetic field. Maxwell went further as he found and predicted that a time-varying electric field also induced a magnetic field as what an electric current did. Meanwhile, he also caught an important and insightful meaning of this finding. That is, an electric field and the associated magnetic field may interact mutually in space and then propagate in the form of an "electromagnetic wave"!

This section introduces wave equations which formulate the principal behaviors of an EM wave. We will try to start from the intuitive viewpoint of Maxwell, and based on his equations we further explain why an EM wave radiates.

A Principles

From Sect. 1.4, Maxwell's equations can be presented by the following four equations:

$$\nabla \times \vec{E} = -\frac{\partial \vec{B}}{\partial t},$$

$$\nabla \times \vec{H} = \vec{J} + \frac{\partial \vec{D}}{\partial t},$$

$$\vec{D} = \in \vec{E},$$

$$\vec{B} = \mu \vec{H}.$$

The last two equations are actually determined by the EM characteristics of the medium. A conductor is a medium which has high conductivity so that a current flows through it when an electric field is applied. On the other hand, a **dielectric medium** is an electrical insulator that has little conductivity. When an electric field or a magnetic field is applied, electric charges do not flow through a dielectric medium, but only slightly shift from their average equilibrium positions causing electric polarization or magnetic polarization as shown in Fig. 2.1. Hence, a dielectric medium can support the formation of electric fields and magnetic fields within itself.

Now, suppose we have a medium having permittivity $\in$ and permeability μ. The relationship between an electric field $\vec{E}$ and the associated electric flux density $\vec{D}$ is given by

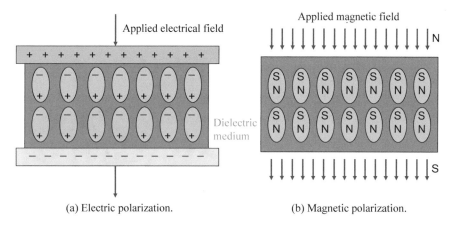

(a) Electric polarization.   (b) Magnetic polarization.

Fig. 2.1 Schematic plot of electric polarization and magnetic polarization

$$\vec{D} = \epsilon \, \vec{E}. \tag{2.1}$$

And the relationship between a magnetic field $\vec{H}$ and the associated magnetic flux density $\vec{B}$ is given by

$$\vec{B} = \mu \vec{H}. \tag{2.2}$$

Therefore, if $\vec{E}$ and $\vec{H}$ are known, $\vec{D}$ and $\vec{B}$ are readily obtained. In the following, we will focus on $\vec{E}$ and $\vec{H}$, and utilize Maxwell's equations to explore the behavior of EM waves.

First, because $\vec{B} = \mu \vec{H}$, Faraday's law can be rewritten as

$$\nabla \times \vec{E} = -\mu \frac{\partial \vec{H}}{\partial t}. \tag{2.3}$$

In Fig. 2.2, suppose a magnetic field $\vec{H}$ changes with time and $-\partial \vec{H}/\partial t$ points to $+\hat{z}$ direction. Then it will induce an electric field $\vec{E}$ on the plane perpendicular to $+\hat{z}$. From Eq. (2.3), the magnitude of $\vec{E}$ is proportional to $-\partial \vec{H}/\partial t$. It means when $\vec{H}$ varies more rapidly, the induced electric field is stronger. Equation (2.3) also reveals an important property: a time-varying magnetic field will generate a corresponding electric field.

On the other hand, because $\vec{D} = \epsilon \, \vec{E}$, Ampere's law can be rewritten as

$$\nabla \times \vec{H} = \vec{J} + \epsilon \frac{\partial \vec{E}}{\partial t}, \tag{2.4}$$

Fig. 2.2 Schematic plot of Faraday's law

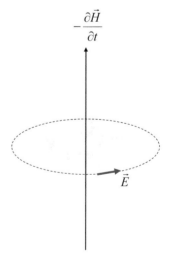

Fig. 2.3 Schematic plot of Ampere's law with current density as source

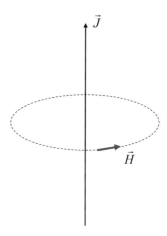

where $\vec{J}$ is the current density. In Eq. (2.4), the second term of the right side is the correction term appended by Maxwell. This term is called *displacement current*, which is the key revealing the existence of *EM radiation*—EM wave can propagate through space.

For conductors or dielectric media, Eq. (2.4) can be rewritten in a simpler form. First, for a conductor, because the conductivity is very good and thus current density $\vec{J}$ is usually much greater than the displacement current $\in \frac{\partial \vec{E}}{\partial t}$. Hence, Eq. (2.4) can be approximated by

$$\nabla \times \vec{H} = \vec{J}. \tag{2.5}$$

In Fig. 2.3, suppose $\vec{J}$ points to $+\hat{z}$-direction. Then a magnetic field $\vec{H}$ is induced on the plane perpendicular to $+\hat{z}$ and the intensity of $\vec{H}$ is proportional to $\vec{J}$. Besides, Eq. (2.5) tells us that once a current exists, there must be an induced magnetic field.

Second, for a dielectric medium, because the conductivity is very weak, the current density $\vec{J} \approx 0$. Hence, Eq. (2.4) can be approximated by

$$\nabla \times \vec{H} = \in \frac{\partial \vec{E}}{\partial t}. \tag{2.6}$$

In Fig. 2.4, suppose an electric field $\vec{E}$ changes with time and $\partial \vec{E}/\partial t$ points to $+\hat{z}$ direction. Then a magnetic field $\vec{H}$ is induced on the plane perpendicular to $+\hat{z}$. The intensity of $\vec{H}$ is proportional to $\partial \vec{E}/\partial t$. When $\vec{E}$ varies more quickly, the induced magnetic field is stronger. Equation (2.6) reveals that a time-varying electric field, being like a conducting current, will generate the corresponding magnetic field.

Fig. 2.4 Schematic plot of
Ampere's law with
time-varying electric field as
source

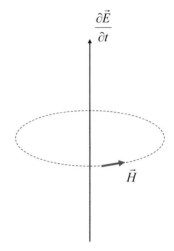

Fig. 2.5 Illustrating
generation of EM waves (I)

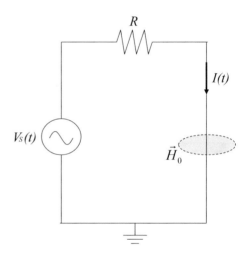

Figures 2.2 and 2.4 clearly show the interaction between electric field $\vec{E}$ and magnetic field $\vec{H}$. The time-varying magnetic field $\vec{H}$ will induce an electric field $\vec{E}$. Similarly, a time-varying electric field $\vec{E}$ will induce a magnetic field $\vec{H}$. In a word, an electric field and a magnetic field induce each other. This fact is the fundamental characteristic of EM fields.

Furthermore, in order to explain why EM waves can radiate, let us arrange a simple experiment as shown in Fig. 2.5. Suppose we have an AC voltage source $V_S(t)$, a resister R, and a conducting line in a loop circuit. The current in the loop is denoted by $I(t)$. According to Eq. (2.5), a magnetic field $\vec{H}_0$ will be induced on the plane perpendicular to the conducting line due to the current flow, and $\vec{H}_0$ will vary with $I(t)$. Then according to Eq. (2.3), this time-varying $\vec{H}_0$ will induce an electric field $\vec{E}_0$ as shown in Fig. 2.6. Next from Eq. (2.6), this induced electric field

Fig. 2.6 Illustrating
generation of EM waves (II)

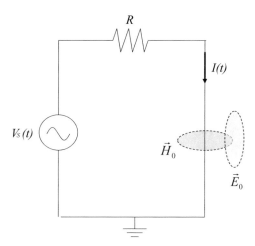

Fig. 2.7 Illustrating
generation of EM waves (III)

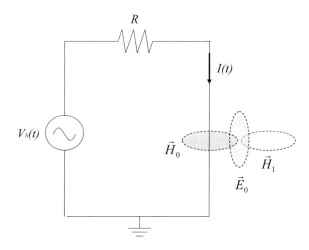

$\vec{E}_0$, which is also time varying with $\vec{H}_0$, will induce another magnetic field $\vec{H}_1$ as
shown in Fig. 2.7. Similarly, this new magnetic field $\vec{H}_1$ will induce another electric
field $\vec{E}_1$. And this induced electric field $\vec{E}_1$ will induce another magnetic field, and
so on. Figure 2.8 shows the most important characteristic of dynamic EM fields: a
time-varying electric field and a time-varying magnetic field will mutually interact
and naturally propagate outward. These dynamic E-fields and H-fields constitute an
EM wave.

When Maxwell proposed the correction term $\epsilon \cdot \partial \vec{E} / \partial t$ for the original Ampere's
law in Eq. (2.4), he already imagined the scenarios of Figs. 2.5, 2.6, 2.7, 2.8 and
predicted the existence of EM wave radiation. This prediction had not been proved
for more than 20 years until a milestone experiment done by the German scientist

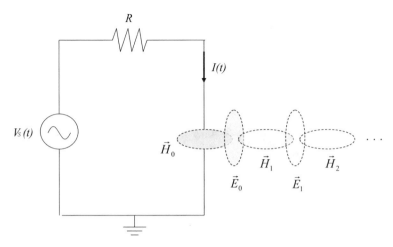

Fig. 2.8 Illustrating generation of EM waves (IV)

H. R. Hertz. This achievement extends electromagnetism from a "static field" to a "dynamic field" in both time and space!

B   Wave Equations

In this section, we utilize Maxwell's equations to derive a mathematical formula describing EM waves in Figs. 2.5, 2.6, 2.7, and 2.8. What we finally get is called **wave equation**. Since EM waves mainly propagate in a dielectric medium, we consider it and derive associated wave equations in the following.

First, we start from Eq. (2.3) and take curl on both sides. We get

$$\nabla \times \left(\nabla \times \vec{E} \right) = \nabla \times \left(-\mu \frac{\partial \vec{H}}{\partial t} \right) = -\mu \frac{\partial}{\partial t} \left(\nabla \times \vec{H} \right). \tag{2.7}$$

For an arbitrary vector field $\vec{A}$, the following equality always holds:

$$\nabla \times \left(\nabla \times \vec{A} \right) = \nabla \left(\nabla \cdot \vec{A} \right) - \nabla^2 \vec{A}, \tag{2.8}$$

where ∇ denotes **gradient** and ∇^2 denotes **Laplace operator**. The mathematical meanings of ∇ and ∇^2 are not our points here, so we skip discussion of their meanings temporarily and utilize Eq. (2.8) directly. Thus, Eq. (2.7) can be rewritten as

$$\nabla \left(\nabla \cdot \vec{E} \right) - \nabla^2 \vec{E} = -\mu \frac{\partial}{\partial t} \left(\nabla \times \vec{H} \right). \tag{2.9}$$

Assume the dielectric medium has current density $\vec{J} = 0$ and charge density $\rho = 0$. From Ampere's law and Gauss's law, we have

$$\nabla \times \vec{H} = \epsilon \frac{\partial \vec{E}}{\partial t}, \tag{2.10}$$

$$\nabla \cdot \vec{D} = \rho = 0 \quad \Rightarrow \quad \nabla \cdot \vec{E} = 0. \tag{2.11}$$

Finally, inserting Eqs. (2.10) and (2.11) into Eq. (2.9), we have

$$\nabla^2 \vec{E} = \mu \epsilon \frac{\partial^2 \vec{E}}{\partial t^2}. \tag{2.12}$$

Equation (2.12) is the **wave equation of electric field** $\vec{E}$. As will be clear later, the solution of Eq. (2.12) reveals the propagation of EM waves. Note that wave equation is a vector equation.

It is worth to mention that when we take curl on both sides of Eq. (2.3), we attain $\nabla \times \vec{H}$ in Eq. (2.7). Then we utilize Eq. (2.10) to replace $\nabla \times \vec{H}$ by $\epsilon \cdot \frac{\partial \vec{E}}{\partial t}$. It results in a partial differential equation containing only electric field $\vec{E}$ as shown in Eq. (2.12). This derivation can be regarded as a mathematical technique to eliminate the unknown vector field $\vec{H}$.

The wave equation in Eq. (2.12) is very important since it completely describes a radiated electric field which may vary with time $\left(\partial^2 \vec{E} / \partial t^2 \right)$ and space $\left(\nabla^2 \vec{E} \right)$. Like Newton's laws of motion, which laid the foundation of classical mechanics, wave equation laid the foundation of dynamic electromagnetism. Readers shall pay more attention to it.

Now, we are going to give the definition of Laplace operator ∇^2 mathematically. Suppose $f(x, y, z)$ is a function of x, y, and z. Then $\nabla^2 f$ is defined as

$$\nabla^2 f = \frac{\partial^2 f}{\partial x^2} + \frac{\partial^2 f}{\partial y^2} + \frac{\partial^2 f}{\partial z^2}. \tag{2.13}$$

Hence, $\nabla^2 f$ is a scalar, which is the sum of all the second-order partial derivatives of $f(x, y, z)$ over x, y, and z.

Furthermore, Laplace operator can be used for vectors. Suppose $\vec{A}$ is a vector given by

$$\vec{A} = A_x \cdot \hat{x} + A_y \cdot \hat{y} + A_z \cdot \hat{z}, \tag{2.14}$$

where A_x, A_y, A_z are three axial components of $\vec{A}$ in space. They are scalars and functions of x, y, and z, denoted by $A_x = A_x(x, y, z)$, $A_y = A_y(x, y, z)$, and $A_z = A_z(x, y, z)$. The Laplace operator of $\vec{A}$ is defined by

$$\nabla^2 \vec{A} = \nabla^2 A_x \cdot \hat{x} + \nabla^2 A_y \cdot \hat{y} + \nabla^2 A_z \cdot \hat{z}. \tag{2.15}$$

Hence, $\nabla^2 \vec{A}$ is still a vector with components $\nabla^2 A_x$, $\nabla^2 A_y$, and $\nabla^2 A_z$ given by

$$\nabla^2 A_x = \frac{\partial^2 A_x}{\partial x^2} + \frac{\partial^2 A_x}{\partial y^2} + \frac{\partial^2 A_x}{\partial z^2}, \tag{2.16}$$

$$\nabla^2 A_y = \frac{\partial^2 A_y}{\partial x^2} + \frac{\partial^2 A_y}{\partial y^2} + \frac{\partial^2 A_y}{\partial z^2}, \tag{2.17}$$

$$\nabla^2 A_z = \frac{\partial^2 A_z}{\partial x^2} + \frac{\partial^2 A_z}{\partial y^2} + \frac{\partial^2 A_z}{\partial z^2}. \tag{2.18}$$

After learning the definition of Laplace operator ∇^2, let us go back to wave equation in Eq. (2.12). First, let $\vec{E}$ be a vector field given by

$$\vec{E} = E_x \cdot \hat{x} + E_y \cdot \hat{y} + E_z \cdot \hat{z}, \tag{2.19}$$

where E_x, E_y, and E_z are functions of x, y, z, and t. Thus, $E_x = E_x(x, y, z, t)$, $E_y = E_y(x, y, z, t)$, and $E_z = E_z(x, y, z, t)$. Hence, they all depend on position and time.

Next, from Eqs. (2.19) and Eq. (2.15), we have

$$\nabla^2 \vec{E} = \nabla^2 E_x \cdot \hat{x} + \nabla^2 E_y \cdot \hat{y} + \nabla^2 E_z \cdot \hat{z}. \tag{2.20}$$

On the other hand, from the definition of second-order partial derivative, we have

$$\frac{\partial^2 \vec{E}}{\partial t^2} = \frac{\partial^2 E_x}{\partial t^2} \cdot \hat{x} + \frac{\partial^2 E_y}{\partial t^2} \cdot \hat{y} + \frac{\partial^2 E_z}{\partial t^2} \cdot \hat{z}. \tag{2.21}$$

Inserting Eqs. (2.20) and (2.21) into Eq. (2.12), the wave equation becomes

$$\nabla^2 E_x \cdot \hat{x} + \nabla^2 E_y \cdot \hat{y} + \nabla^2 E_z \cdot \hat{z} = \mu \in \left(\frac{\partial^2 E_x}{\partial t^2} \cdot \hat{x} + \frac{\partial^2 E_y}{\partial t^2} \cdot \hat{y} + \frac{\partial^2 E_z}{\partial t^2} \cdot \hat{z} \right). \tag{2.22}$$

Because the above vector equation holds along each axis, we obtain the following three independent equations:

$$\nabla^2 E_x = \mu \in \frac{\partial^2 E_x}{\partial t^2}, \tag{2.23}$$

$$\nabla^2 E_y = \mu \in \frac{\partial^2 E_y}{\partial t^2}, \tag{2.24}$$

$$\nabla^2 E_z = \mu \in \frac{\partial^2 E_z}{\partial t^2}. \tag{2.25}$$

Because Eqs. (2.23), (2.24), and (2.25) are independent equations, each component electric field E_x, E_y, and E_z can be derived separately in each equation.

Finally, since Eqs. (2.23), (2.24), (2.25) are scalar equations, their solutions shall be simpler than that of Eq. (2.12). However, these equations are still second-order partial differential equations and it may take much effort to derive the associated solutions. In the following section, we will utilize a mathematical technique to simplify wave equations. The technique is based on phasor concept, which is widely used in electrical engineering. Utilizing phasor concept, the derivation of solutions in Eqs. (2.23), (2.24), (2.25) will become much simpler.

Example 2.1

Suppose a scalar function $f(x, y, z) = 3x^2y + 4z^3$. Please derive $\nabla^2 f$ at the point $(2, 3, -1)$.

Solution

From Eq. (2.13), we have

$$\nabla^2 f = \frac{\partial^2 f}{\partial x^2} + \frac{\partial^2 f}{\partial y^2} + \frac{\partial^2 f}{\partial z^2} = 6y + 24z.$$

Hence, $\nabla^2 f$ at the point $(2, 3, -1)$ is given by

$$\nabla^2 f = 6 \cdot 3 + 24 \cdot (-1) = -6.$$

∎

Example 2.2

Suppose a vector field $\vec{A} = 4xy^2 \cdot \hat{x} + 2y^3 \cdot \hat{y} + (z^2 + xy) \cdot \hat{z}$. Please derive $\nabla^2 \vec{A}$ at the point $(1, 3, 2)$.

Solution

First, we have

$$A_x = 4xy^2,$$

$$A_y = 2y^3,$$

$$A_z = z^2 + xy.$$

From Eqs. (2.16), (2.17), (2.18), we have

$$\nabla^2 A_x = \frac{\partial^2 A_x}{\partial x^2} + \frac{\partial^2 A_x}{\partial y^2} + \frac{\partial^2 A_x}{\partial z^2} = 8x,$$

$$\nabla^2 A_y = \frac{\partial^2 A_y}{\partial x^2} + \frac{\partial^2 A_y}{\partial y^2} + \frac{\partial^2 A_y}{\partial z^2} = 12y,$$

$$\nabla^2 A_z = \frac{\partial^2 A_z}{\partial x^2} + \frac{\partial^2 A_z}{\partial y^2} + \frac{\partial^2 A_z}{\partial z^2} = 2.$$

At the point $(1, 3, 2)$, we get

$$\nabla^2 A_x = 8, \quad \nabla^2 A_y = 36, \quad \nabla^2 A_z = 2.$$

Finally, from Eq. (2.15), we get

$$\nabla^2 \vec{A} = \nabla^2 A_x \cdot \hat{x} + \nabla^2 A_y \cdot \hat{y} + \nabla^2 A_z \cdot \hat{z}$$
$$= 8\hat{x} + 36\hat{y} + 2\hat{z}.$$

■

Example 2.3

Suppose an electric field $\vec{E} = E_x \cdot \hat{x}$, where $E_x = A \cdot \sin(\omega t - kz)$. Let the permeability and permittivity be μ and ϵ, respectively. Prove that $k^2 = \omega^2 \mu \, \epsilon$.

Solution

First, E_x must satisfy the wave equation in Eq. (2.23). Thus

$$\nabla^2 E_x = \mu \, \epsilon \, \frac{\partial^2 E_x}{\partial t^2},$$

Because

$$\nabla^2 E_x = \frac{\partial^2 E_x}{\partial x^2} + \frac{\partial^2 E_x}{\partial y^2} + \frac{\partial^2 E_x}{\partial z^2} = -Ak^2 \cdot \sin(\omega t - kz)$$

and

$$\frac{\partial^2 E_x}{\partial t^2} = -A\omega^2 \sin(\omega t - kz),$$

we finally get

$$-Ak^2 \sin(\omega t - kz) = -A\omega^2 \mu \in \cdot \sin(\omega t - kz)$$
$$\Rightarrow k^2 = \omega^2 \mu \in$$

It completes the proof.

■

Example 2.4

Assume $\vec{E} = E_x \cdot \hat{x}$ and $k^2 = \omega^2 \mu \in$. Prove that $E_x = A \cdot \cos(\omega t - kz + \theta)$ and $E_x = B \cdot \cos(\omega t + kz + \phi)$ both are solutions of wave equation.

Solution

First, when $E_x = A \cdot \cos(\omega t - kz + \theta)$, we have

$$\nabla^2 E_x = \frac{\partial^2 E_x}{\partial z^2} = -k^2 A \cos(\omega t - kz + \theta)$$
$$= -k^2 E_x.$$

$$\frac{\partial^2 E_x}{\partial t^2} = -\omega^2 A \cos(\omega t - kz + \theta) = -\omega^2 E_x$$

Since $k^2 = \omega^2 \mu \in$, we have

$$\nabla^2 E_x = -\omega^2 \mu \in E_x = \mu \in \frac{\partial^2 E_x}{\partial t^2}.$$

Hence, from Eq. (2.23), $E_x = A \cdot \cos(\omega t - kz + \theta)$ is the solution of wave equation.

Next, when $E_x = B \cdot \cos(\omega t + kz + \phi)$, we have

$$\nabla^2 E_x = \frac{\partial^2 E_x}{\partial z^2} = -k^2 B \cos(\omega t + kz + \phi)$$
$$= -k^2 E_x.$$

$$\frac{\partial^2 E_x}{\partial t^2} = -\omega^2 B \cos(\omega t + kz + \phi) = -\omega^2 E_x.$$

Since $k^2 = \omega^2 \mu \in$, we have

$$\nabla^2 E_x = -\omega^2 \mu \in E_x = \mu \in \frac{\partial^2 E_x}{\partial t^2}.$$

Hence, from Eq. (2.23), $E_x = B \cdot \cos(\omega t + kz + \phi)$ is the solution of wave equation.

∎

2.2   Phasor Representation

In electrical engineering, complex numbers are very useful in many areas such as electronics, circuitry, control, communications, signal processing, and EM waves. Based on complex numbers, we will introduce a useful method, called phasor representation, in this section. The method is used in EM field to represent a signal consisting of sinusoids. Here we will start from the simplest one: a sinusoid.

A   Basic Concepts

First, we show a principle result coming from linear analysis:

When a sinusoidal signal $A \cos(\omega t + \phi)$ goes through a linear system, the output is still a sinusoidal signal with identical frequency. Only the amplitude (A) and the phase (ϕ) may change after passing through the system.

Let us consider a circuit as a linear system, and an example is provided in Fig. 2.9. Suppose we have an input signal $A \cos(\omega t + \phi)$, and the output signal can be represented by $B \cos(\omega t + \theta)$. In this example, the amplitude changes from A to B and the phase changes from ϕ to θ. But the frequency ω keeps the same.

The above result is very useful for an electrical engineer because most electronic circuits are linear. Furthermore, this result naturally gives us an idea that

When a sinusoid goes into a linear circuit, we only need to focus on the amplitude and phase. The frequency may be completely ignored since it keeps the same.

This idea enables us to simplify the analysis of electronic circuits. The way to realize this idea is using phasor representation in circuit analysis. In the following context, we start from the principles of phasors and then apply the principles in circuit analysis.

First, for a complex number e^{jx}, Euler's formula tells us that it can be represented by

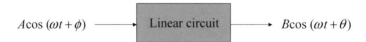

Fig. 2.9   A linear circuit with sinusoidal input signal and output signal

$$e^{jx} = \cos x + j \sin x, \tag{2.26}$$

where x is a real number and $j = \sqrt{-1}$. Note that the relationship between exponential function and trigonometric functions is established through Eq. (2.26). In Eq. (2.26), let $x = \omega t$, where ω is the frequency and t is the time. Then we have

$$e^{j\omega t} = \cos \omega t + j \sin \omega t. \tag{2.27}$$

The real part of $e^{j\omega t}$ can be represented by

$$\mathrm{Re}\{e^{j\omega t}\} = \cos \omega t. \tag{2.28}$$

Furthermore, suppose $x = \omega t + \theta$. Then we have

$$e^{j(\omega t+\theta)} = \cos(\omega t + \theta) + j \sin(\omega t + \theta). \tag{2.29}$$

Thus, the real part of $e^{j(\omega t+\theta)}$ can be represented by

$$\mathrm{Re}\{e^{j(\omega t+\theta)}\} = \cos(\omega t + \theta). \tag{2.30}$$

Equation (2.30) implies an important idea: we can represent a sinusoidal signal $\cos(\omega t + \theta)$ by a complex number $e^{j(\omega t+\theta)}$. The signal can be obtained by simply taking the real part of $e^{j(\omega t+\theta)}$.

Suppose $y(t)$ is a sinusoid given by

$$y(t) = A \cos(\omega t + \theta), \tag{2.31}$$

where A is the amplitude, ω is the frequency, and θ is the phase. Utilizing Eq. (2.30), $y(t)$ can be rewritten as

$$y(t) = \mathrm{Re}\{Ae^{j(\omega t+\theta)}\} = \mathrm{Re}\{Ae^{j\theta} \cdot e^{j\omega t}\} = \mathrm{Re}\{Y \cdot e^{j\omega t}\}, \tag{2.32}$$

where

$$Y = Ae^{j\theta}. \tag{2.33}$$

In Eq. (2.33), Y is a complex number and is fully determined by the amplitude A and the phase θ of $y(t)$. Remember when going through a linear system, only amplitude and phase of a sinusoidal signal may change. Thus, Y has all the things we need to characterize when $y(t)$ goes through a linear system.

Now, we define $Y = Ae^{j\theta}$ as the **phasor** of $y(t)$. In the representation, the phasor Y is a complex number while A is the amplitude of $y(t)$ and θ is the phase of $y(t)$. For an arbitrary sinusoidal signal $y(t) = A\cos(\omega t + \theta)$, it has a corresponding phasor Y, and Y is solely determined by the amplitude and phase of $y(t)$.

Example 2.5

Suppose we have a sinusoid $y(t) = 5 \cdot \cos(\omega t + 30°)$. Please derive the phasor of $y(t)$.

Solution

From the amplitude and phase of $y(t)$, we get the phasor given by

$$Y = 5e^{j30°}.$$

■

Example 2.6

Suppose we have a sinusoid $y(t) = 5 \cdot \sin(\omega t + 20°)$. Please derive the phasor of $y(t)$.

Solution

Because the phasor is defined on cosine function but not sine function, we have to convert $\sin \omega t$ into $\cos \omega t$. From the trigonometric formula $\sin \phi = \cos(\phi - 90°)$, we have

$$\sin(\omega t + 20°) = \cos(\omega t + 20° - 90°) = \cos(\omega t - 70°).$$

Then

$$y(t) = 5 \cdot \sin(\omega t + 20°) = 5 \cdot \cos(\omega t - 70°).$$

Hence, the phasor of $y(t)$ is given by

$$Y = 5 \cdot e^{-j70°}.$$

■

Example 2.7

Suppose we have a signal $y(t) = 5 \cdot \cos(\omega t + 70°) + 3 \cdot \cos(\omega t - 40°)$. Please derive the phasor of Y.

Solution

First, $y(t)$ consists of two sinusoids with the corresponding phasors

$$5 \cdot \cos(\omega t + 70°) \xrightarrow{\text{phasor}} 5 \cdot e^{j70°},$$

$$3 \cdot \cos(\omega t - 40°) \xrightarrow{\text{phasor}} 3 \cdot e^{-j40°}.$$

Because the phasor operation allows addition, we obtain

$$Y = 5e^{j70°} + 3 \cdot e^{-j40°}.$$

∎

From the above examples, if $y(t)$ is a sinusoid, the corresponding phasor can be readily derived. On the other hand, if the phasor Y is given, the corresponding signal $y(t)$ can be easily derived. The examples are provided below.

Example 2.8

Suppose $y(t)$ is a sinusoid with frequency ω and its phasor is given as $Y = 3e^{j80°}$. Please derive $y(t)$.

Solution

From $Y = 3e^{j80°}$, we know the amplitude of $y(t)$ is 3 and the phase is 80°. Hence, we get

$$y(t) = 3 \cdot \cos(\omega t + 80°).$$

∎

Example 2.9

Suppose $y(t)$ is a sinusoid with frequency ω and its phasor is given by $Y = 2e^{j50°} - 3e^{-j130°}$. Please derive $y(t)$.

Solution

From $Y = 2e^{j50°} - 3e^{-j130°}$, we know $y(t)$ consists of two sinusoids given by

$$2 \cdot e^{j50°} \xrightarrow{\text{phasor}} 2 \cdot \cos(\omega t + 50°),$$

$$3 \cdot e^{-j130^\circ} \quad \xrightarrow{\text{phasor}} \quad 3 \cdot \cos(\omega t - 130^\circ).$$

Hence, $y(t)$ is given by

$$y(t) = 2 \cdot \cos(\omega t + 50^\circ) - 3 \cdot \cos(\omega t - 130^\circ).$$

∎

From the above examples, we know $y(t)$ and Y are one-to-one mapping. That means, if we get one of them, the other one can be got immediately. In other words, each one of them can represent the other.

B   Differentiation of Phasors

Because of the promising property of phasor in differentiation, phasor representation is very useful in signal analysis. Suppose we have a sinusoid $y(t) = A \cos(\omega t + \theta)$ represented by

$$y(t) = \text{Re}\{Ye^{j\omega t}\}, \tag{2.34}$$

where $Y = Ae^{j\theta}$ is the phasor of $y(t)$. When we take the differentiation of Eq. (2.34) with time, we have

$$\frac{dy(t)}{dt} = \frac{d}{dt}\text{Re}\{Ye^{j\omega t}\} = \text{Re}\left\{\frac{d}{dt}Ye^{j\omega t}\right\}. \tag{2.35}$$

Because Y is not a function of t, we get

$$\frac{d}{dt}Ye^{j\omega t} = j\omega Y \cdot e^{j\omega t}. \tag{2.36}$$

Hence, Eq. (2.35) becomes

$$\frac{dy(t)}{dt} = \text{Re}\{j\omega Y \cdot e^{j\omega t}\}. \tag{2.37}$$

Equation (2.37) actually reveals an important result:

If the phasor of $y(t)$ is Y, then the phasor of its differentiation, $\frac{dy(t)}{dt}$, is $j\omega Y$.
Based on this result, we can obtain the following:

$$y(t) \quad \xrightarrow{\text{phasor}} \quad Y,$$

$$\frac{dy(t)}{dt} \xrightarrow{\text{phasor}} j\omega Y,$$

$$\frac{d^2 y(t)}{dt^2} = \frac{d}{dt}(\frac{dy(t)}{dt}) \xrightarrow{\text{phasor}} j\omega \cdot (j\omega Y) = (j\omega)^2 Y,$$

$$\frac{d^3 y(t)}{dt^3} = \frac{d}{dt}(\frac{d^2 y(t)}{dt^2}) \xrightarrow{\text{phasor}} j\omega \cdot (j\omega)^2 Y = (j\omega)^3 Y,$$

$$\ldots\ldots$$

and so on. Thus, we get a useful formula

$$\frac{d^n y(t)}{dt^n} \xrightarrow{\text{phasor}} (j\omega)^n Y, \ n = 1, \ 2, \ 3, \ldots \tag{2.38}$$

Equation (2.38) tells us that the phasor of the n-th-order differentiation of $y(t)$ simply equals to $(j\omega)^n Y$. This property can greatly simplify signal analysis.

Example 2.10

Suppose we have a sinusoid $y(t) = 5\cos(\omega t - 30°)$. Please derive the phasor of its differentiation, $\frac{dy(t)}{dt}$.

Solution

First, the phasor of $y(t)$ is given by

$$Y = 5e^{-j30°}.$$

Let Y_1 be the phasor of $\frac{dy(t)}{dt}$. Then from Eq. (2.38), we have

$$Y_1 = (j\omega) \cdot Y = j\omega \cdot 5e^{-j30°}.$$

Because $j = e^{j90°}$, Y_1 can also be represented by

$$Y_1 = e^{j90°} \cdot 5\omega \cdot e^{-j30°} = 5\omega \cdot e^{j60°}.$$

∎

Example 2.11

Suppose $y(t) = 3\sin(\omega t + 130°)$. Please derive the phasor of $\frac{d^2 y(t)}{dt^2}$.

Fig. 2.10 Illustrating a
linear (RC) circuit

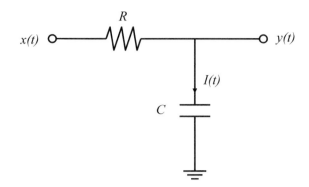

Solution

First, we convert $\sin \omega t$ into $\cos \omega t$. Thus

$$y(t) = 3\sin(\omega t + 130°) = 3\cos(\omega t + 130° - 90°) = 3\cos(\omega t + 40°).$$

Hence, the phasor of $y(t)$ is given by

$$Y = 3e^{j40°}.$$

Let Y_2 be the phasor of $\frac{d^2 y(t)}{dt^2}$. Then from Eq. (2.38), we have

$$Y_2 = (j\omega)^2 Y = -\omega^2 Y = -3\omega^2 \cdot e^{j40°}.$$

∎

C   Applications of Phasors

In this subsection, we provide two examples to illustrate the applications of phasors.
In Fig. 2.10, we have a resistor–capacitor (RC) circuit. The input voltage is a sinusoid
given by

$$x(t) = A \cdot \cos(\omega t + \phi). \tag{2.39}$$

The corresponding phasor of $x(t)$ is

$$X = Ae^{j\phi}, \tag{2.40}$$

where A is the amplitude and ϕ is the phase of $x(t)$.

In Fig. 2.10, the output voltage $y(t)$ is also a sinusoid with the same frequency as $x(t)$. Now, how do we derive $y(t)$? Basically we have two approaches to resolve this problem.

Approach 1

Given $x(t)$, we can derive $y(t)$ directly from the corresponding differential equation of the RC circuit.

Approach 2

Suppose the phasor of $x(t)$ is X and the phasor of $y(t)$ is Y. Then we can convert our problem from

$$x(t) \xrightarrow{\text{RC circuit}} y(t)$$

to a simple form:

$$X \xrightarrow{\text{RC circuit}} Y.$$

In other words, we convert the problem of "deriving $y(t)$ from $x(t)$" to the problem of "deriving phasor Y from phasor X".

Apparently, Approach 2 will be much simpler than Approach 1 because

- In Approach 1, both $x(t)$ and $y(t)$ are functions of time. On the other hand, in Approach 2 both **phasor X and phasor Y are not functions of time.**
- By utilizing the simple differentiation property of phasor in Eq. (2.38), we can easily derive Y from X.

In order to clarify the above points, we deal with the problem separately using Approach 1 and Approach 2 in the following. Readers can compare their difference to learn the convenience of using phasors.

Approach 1

We apply Kirchhoff's law in Fig. 2.10 to have

$$\frac{x(t) - y(t)}{R} = I(t) = C\frac{dy(t)}{dt}. \tag{2.41}$$

Then we have

$$RC\frac{dy(t)}{dt} + y(t) = x(t). \tag{2.42}$$

Next, we insert $x(t) = A \cdot \cos(\omega t + \phi)$ into Eq. (2.42) to obtain

$$RC\frac{dy(t)}{dt} + y(t) = A\cos(\omega t + \phi). \tag{2.43}$$

Equation (2.43) is a first-order differential equation. Although $y(t)$ can be derived in Eq. (2.43), the process is somewhat complicated and is skipped here.

Approach 2

Like Approach 1, from Kirchhoff's law, we have

$$RC\frac{dy(t)}{dt} + y(t) = x(t). \tag{2.44}$$

Next, we convert Eq. (2.44) into phasor representation by applying

$$x(t) \xrightarrow{phasor} X,$$

$$y(t) \xrightarrow{phasor} Y,$$

$$\frac{dy(t)}{dt} \xrightarrow{phasor} j\omega Y.$$

Then we have

$$RC(j\omega Y) + Y = X. \tag{2.45}$$

From Eq. (2.45), it is easy to get the following:

$$Y = \frac{X}{1 + j\omega RC} = \frac{Ae^{j\phi}}{1 + j\omega RC}. \tag{2.46}$$

Since a complex number $z = a + jb$ can be represented by

$$z = \sqrt{a^2 + b^2} \cdot e^{j\varphi}, \tag{2.47}$$

where φ is given by

$$\varphi = \tan^{-1}(\frac{b}{a}). \tag{2.48}$$

Utilizing Eqs. (2.47) and (2.48), we have

$$1 + j\omega RC = \sqrt{1 + (\omega RC)^2} \cdot e^{j\alpha}, \tag{2.49}$$

where α is given by

$$\alpha = \tan^{-1}(\omega RC). \tag{2.50}$$

Fig. 2.11 An RLC circuit for Example 2.12

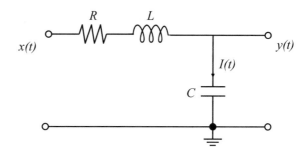

Hence, Eq. (2.46) can be rewritten as

$$Y = \frac{A}{\sqrt{1 + (\omega RC)^2}} \cdot e^{j(\phi - \alpha)}. \tag{2.51}$$

We can represent Y by

$$Y = B e^{j\theta}, \tag{2.52}$$

where

$$B = \frac{A}{\sqrt{1 + (\omega RC)^2}}, \tag{2.53}$$

$$\theta = \phi - \alpha = \phi - \tan^{-1}(\omega RC). \tag{2.54}$$

Finally, the output signal $y(t)$ is readily given by

$$y(t) = \mathrm{Re}\{Y e^{j\omega t}\} = B \cos(\omega t + \theta). \tag{2.55}$$

From the above example, we can easily derive the phasor Y from the phasor X. Then $y(t)$ is readily obtained from Y. Notice that we do not need to solve the differential equation in Approach 2 as we do need in Approach 1. Besides, Fig. 2.10 is just a simple RC circuit. As circuits become more complicated, the effectiveness and convenience of phasor will become more significant.

Example 2.12

In Fig. 2.11, we have an RLC circuit with an input voltage $x(t) = A \cdot \cos(\omega t + \phi)$. Please use phasors to derive the output voltage $y(t)$.

Solution

First, from Kirchhoff's law, we have

$$x(t) = R \cdot I(t) + L\frac{dI(t)}{dt} + y(t),$$

where $I(t)$ is the current given by

$$I(t) = C\frac{dy(t)}{dt}.$$

From the above two equations, we have

$$x(t) = RC\frac{dy(t)}{dt} + LC\frac{d^2y(t)}{dt^2} + y(t).$$

Next, we convert the above differential equation through phasor representation:

$$x(t) \quad \xrightarrow{\text{phasor}} \quad X$$

$$y(t) \quad \xrightarrow{\text{phasor}} \quad Y$$

$$\frac{dy(t)}{dt} \quad \xrightarrow{\text{phasor}} \quad j\omega Y$$

$$\frac{d^2y(t)}{dt^2} \quad \xrightarrow{\text{phasor}} \quad (j\omega)^2 Y = -\omega^2 Y.$$

Then, we get the following phasor equation:

$$X = RC(j\omega Y) + LC(-\omega^2 Y) + Y = [(1 - \omega^2 LC) + j\omega RC] \cdot Y.$$

Hence

$$Y = \frac{X}{(1 - \omega^2 LC) + j\omega RC} = \frac{Ae^{j\phi}}{(1 - \omega^2 LC) + j\omega RC}.$$

The denominator can be rewritten as

$$(1 - \omega^2 LC) + j\omega RC = \sqrt{(1 - \omega^2 LC)^2 + (\omega RC)^2} \cdot e^{j\alpha},$$

where

$$\alpha = \tan^{-1}\left(\frac{\omega RC}{1 - \omega^2 LC}\right).$$

Suppose $Y = Be^{j\theta}$, then

$$B = \frac{A}{\sqrt{(1 - \omega^2 LC)^2 + (\omega RC)^2}},$$

$$\theta = \phi - \alpha = \phi - \tan^{-1}\left(\frac{\omega RC}{1 - \omega^2 LC}\right).$$

Since Y is derived, we readily obtain $y(t)$ as

$$y(t) = \text{Re}\{Ye^{j\omega t}\} = B\cos(\omega t + \theta).$$

∎

From Example 2.12, we can see when the number of circuit components increases, the differential equation of circuit analysis will become more complicated. In this case, it is difficult to derive the solution from Approach 1. Fortunately, with the aid of phasors, it becomes much easier to derive the solution. In the following example, we will take the first step to see how phasor representation can help us simplify EM analysis.

Example 2.13

We have learned that Maxwell's equations are given by

$$\nabla \times \vec{E} = -\frac{\partial \vec{B}}{\partial t}, \tag{2.56}$$

$$\nabla \times \vec{H} = \vec{J} + \frac{\partial \vec{D}}{\partial t}, \tag{2.57}$$

$$\nabla \cdot \vec{D} = \rho, \tag{2.58}$$

$$\nabla \cdot \vec{B} = 0. \tag{2.59}$$

Please represent Maxwell's equations using phasors.

Solution

Since the left side of these four equations does not involve the differentiation in time, we only need to take care of the right side in Eq. (2.56) and Eq. (2.57). We convert the following two components through phasor representation:

$$\frac{\partial \vec{B}}{\partial t} \xrightarrow{phasor} j\omega\vec{B},$$

$$\frac{\partial \vec{D}}{\partial t} \xrightarrow{\text{phasor}} j\omega\vec{D}.$$

Thus, we have phasor representation of Maxwell's equations given by

$$\nabla \times \vec{E} = -j\omega\vec{B}, \tag{2.60}$$

$$\nabla \times \vec{H} = \vec{J} + j\omega\vec{D}, \tag{2.61}$$

$$\nabla \cdot \vec{D} = \rho, \tag{2.62}$$

$$\nabla \cdot \vec{B} = 0. \tag{2.63}$$

∎

As we compare Eqs. (2.56)–(2.59) with Eqs. (2.60)–(2.63), we can find the equations are greatly simplified because differentiation in time disappears. The advantage of phasor expression can be readily appreciated in the simplification of Maxwell's equations.

Finally, before ending this section, readers may have a question in mind:

Q: Phasor may be an efficient method when the input signal is a sinusoid. What if the input signal is not a sinusoid?

Of course, phasor representation only applies to sinusoids. Fortunately, from Fourier transform, it can be proved that all the signals can be represented by a sum of sinusoids. For example, suppose we have a square signal getting into a circuit. Then from Fourier transform, this signal can be equivalently regarded as a combination of sinusoids. Because each sinusoid can be represented by a respective phasor, the corresponding output can be individually derived. Finally, we sum the outputs of all the sinusoids to get the resulting output signal. Therefore, phasors and Fourier transform are very useful analytical tools for all the realistic signals.

In this section, we mainly focus on the applications of phasors in linear circuits. In the next section, we will see its applications in electromagnetics. Specifically, we use it to derive the solutions of wave equations and then explore the insight of EM waves.

2.3 Uniform Plane Wave

Based on Faraday's law and Ampere's law, we understand that EM wave will propagate. But we still don't get the details like: how do electric fields or magnetic fields vary with time and with space? How fast can they propagate? How do they interact mutually? … etc. The solutions of wave equations in Sect. 2.1 will provide answers for these questions. In this section, we will solve wave equations in order to understand the properties of EM wave.

A Uniform Plane Wave

First, from Sect. 2.1, the wave equation of a propagating electric field is given by

$$\nabla^2 \vec{E} = \mu \in \frac{\partial^2 \vec{E}}{\partial t^2}. \tag{2.64}$$

In Eq. (2.64), the electric field $\vec{E}$ is denoted by

$$\vec{E} = E_x \cdot \hat{x} + E_y \cdot \hat{y} + E_z \cdot \hat{z}, \tag{2.65}$$

where E_x, E_y, E_z are axial components along x, y, and z, respectively. They are functions of space coordinate x, y, z, and time t. Hence, they can be represented by $E_x = E_x(x, y, z, t)$, $E_y = E_y(x, y, z, t)$, and $E_z = E_z(x, y, z, t)$.

As we take a look at Eq. (2.64), it is a second-order linear partial differential equation. Besides, $\vec{E}$ has three components in space. Hence, it is difficult to solve Eq. (2.64). In order to simplify the problem, we make two assumptions as follows:

1. The EM wave of concern propagates along z-axis, and the electric field only has y-component. That is, $\vec{E} = E_y \cdot \hat{y}$.
2. E_y depends only on z and t. That is, $E_y = E_y(z, t)$.

Under the above assumptions, it will be much easier to solve wave equation since the associated EM waves can be formulated in a simple form: its electric field only has y-component and E_y is a function of z and t. Because E_y does not depend on x or y, the electric fields of each point on an xy-plane are all identical. This kind of wave is called **uniform plane wave**. Uniform plane wave is the most important and fundamental EM wave which can be applied to analyze complicated EM waves. For convenience, uniform plane wave will be abbreviated as plane wave throughout the context.

Figure 2.12 is a schematic plot of plane wave. For each xy-plane, the wave propagates at a fixed speed along z-axis. Hence, for a plane at $z = z_1$, all the electric fields have identical magnitude and phase on this plane. However, two electric fields on different planes, e.g., $z = z_1$ plane and $z = z_2$ plane, may differ. Readers may keep Fig. 2.12 in mind because it provides a good example to understand the propagation of EM wave. Now, we start to solve the wave equation.

Because $\vec{E} = E_y \cdot \hat{y}$, Eq. (2.64) can be simplified as

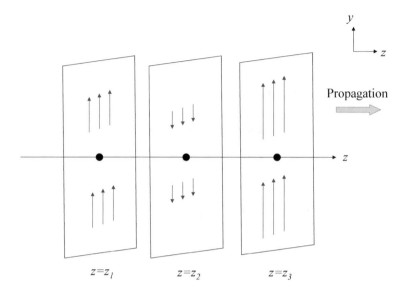

Fig. 2.12 Schematic plot of plane wave

$$\nabla^2 E_y = \mu \in \frac{\partial^2 E_y}{\partial t^2}. \tag{2.66}$$

Since E_y does not depend on x or y, we have

$$\nabla^2 E_y = \frac{\partial^2 E_y}{\partial x^2} + \frac{\partial^2 E_y}{\partial y^2} + \frac{\partial^2 E_y}{\partial z^2} = \frac{\partial^2 E_y}{\partial z^2}. \tag{2.67}$$

Inserting Eq. (2.67) into Eq. (2.66), we have

$$\frac{\partial^2 E_y}{\partial z^2} = \mu \in \frac{\partial^2 E_y}{\partial t^2}. \tag{2.68}$$

Equation (2.68) is a scalar equation which is simpler than the vector one in Eq. (2.64). However, it is still a second-order partial differential equation and deriving the solution is a complicated task. Then, how can we further simplify the equation?

We use the phasor method! Suppose E_y is a sinusoidal electric field with frequency ω. The corresponding phasor is represented by $\hat{E}_y$. Note that although E_y is a function of z and t, **the corresponding phasor $\hat{E}_y$ is only a function of z.** Hence, $\hat{E}_y$ is not a function of t and can be simply represented by $\hat{E}_y = \hat{E}_y(z)$. This is why we can further simplify our analysis when solving wave equation.

Next, for Eq. (2.68), we utilize the differential property of phasors as follows:

$$E_y \xrightarrow{\;phasor\;} \hat{E}_y,$$

$$\frac{\partial^2 E_y}{\partial t^2} \xrightarrow{phasor} (j\omega)^2 \hat{E}_y.$$

Hence, Eq. (2.68) can be represented by

$$\frac{\partial^2 \hat{E}_y}{\partial z^2} = \mu \in (j\omega)^2 \hat{E}_y = -\omega^2 \mu \in \hat{E}_y. \tag{2.69}$$

The corresponding wave equation of $\hat{E}_y$ is given by

$$\frac{d^2 \hat{E}_y}{dz^2} + k^2 \hat{E}_y = 0, \tag{2.70}$$

where

$$k^2 = \omega^2 \mu \in . \tag{2.71}$$

Obviously, Eq. (2.70) is much simpler than Eq. (2.68) because $\hat{E}_y$ is merely a function of z.

B Solutions of Wave Equations

For the second-order linear differential equation in Eq. (2.70), we have two independent solutions. The first one is given by

$$\hat{E}_y(z) = E_a e^{-jkz}, \tag{2.72}$$

where E_a is a constant corresponding to $\hat{E}_y$ at $z = 0$, and $j = \sqrt{-1}$. In Eq. (2.72), the electric field $\hat{E}_y$ varies with e^{-jkz} along z. As will become clear later, Eq. (2.72) represents a plane wave called **forward wave** which propagates toward $+z$ direction.

Next, the second solution of Eq. (2.70) is given by

$$\hat{E}_y(z) = E_b e^{jkz}, \tag{2.73}$$

where E_b is a constant corresponding to $\hat{E}_y$ at $z = 0$. In Eq. (2.73), the electric field $\hat{E}_y$ varies with e^{jkz} along z. It will be shown later that Eq. (2.73) is a plane wave called **backward wave** which propagates toward $-z$ direction.

From above, a wave equation has two independent solutions in general. These two solutions represent forward wave and backward wave, respectively. **The forward wave propagates toward +z direction, while the backward wave travels toward**

−z direction. Note that a forward wave varies with e^{-jkz}, and a backward wave varies with e^{jkz} along z.

In Eq. (2.72) or Eq. (2.73), the solution $\hat{E}_y$ is a phasor and the corresponding time-dependent electric field $E_y(z, t)$ can be easily got from $\hat{E}_y$. In Eq. (2.72), E_a is the phasor of a forward wave at $z = 0$ and it is, in general, a complex number. Suppose

$$E_a = Ae^{j\theta}, \tag{2.74}$$

where $A = |E_a|$ denotes the **magnitude** of E_a and θ denotes the **phase** of E_a. From Eq. (2.72) and Eq. (2.74), we have

$$\hat{E}_y(z) = E_a e^{-jkz} = Ae^{j\theta} \cdot e^{-jkz}. \tag{2.75}$$

From Sect. 2.2, we learned that the time-dependent electric field $E_y(z, t)$ and the corresponding phasor $\hat{E}_y(z)$ have the following relationship:

$$E_y(z, t) = \mathrm{Re}\left\{\hat{E}_y(z)e^{j\omega t}\right\}. \tag{2.76}$$

Inserting Eq. (2.75) into Eq. (2.76), we have

$$\begin{aligned}
E_y(z, t) &= \mathrm{Re}\{Ae^{j\theta} \cdot e^{-jkz} \cdot e^{j\omega t}\} = \mathrm{Re}\left\{Ae^{j(\omega t - kz + \theta)}\right\} \\
&= A\cos(\omega t - kz + \theta).
\end{aligned} \tag{2.77}$$

Because $\vec{E}(z, t) = E_y(z, t) \cdot \hat{y}$, we finally get

$$\vec{E}(z, t) = A\cos(\omega t - kz + \theta) \cdot \hat{y}. \tag{2.78}$$

The above example tells us that the electric field $E_y(z, t)$ can be easily derived from its phasor $\hat{E}_y(z)$.

Similarly, in Eq. (2.73), E_b is the phasor of a backward wave at $z = 0$. Suppose

$$E_b = Be^{j\phi}, \tag{2.79}$$

where $B = |E_b|$ denotes the magnitude of E_b and ϕ denotes the associated phase. From Eq. (2.73) and Eq. (2.79), we have

$$\hat{E}_y = E_b e^{jkz} = Be^{j\phi} \cdot e^{jkz}. \tag{2.80}$$

Therefore, the time-dependent electric field $E_y(z, t)$ is given by

$$E_y(z, t) = \mathrm{Re}\left\{\hat{E}_y(z) \cdot e^{j\omega t}\right\} = B\cos(\omega t + kz + \phi). \tag{2.81}$$

Fig. 2.13 Illustrating EM
waves for a given time

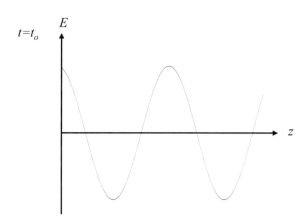

Finally, we have

$$\vec{E}(z, t) = B \cos(\omega t + kz + \phi) \cdot \hat{y}. \tag{2.82}$$

From above, once the associated phasor is known, the corresponding time-dependent electric field is readily derived. Hence, phasor representation is a useful tool for solving electromagnetic problems. In the remaining context, readers will find that when analyzing EM wave, we usually use phasors representing electric fields or magnetic fields. By doing this, we not only simplify our analysis, but also help ourselves access the key of EM waves. Therefore, it is important for readers to practice and be familiar with phasors.

Next, in order to help readers understand the waveforms of forward wave and backward wave, we consider the formulas in Eq. (2.78) and Eq. (2.82), respectively. First, we consider backward wave in Eq. (2.82). Imagine what the EM wave will look like when the time freezes. Let $t = t_0$ in Eq. (2.82), and then we have

$$\vec{E}(z, t_0) = B \cos(kz + \omega t_0 + \phi) \cdot \hat{y}. \tag{2.83}$$

Because t is fixed, the electric field is simply a function of z. Thus, the EM wave forms a sinusoid as shown in Fig. 2.13. On the other hand, when we observe this EM wave at a fixed point $z = z_0$ in space, from Eq. (2.82), the EM wave can be represented by

$$\vec{E}(z_0, t) = B \cos(\omega t + kz_0 + \phi) \cdot \hat{y}. \tag{2.84}$$

Because z is fixed, the electric field is simply a function of time t. If we stand at $z = z_0$ and observe the wave for a while, we will find the waveform is also a sinusoid as shown in Fig. 2.14.

Similarly, from Eq. (2.78), the forward wave at a fixed time $t = t_0$ is given by

Fig. 2.14 Illustrating EM waves for a given position

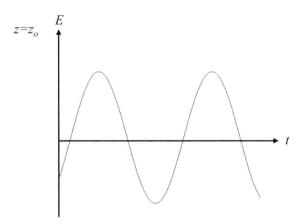

$$\vec{E}(z, t_0) = A \cos(\omega t_0 - kz + \theta) \cdot \hat{y}. \tag{2.85}$$

And the forward wave at a fixed point $z = z_0$ is given by

$$\vec{E}(z_0, t) = A \cos(\omega t - kz_0 + \theta) \cdot \hat{y}. \tag{2.86}$$

Hence, in both cases, the forward wave forms a sinusoid too.

From the above results, we know that for a single-frequency plane wave, the waveform at any time or at any point is a sinusoid. This feature reveals the usefulness of phasor representation since all the practical EM waves are composed of plane waves.

Finally, it is worth to mention that from Maxwell's equations we obtain the wave equation in Eq. (2.64). Then, we get plane waves as represented in Eq. (2.78) and Eq. (2.82) by solving wave equation. From these mathematical representations, we can see the properties of EM wave: how it distributes in space and how it varies with time. Without Maxwell's equations, it is hard to imagine the waveform of EM waves!

Example 2.14

Prove that (a). $\hat{E}_y = ae^{-jkz}$ (b). $\hat{E}_y = be^{jkz}$ (c). $\hat{E}_y = ae^{-jkz} + be^{jkz}$, all are solutions for Eq. (2.70), where a and b are arbitrary complex numbers.

Solution

(a). Let $\hat{E}_y = ae^{-jkz}$.

Then

$$\frac{d^2 \hat{E}_y}{dz^2} = (-jk)^2 \cdot ae^{-jkz} = -k^2 \hat{E}_y.$$

Hence

$$\frac{d^2 \hat{E}_y}{dz^2} + k^2 \hat{E}_y = -k^2 \hat{E}_y + k^2 \hat{E}_y = 0.$$

It means that $\hat{E}_y = ae^{-jkz}$ is a solution of Eq. (2.70).

(b). Let $\hat{E}_y = be^{jkz}$.

Then

$$\frac{d^2 \hat{E}_y}{dz^2} = (jk)^2 \cdot be^{jkz} = -k^2 \hat{E}_y.$$

Hence

$$\frac{d^2 \hat{E}_y}{dz^2} + k^2 \hat{E}_y = -k^2 \hat{E}_y + k^2 \hat{E}_y = 0.$$

It means that $\hat{E}_y = be^{jkz}$ is a solution of Eq. (2.70).

(c). Let $\hat{E}_y = ae^{-jkz} + be^{jkz}$.

Then

$$\frac{d^2 \hat{E}_y}{dz^2} = (-jk)^2 \cdot ae^{-jkz} + (jk)^2 \cdot be^{jkz} = -k^2 \hat{E}_y.$$

Hence

$$\frac{d^2 \hat{E}_y}{dz^2} + k^2 \hat{E}_y = -k^2 \hat{E}_y + k^2 \hat{E}_y = 0.$$

It means that $\hat{E}_y = ae^{-jkz} + be^{jkz}$ is a solution of Eq. (2.70).

∎

Example 2.15

A forward plane wave has frequency ω and its phasor of electric field is $\vec{E}(z) = E_a e^{-jkz} \cdot \hat{y}$, where E_a is the phasor of the electric field at $z = 0$ and $k = \frac{2\pi}{9}$ (1/m). Please derive the electric field at $z = 15$ m in the following three cases:

(a). $E_a = 5$.
(b). $E_a = 5 \cdot e^{j\frac{\pi}{3}}$
(c). $E_a = 10 \cdot e^{-j\frac{\pi}{4}}$.

Solution

(a).   The phasor of E-field at $z = 15$ m is

$$\hat{E}(z) = E_a \cdot e^{-j\frac{2}{9}\pi \times 15} = 5 \cdot e^{-j\frac{10}{3}\pi} = 5 \cdot e^{j\frac{2}{3}\pi}.$$

Then the real E-field is

$$\vec{E}(z, t) = \text{Re}\{\hat{E}(z)e^{j\omega t}\} \cdot \hat{y} = 5\cos(\omega t + \frac{2}{3}\pi) \cdot \hat{y}.$$

(b).   The phasor of E-field at $z = 15$ m is

$$\hat{E}(z) = E_a \cdot e^{-j\frac{2}{9}\pi \times 15} = 5e^{j\frac{\pi}{3}} \cdot e^{-j\frac{10}{3}\pi} = 5 \cdot e^{-j\pi} = -5.$$

Then the real E-field is

$$\vec{E}(z, t) = \text{Re}\{\hat{E}(z)e^{j\omega t}\} \cdot \hat{y} = -5\cos\omega t \cdot \hat{y}.$$

(c).   The phasor of E-field at $z = 15$ m is

$$\hat{E}(z) = E_a \cdot e^{-j\frac{2}{9}\pi \times 15} = 10e^{-j\frac{\pi}{4}} \cdot e^{-j\frac{10}{3}\pi} = 10 \cdot e^{-j\frac{43}{12}\pi} = 10 \cdot e^{j\frac{5}{12}\pi}.$$

Then the real E-field is

$$\vec{E}(z, t) = \text{Re}\{\hat{E}(z)e^{j\omega t}\} \cdot \hat{y} = 10\cos(\omega t + \frac{5}{12}\pi) \cdot \hat{y}.$$

∎

Example 2.16

A backward plane wave has frequency ω and the phasor of its electric field is $\vec{E}(z) = E_b e^{jkz} \cdot \hat{y}$, where E_b is the phasor of the electric field at $z = 0$ and $k = \frac{2\pi}{5}$ (1/m). Please derive the electric field at $z = 18$ m in the following three cases:

(a). $E_b = 3$.
(b). $E_b = 3 \cdot e^{j\frac{\pi}{6}}$.
(c). $E_b = 4 \cdot e^{-j\frac{\pi}{2}}$.

Solution

(a). The phasor of E-field at $z = 18$ m is

$$\hat{E}(z) = E_b \cdot e^{j\frac{2}{5}\pi \times 18} = 3 \cdot e^{j\frac{36}{5}\pi} = 3 \cdot e^{j\frac{6}{5}\pi}.$$

Then the real E-field is

$$\vec{E}(z, t) = \text{Re}\{\hat{E}(z)e^{j\omega t}\} \cdot \hat{y} = 3\cos(\omega t + \frac{6}{5}\pi) \cdot \hat{y}.$$

(b). The phasor of E-field at $z = 18$ m is

$$\hat{E}(z) = E_b \cdot e^{j\frac{2}{5}\pi \times 18} = 3e^{j\frac{\pi}{6}} \cdot e^{j\frac{36}{5}\pi} = 3 \cdot e^{j\frac{221}{30}\pi} = 3 \cdot e^{j\frac{41}{30}\pi}.$$

Then the real E-field is

$$\vec{E}(z, t) = \text{Re}\{\hat{E}(z)e^{j\omega t}\} \cdot \hat{y} = 3\cos(\omega t + \frac{41}{30}\pi) \cdot \hat{y}.$$

(c). The phasor of E-field at $z = 18$ m is

$$\hat{E}(z) = E_b \cdot e^{j\frac{2}{5}\pi \times 18} = 4e^{-j\frac{\pi}{2}} \cdot e^{j\frac{36}{5}\pi} = 4 \cdot e^{j\frac{67}{10}\pi} = 4 \cdot e^{j\frac{7}{10}\pi}.$$

Then the real E-field is

$$\vec{E}(z, t) = \text{Re}\{\hat{E}(z)e^{j\omega t}\} \cdot \hat{y} = 4\cos(\omega t + \frac{7}{10}\pi) \cdot \hat{y}.$$

∎

Example 2.17

An EM wave consists of a forward wave and a backward wave having frequency ω. The phasor of its electric field is given by $\vec{E}(z) = (E_a e^{-jkz} + E_b e^{jkz}) \cdot \hat{y}$, where $k = \frac{2\pi}{9}$ (1/m). Please derive the electric field at $z = 3$ m in the following cases:

(a). $E_a = 3$ and $E_b = 5$.

(b). $E_a = 6 \cdot e^{j\frac{\pi}{5}}$ and $E_b = 2 \cdot e^{-j\frac{\pi}{6}}$.

Solution

(a).   The phasor of the forward wave at $z = 3$ m is

$$E^+(z) = E_a \cdot e^{-j\frac{2}{9}\pi \times 3} = 3 \cdot e^{-j\frac{2}{3}\pi}.$$

The phasor of backward wave at $z = 3$ m is

$$E^-(z) = E_b \cdot e^{j\frac{2}{9}\pi \times 3} = 5 \cdot e^{j\frac{2}{3}\pi}.$$

Then the real E-field at $z = 3$ m is

$$\vec{E}(z,t) = \text{Re}\{[E^+(z) + E^-(z)]e^{j\omega t}\} \cdot \hat{y} = [3\cos(\omega t - \frac{2}{3}\pi) + 5\cos(\omega t + \frac{2}{3}\pi)] \cdot \hat{y}.$$

(b).   The phasor of forward wave at $z = 3$ m is

$$E^+(z) = E_a \cdot e^{-j\frac{2}{9}\pi \times 3} = 6e^{j\frac{\pi}{5}} \cdot e^{-j\frac{2}{3}\pi} = 6 \cdot e^{-j\frac{7}{15}\pi}.$$

The phasor of backward wave at $z = 3$ m is

$$E^-(z) = E_b \cdot e^{j\frac{2}{9}\pi \times 3} = 2e^{-j\frac{\pi}{6}} \cdot e^{j\frac{2}{3}\pi} = 2 \cdot e^{j\frac{\pi}{2}}.$$

Then the real E-field at $z = 3$ m is

$$\vec{E}(z,t) = \text{Re}\{[E^+(z) + E^-(z)]e^{j\omega t}\} \cdot \hat{y} = [6\cos(\omega t - \frac{7}{15}\pi) + 2\cos(\omega t + \frac{\pi}{2})] \cdot \hat{y}.$$

∎

2.4   Propagation of Plane Waves

In Sect. 2.3, we use phasors to derive the plane wave solution from wave equations. Plane waves are very useful because all the practical EM waves can be regarded as a combination of plane waves. By learning plane waves, we can catch the critical characteristics of EM waves, which exist everywhere but can not be seen or touched.

In this section, we will study how plane waves propagate. Because EM waves propagate mainly in an insulator (dielectric medium), we will focus on this medium to get the fundamental properties of EM waves.

A *Propagation of Plane Waves in Free Space*

Different from acoustic waves, EM waves can propagate even in free space (a perfect vacuum, i.e., a space free of matter). Let us consider a forward plane wave propagating along $+z$ direction while the associated electric field has only y-component, i.e., $\vec{E} = E_y \cdot \hat{y}$, where E_y is a function of (z, t) and is denoted by $E_y = E_y(z, t)$. Obviously, E_y is independent of x and y. This forward wave is shown in Fig. 2.15, where the waves in each plane travel toward $+z$ direction with a constant speed. For the wave at the plane of $z = z_1$, the electric field is given by $\vec{E} = E_y(z_1, t) \cdot \hat{y}$; for the wave at the plane of $z = z_2$, the electric field is given by $\vec{E} = E_y(z_2, t) \cdot \hat{y}$. Apparently, the electric field varies with time and position. But for those points on the same plane, they have identical electric fields. This is an important feature of plane waves.

From the previous section, we know that a forward wave has a phasor representation given by

$$\hat{E}_y(z) = E_a \cdot e^{-jkz}, \tag{2.87}$$

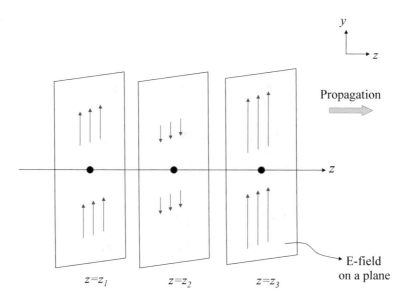

Fig. 2.15 A forward plane wave

where E_a denotes the electric field at $z = 0$. The parameter k is given by

$$k^2 = \omega^2 \mu \in \; \Rightarrow \; k = \omega\sqrt{\mu \in}. \tag{2.88}$$

From Eq. (2.88), k depends on the wave frequency ω and its meaning will become clear later.

In Eq. (2.87), E_a is a complex number. Let

$$E_a = Ae^{j\theta}. \tag{2.89}$$

Then the electric field of the plane wave can be represented by

$$E_y(z, t) = \text{Re}\{\hat{E}_y e^{j\omega t}\} = A\cos(\omega t - kz + \theta), \tag{2.90}$$

where A is the amplitude and θ is the phase.

Next, we consider that this plane wave propagates in free space. For free space, the permittivity and the permeability are given by

$$\in \; = \; \in_0 \; = \; \frac{1}{36\pi} \times 10^{-9} \; (\text{Farad/m}), \tag{2.91}$$

$$\mu = \mu_0 = 4\pi \times 10^{-7} \; (\text{Henry/m}), \tag{2.92}$$

respectively. Hence, the parameter k in Eq. (2.88) is given by

$$k = k_0 = \omega\sqrt{\mu_0 \in_0}, \tag{2.93}$$

where k_0 specifies this parameter in free space. Note that in the following context, the permittivity and the permeability in free space are denoted by $\in_0$ and μ_0, respectively.

To simplify our analysis, we let the phase $\theta = 0$ in Eq. (2.90). Thus, the electric field of the forward wave becomes

$$E_y(z, t) = A\cos(\omega t - k_0 z). \tag{2.94}$$

Now, we observe the forward wave in Eq. (2.94) and see if this wave travels toward $+z$. When $t = 0$, we have

$$E_y(z) = A\cos(-k_0 z) = A\cos(k_0 z). \tag{2.95}$$

As we plot E_y along z in Fig. 2.16, the maximum (amplitude $= A$) occurs at $k_0 z = 2m\pi$ and the minimum (amplitude $= -A$) occurs at $k_0 z = (2m + 1)\pi$, where m is an integer. Furthermore, as we observe Fig. 2.16 carefully, we find the waveform repeats every distance of $\frac{2\pi}{k_0}$. This distance is called **wavelength** of the EM wave. Hence, in free space, the wavelength λ_0 and the parameter k_0 have the following relationship:

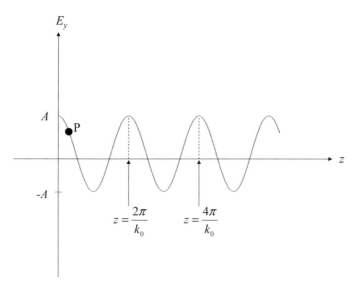

Fig. 2.16 Explaining propagation of forward wave (I)

$$\lambda_0 = \frac{2\pi}{k_0}. \tag{2.96}$$

From Eq. (2.96), the wavelength λ_0 is inversely proportional to the parameter k_0. The greater the parameter k_0, the shorter the wavelength λ_0. Note that for wave-like phenomena such as acoustic wave, water wave, or EM wave, the definition of their wavelengths is identical—the spatial distance over which the waveform repeats is defined as the wavelength.

Next, when $t = \frac{\pi}{2\omega}$ and thus $\omega t = \frac{\pi}{2}$, Eq. (2.94) becomes

$$E_y(z) = A \cos\left(\frac{\pi}{2} - k_0 z\right) = A \sin(k_0 z). \tag{2.97}$$

Along z-axis, the electric field E_y can be plotted as shown in Fig. 2.17. Notice that the waveform is similar to that in Fig. 2.16, but the corresponding electric field at each point already changes. Meanwhile, in Fig. 2.17, the waveform still repeats every distance of $\frac{2\pi}{k_0}$. Hence, the wavelength does not change with time.

As time goes on, when $t = \frac{\pi}{\omega}$ and thus $\omega t = \pi$. Then Eq. (2.94) becomes

$$E_y(z) = A \cos(\pi - k_0 z) = -A \cos k_0 z. \tag{2.98}$$

The corresponding electric field is shown in Fig. 2.18. Like Fig. 2.17, the waveform remains sinusoidal, but the electric field at each point along z-axis changes. Similarly, when $t = \frac{3\pi}{2\omega}$ and $\omega t = \frac{3\pi}{2}$, the corresponding waveform is shown in Fig. 2.19. (We

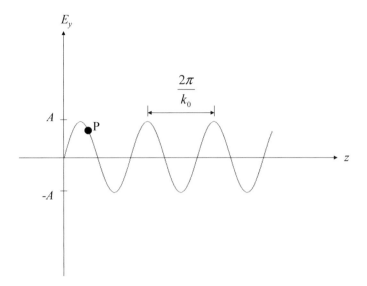

Fig. 2.17   Explaining propagation of forward wave (II)

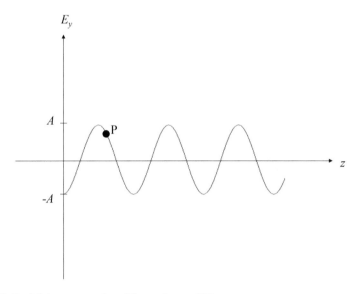

Fig. 2.18   Explaining propagation of forward wave (III)

encourage readers draw these figures by themselves. It will help you understand the propagation of EM waves.)

Now, we can explain why we name it forward wave in Eq. (2.94) by using the above figures. First, please keep an eye on the point P in Fig. 2.16. Then trace it in Figs. 2.17, 2.18, and 2.19. Do you notice that as time goes on, the point P moves

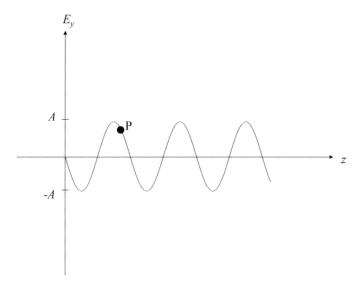

Fig. 2.19 Explaining propagation of forward wave (IV)

toward $+z$ direction? If you do, you shall understand why it is a "forward wave". When $t = t_1$, the point P is at $z = z_1$. After Δt, it moves to $z = z_2$ and $z_2 > z_1$. Hence, the EM wave "moves forward" and therefore is called forward wave.

From above, we explain why $E_y = A\cos(\omega t - k_0 z)$ is a forward wave. On the other hand, if we have the electric field $E_y = B\cos(\omega t + k_0 z)$, the wave will move toward $-z$ direction and called backward wave. Readers may draw figures similar to Figs. 2.16–2.19 to see if it moves backward!

B   Phase Velocity

When we see a moving object like a vehicle, its speed is of interest. Similarly, for a "moving" EM wave, we are also interested in its velocity. The question is: how do we derive the velocity of an EM wave?

First, we consider a forward wave and rewrite Eq. (2.94) as

$$E_y(z, t) = A\cos(\omega t - k_0 z) = A\cos\Omega, \qquad (2.99)$$

where

$$\Omega = \omega t - k_0 z. \qquad (2.100)$$

In Eq. (2.99), the angle Ω is the phase of E_y and it varies with t and z. Next, we try to measure a "moving distance" of an EM wave for a specific time period as shown

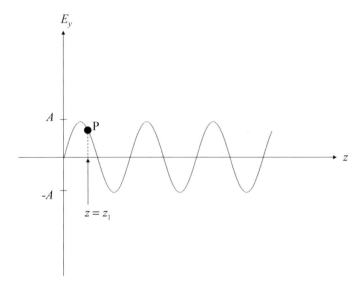

Fig. 2.20 Explaining phase velocity (I)

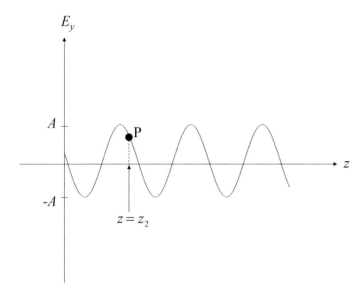

Fig. 2.21 Explaining phase velocity (II)

in Figs. 2.20 and 2.21. In Fig. 2.20, when $t = t_1$, the phase of point P at $z = z_1$ is given by

$$\Omega_P = \omega t_1 - k_0 z_1. \tag{2.101}$$

In Fig. 2.21, it shows that when $t = t_2$ and $t_2 > t_1$, the point P moves to z_2 and its phase Ω_P remains. Hence Ω_P is given by

$$\Omega_P = \omega t_2 - k_0 z_2. \tag{2.102}$$

Between the time interval $\Delta t = t_2 - t_1$, the moving distance of P is $\Delta z = z_2 - z_1$. Thus, the velocity v can be derived by

$$v = \frac{\Delta z}{\Delta t} = \frac{z_2 - z_1}{t_2 - t_1}. \tag{2.103}$$

From Eq. (2.101) and Eq. (2.102), we obtain

$$\omega t_1 - k_0 z_1 = \omega t_2 - k_0 z_2 \quad \Rightarrow \quad k_0(z_2 - z_1) = \omega(t_2 - t_1). \tag{2.104}$$

Inserting Eq. (2.104) into Eq. (2.103), we finally get

$$v = \frac{\omega}{k_0}. \tag{2.105}$$

Because v represents the velocity of the phase point P (a point of particular phase) of the EM wave, it is called the **phase velocity**. The physical meaning of v may be interpreted as follows: when we fix our eyes at a point with a specific phase of a traveling EM wave, the moving speed of this point is regarded as the phase velocity.

As we take a look at Eq. (2.105), we may think the phase velocity v is proportional to the frequency ω. But it is not true! In free space, because $k_0 = \omega\sqrt{\mu_0 \epsilon_0}$, we have

$$v = \frac{\omega}{k_0} = \frac{1}{\sqrt{\mu_0 \epsilon_0}}, \tag{2.106}$$

where

$$\frac{1}{\sqrt{\mu_0 \epsilon_0}} = \frac{1}{\sqrt{(4\pi \times 10^{-7}) \cdot (\frac{1}{36\pi} \times 10^{-9})}} = 3 \times 10^8 = c. \tag{2.107}$$

Here $c = 3 \times 10^8 \, (m/s)$ denotes the speed of light. Hence, we have

$$v = c. \tag{2.108}$$

Equation (2.108) tells us that the speed of an EM wave in free space is independent of frequency and is equal to the speed of light. This implies that light might be a kind of EM wave, and it is true! Hence, the behavior of light also follows Maxwell's equations.

Faraday and Maxwell are pioneers predicting that light is an EM wave. Since then, the conventional perspective of sunlight changes. It is actually an EM radiation consisting of time-varying electric field and magnetic field. This change also gives us an advanced understanding of sunlight. For example, conventionally we know that white light is dispersed by a prism into seven visible colors from red to violet. Now, we understand that white light is made up of EM waves having different frequencies. The seven visible colors actually represent a group of EM waves having different frequencies. Because the frequencies are different, the respective refraction angles through a prism are different and thus lead to various spectral colors.

On the other hand, because the sunlight provides heat according to our daily experience, scientists intuitively consider EM radiation as a transfer of energy. This perspective will be discussed in Chap. 3.

C   Wave Number

From previous results, we learn that k_0 is an important parameter. It is closely related to the propagation of EM waves. In free space, because $v = c$, from Eq. (2.105), we have

$$k_0 = \frac{\omega}{c}. \tag{2.109}$$

Hence, k_0 is proportional to ω. The higher the frequency, the larger the parameter k_0.

Suppose an EM wave has the frequency f and the relationship between f and radial frequency ω is given by

$$\omega = 2\pi f. \tag{2.110}$$

It is well known that the wave velocity = frequency × wavelength, we have

$$c = f \cdot \lambda_0, \tag{2.111}$$

where λ_0 is the wavelength in free space. From Eqs. (2.109)–(2.111), we get

$$k_0 = \frac{2\pi}{\lambda_0}. \tag{2.112}$$

Equation (2.112) is equivalent to Eq. (2.96). Notice that Eq. (2.96) is derived from the time-varying and space-varying behavior of EM waveform as shown in Fig. 2.16, while Eq. (2.112) is derived from the definition of velocity. In addition, Eq. (2.112) shows that the parameter k_0 is inversely proportional to the wavelength λ_0. The larger the wavelength, the smaller the parameter k_0.

The parameter k_0 is called **wave number**, whose physical meaning is interpreted in the following. Because the unit of the wavelength λ_0 is m, intuitively $1/\lambda_0$ tells us how many wavelengths within a meter. Since the parameter k_0 is proportional to

$1/\lambda_0$, it is called "wave number". For an EM wave, if the wave number k_0 is larger, the more repeating waveforms occur within a unit length.

Because the wave number k_0 is an important parameter, we need to further explore its physical meaning. In Eq. (2.94), we have the electric field $E_y(z, t) = A\cos(\omega t - k_0 z)$, which shows that the electric field varies with time t and position z. The time-varying rate is determined by ω and the space-varying rate is determined by k_0. The relationship between ω and k_0 is described as follows:

1. When we consider an electric field E_y in time domain, as the frequency ω grows up, the waveform shall vary with time more quickly. In other words, the higher the frequency, the more the occurrences of repeating waveform per unit time. It is consistent with our intuition.
2. When we consider the electric field in space domain, from Eq. (2.94), $k_0 z$ actually plays the same role as ωt. The physical meaning of k_0 is similar to that of ω: the greater the wave number k_0, the more the repeating waveform per unit length. Thus, we can regard k_0 as the "frequency" in space domain. As the wave number k_0 grows up, the waveform shall vary with position more quickly.

From Eq. (2.109), the wave number k_0 is proportional to the frequency ω. It means that when time-varying rate of an electric field increases, the space-varying rate also increases, and vice versa. Therefore, a signal having higher frequency varies more rapidly both in time and in space than that of a signal having lower frequency.

Example 2.18

Suppose a plane wave propagates in free space with frequency $f = 1\,\text{MHz}$. Please derive the associated wavelength, wave number, and phase velocity.

Solution

First, from Eq. (2.111), the wavelength of the plane wave is given by

$$\lambda_0 = \frac{c}{f} = \frac{3 \times 10^8}{1 \times 10^6} = 300\,(\text{m}).$$

Next, from Eq. (2.112), the wave number is given by

$$k_0 = \frac{2\pi}{\lambda_0} = \frac{\pi}{150}\,(\text{m}^{-1}).$$

Finally, the phase velocity is given by

$$v = c = f \cdot \lambda_0 = 3 \times 10^8\,(\text{m/s}).$$

∎

D   Plane Waves in Dielectric Media

After learning the propagation of plane waves in free space, we extend our learning to dielectric media. First, suppose we have a dielectric medium with permittivity ϵ and permeability μ. For simplicity, we consider a non-magnetic medium and thus $\mu = \mu_0$. The permittivity ϵ can be represented by

$$\epsilon = \epsilon_r \cdot \epsilon_0, \tag{2.113}$$

where ϵ_0 denotes the permittivity in free space and ϵ_r denotes the **relative permittivity** or **dielectric constant**. Note that free space has the least permittivity. Thus, $\epsilon_r \geq 1$. For the most popular propagation medium, air, in wireless communications, it has the relative permittivity $\epsilon_r = 1.00059$. Hence, air is usually regarded as free space for EM wave propagation.

Because $\epsilon = \epsilon_r \epsilon_0$ and $\mu = \mu_0$, the wave number k of a dielectric medium is given by

$$k = \omega\sqrt{\epsilon\mu} = \omega\sqrt{\epsilon_r \cdot \epsilon_0\,\mu_0} = \sqrt{\epsilon_r} \cdot \frac{\omega}{c}. \tag{2.114}$$

Next, we define the **refractive index** by

$$n = \sqrt{\epsilon_r}, \tag{2.115}$$

where $n \geq 1$ because $\epsilon_r \geq 1$. This index determines how much the light is bent, or refracted, when entering the medium and is thus named. Using Eq. (2.115) and Eq. (2.114), we can rewrite k as

$$k = \frac{n\omega}{c}. \tag{2.116}$$

From Eq. (2.116), the wave number k is proportional to the refractive index n. When the refractive index n increases, the associated wave number k increases as well. The refractive index n is an important parameter and closely related to the behavior of EM waves.

From Eqs. (2.109) and (2.116), we have

$$k = nk_0. \tag{2.117}$$

Hence, the wave number k of a dielectric medium is greater than the wave number k_0 of free space since $n \geq 1$ ($n = 1$ holds only for free space). This grants EM waves two major characteristic differences between dielectric medium and free space:

1. Phase velocity

Following the same process as in deriving Eq. (2.106), we can get the phase velocity of an EM wave in a dielectric medium. It is given by

$$v = \frac{\omega}{k} = \frac{c}{n}. \tag{2.118}$$

Because $n \geq 1$, the phase velocity v in a dielectric medium is less than that in free space. The larger the refractive index, the less the phase velocity.

2. Wavelength

Suppose an EM wave has the wavelength λ in a dielectric medium. From Eq. (2.118) and because the wave velocity = frequency $\times$ wavelength, we have

$$v = f \cdot \lambda = \frac{c}{n}. \tag{2.119}$$

Thus

$$\lambda = \frac{c}{nf} = \frac{\lambda_0}{n}, \tag{2.120}$$

where λ_0 is the associated wavelength in free space. Because $n \geq 1$, the wavelength in a dielectric medium is smaller than that in free space. The larger the refractive index, the smaller the wavelength.

From the previous discussions, an EM wave will have different phase velocities and wavelengths when traveling in different dielectric media. In addition, from Eqs. (2.117) and (2.120), we have

$$k = n \cdot \frac{2\pi}{\lambda_0} = \frac{2\pi}{\lambda}. \tag{2.121}$$

Readers may notice the similarity between Eqs. (2.121) and (2.112). Thus, no matter in free space or in dielectric medium, the relationship between wave number and the wavelength keeps the same mathematical form.

Example 2.19

A plane wave propagates in a dielectric medium with the relative permittivity $\epsilon_r = 9$. Suppose the frequency $f = 1$ GHz. Please derive the associated wavelength, wave number, and phase velocity.

Solution

From Eq. (2.115), the refractive index is given by

$$n = \sqrt{\epsilon_r} = 3.$$

From Eq. (2.111), the wavelength in free space is given by

$$\lambda_0 = \frac{c}{f} = \frac{3 \times 10^8}{1 \times 10^9} = 0.3 \quad (\text{m}).$$

Hence, from Eq. (2.120), the wavelength in this dielectric medium is given by

$$\lambda = \frac{\lambda_0}{n} = 0.1 \quad (\text{m}).$$

From Eq. (2.121), the associated wave number is given by

$$k = \frac{2\pi}{\lambda} = \frac{2\pi}{0.1} = 20\pi \quad (\text{m}^{-1}).$$

Finally, from Eq. (2.118), the phase velocity is given by

$$v = \frac{c}{n} = \frac{3 \times 10^8}{3} = 1 \times 10^8 \quad (\text{m/s}).$$

■

Summary

2.1. Wave Equations

Based on Maxwell's equations, we learn the interaction between electric field and magnetic field. Then, we derive the associated wave equations whose solutions represent the propagating EM waves.

2.2. Phasor

We learn the concept and applications of phasors.

2.3. Plane waves

We consider plane waves and learn how to derive the associated solutions of the wave equations. The formulas of the resulting forward waves and backward waves are also provided.

2.4. Propagation of Plane waves

We learn how the electric field varies with time and space when forward waves and backward waves propagate. The associated wavelength, wave number, and phase velocity are derived.

In this chapter, we learn the fundamental concept and behavior of EM waves. In Chap. 3, we will further learn the characteristics of EM waves in order to get more insights.

Exercises

1. Suppose $f(x, y, z)$ is a scalar function. Then the gradient of $f(x, y, z)$ is a vector and defined as follows:

$$\nabla f = \frac{\partial f}{\partial x} \cdot \hat{x} + \frac{\partial f}{\partial y} \cdot \hat{y} + \frac{\partial f}{\partial z} \cdot \hat{z}.$$

Please prove the following equality holds:

$$\nabla \times (\nabla \times \vec{A}) = \nabla(\nabla \cdot \vec{A}) - \nabla^2 \vec{A},$$

where $\vec{A}$ is a vector field given by

$$\vec{A} = A_x \cdot \hat{x} + A_y \cdot \hat{y} + A_z \cdot \hat{z}.$$

[Hint: First, prove the equality holds for x-component.]

2. Similar to how we derived wave equations of electric fields in Sect. 2.1, please derive wave equations of magnetic fields.
3. Suppose we have a scalar function $f = 2x^2 y + z^3 x$. Please derive $\nabla^2 f$ at the point $(2, 1, 2)$. (Hint: Refer to Example 2.1.)
4. Suppose we have a vector field $\vec{A} = A_x \cdot \hat{x} = (xy^2 + z^2) \cdot \hat{x}$. Please derive $\nabla^2 \vec{A}$ at the point $(-5, 2, 0)$. (Hint: Refer to Example 2.2.)
5. Suppose we have a vector field $\vec{A} = 2xy \cdot \hat{x} + z^3 \cdot \hat{y} + (z + x^2 y) \cdot \hat{z}$. Please derive $\nabla^2 \vec{A}$ at the point $(2, 5, 1)$.
6. Suppose we have a sinusoid $x(t) = 3 \cdot \cos(\omega t + 30°)$. Please represent $x(t)$ by phasor. (Hint: Example 2.5.)
7. Suppose we have a sinusoid $x(t) = 5 \cdot \sin(\omega t + 40°)$. Please represent $x(t)$ by phasor.
8. Suppose we have a signal $x(t) = 2 \cdot \cos(\omega t - 20°) + 4 \sin(\omega t - 45°)$. Please represent $x(t)$ by phasors. (Hint: Example 2.7.)
9. Suppose we have a combination of two sinusoids $x(t) = 6 \cdot \cos(\omega t - \frac{\pi}{6}) - 3 \sin(\omega t - \frac{\pi}{5})$. Please represent $x(t)$ by phasors.
10. Suppose we have a signal $x(t) = 2 \cdot \cos(\omega t + \frac{\pi}{3})$. Please represent $x(t)$, $\frac{dx(t)}{dt}$, and $\frac{d^2 x(t)}{dt^2}$ by phasors. (Hint: Example 2.10 and Example 2.11.)
11. Suppose we have a signal $x(t) = 5 \cdot \cos(\omega t + \frac{\pi}{6}) - 3 \cdot \sin(\omega t - \frac{\pi}{3})$. Please represent $x(t), \frac{dx(t)}{dt}$, and $\frac{d^2 x(t)}{dt^2}$ by phasors.

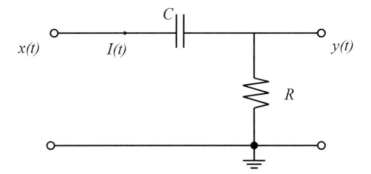

Fig. 2.22 Illustrating Exercise 12

12. In Fig. 2.22, if the input voltage is provided by $x(t) = a\cos(\omega t + \theta)$. Please derive the output voltage $y(t)$ using phasor method. (Hint: Example 2.12.)

13. Suppose we have a signal $x(t) = \sin(\omega t + \frac{\pi}{4})$. Please represent $\frac{d^3 x(t)}{dt^3}$ and $\frac{d^4 x(t)}{dt^4}$ by phasors.

14. Similar to how we derive the wave equation of electric field in Eq. (2.98) of Sect. 2.1, i.e.,

$$\nabla^2 \vec{E} = \mu \in \frac{\partial^2 \vec{E}}{\partial t^2},$$

please derive the wave equation of magnetic field in a dielectric medium. In addition, compare the derived equation with Eq. (2.98) and notice the symmetry of these two equations.

14. A forward plane wave has frequency ω and its phasor of electric field is $\vec{E}(z) = E_a e^{-jkz} \cdot \hat{y}$, where E_a is the phasor of the electric field at $z = 0$ and the wave number $k = \frac{\pi}{8}$ (1/m). Please derive the electric field at $z = 10$ m in the following three cases:

(a). $E_a = 7$,
(b). $E_a = 7 \cdot e^{j\frac{\pi}{4}}$,
(c). $E_a = 6 \cdot e^{-j\frac{\pi}{3}}$.

    (Hint: Example 2.15.)

15. A backward plane wave has frequency ω and the phasor of its electric field is $\vec{E}(z) = E_b e^{jkz} \cdot \hat{y}$, where E_b is the phasor of the electric field at $z = 0$ and the wave number $k = \frac{\pi}{3}$ (1/m). Please derive the electric field at $z = 5$ m in the following three cases:

(a). $E_b = 5$,
(b). $E_b = 5 \cdot e^{j\frac{\pi}{3}}$,
(c). $E_b = 7 \cdot e^{-j\frac{\pi}{4}}$.

    (Hint: Example 2.16.)

16. Suppose an EM wave consists of a forward wave and a backward wave having frequency ω. The phasor of its electric field is given by $\vec{E}(z) = (E_a e^{-jkz} + E_b e^{jkz}) \cdot \hat{y}$,

    where $k = \frac{\pi}{8}$ (1/m). Please derive the electric field at $z = 6$ m in the following two cases:

    (a). $E_a = 7$ and $E_b = 5$.
    (b). $E_a = 4 \cdot e^{j\frac{\pi}{3}}$ and $E_b = 11 \cdot e^{-j\frac{\pi}{6}}$.

    (Hint: Example 2.17.)

17. Suppose the electric field of an EM wave is $\vec{E} = E_y \cdot \hat{y}$, where $E_y = 3 \cdot \cos(\omega t - k_0 z + \frac{\pi}{4})$.

    (a). Please plot the waveform of the electric field between $z = 0$ and $z = \frac{4\pi}{k_0}$ when $t = 0$, $t = \frac{\pi}{2\omega}$, $t = \frac{\pi}{\omega}$, and $t = \frac{3\pi}{2\omega}$.
    (b). From the results in (a), please explain why the wave is a forward wave.

    (Hint: Refer to Sect. 2.4.)

18. Suppose the electric field of an EM wave is $\vec{E} = E_y \cdot \hat{y}$, where $E_y = 5 \cdot \cos(\omega t + k_0 z)$.

    (a). Please plot the waveform of the electric field between $z = 0$ and $z = \frac{4\pi}{k_0}$ when $t = 0$, $t = \frac{\pi}{2\omega}$, $t = \frac{\pi}{\omega}$, and $t = \frac{3\pi}{2\omega}$.
    (b). From the results in (a), explain why this wave is a backward wave.

19. A plane wave propagates along $+z$ in free space and its electric field is $\vec{E} = E_x \cdot \hat{x}$, where $E_x = 5 \cdot \cos(\omega t - k_0 z)$ and the wave number $k_0 = \frac{2\pi}{3}$ (1/m). Please derive the following parameters:

    (a). Frequency ω.
    (b). Wavelength λ_0.
    (c). Phase velocity v_p.

    (Hint: In free space, phase velocity is equal to light speed.)

20. (Continued from Exercise 19)

    (a). Plot E_x at $0 \le z \le 6$ m when $t = 0$.
    (b). Plot E_x at $0 \le z \le 6$ m when $t = 2$ ns. (Note: 1 ns= 10^{-9}s.)
    (c). Plot E_x at $0 \le z \le 6$ m when $t = 5$ ns.

20. (Continued from Exercise 19)

    (a). At $z = 0$, plot E_x when $0 \le t \le 20$ ns.
    (b). At $z = 2$ m, plot E_x when $0 \le t \le 20$ ns.

21. A plane wave propagates in a dielectric medium along $+z$ and its electric field is $\vec{E} = E_x \cdot \hat{x}$, where $E_x = 3 \cdot \cos(\omega t - kz)$ and the frequency $\omega = 2\pi \times 10^7$ (rad/s).

Suppose the refractive index of this medium is $n = 4$. Please derive the following parameters:

(a).   Wave number k.
(b).   Wavelength λ.
(c).   Phase velocity v_p.

(Hint: Example 2.18.)

Chapter 3
Characteristics of EM Waves

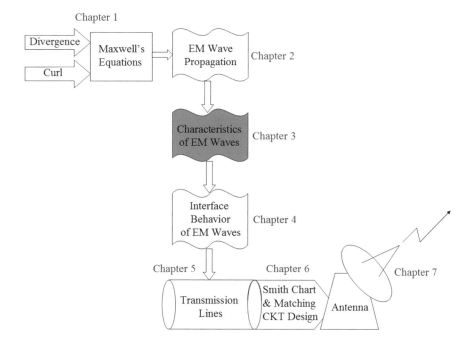

© Springer Nature Switzerland AG 2020

M.-S. Kao and C.-F. Chang, *Understanding Electromagnetic Waves*,

https://doi.org/10.1007/978-3-030-45708-2_3

Abstract When applying EM waves, we need to understand four important characteristics in advance:

1. Wave impedance.
2. Phase velocity and group velocity.
3. Poynting vector.
4. Skin effect and skin depth.

   In order to help readers understand these properties, we adopt a math-friendly approach by using many examples to explain their physical meanings step by step. For instance, we start from a simple case of two waves to derive the group velocity and then extend it to general cases. Furthermore, an interesting example of a shark chasing a group of small fishes in the ocean is provided to help readers catch the core idea and understand the meaning of group velocity. After learning this chapter, readers will have a solid background for advanced study of EM waves.

Keywords Wave impedance · Intrinsic impedance · Forward wave · Backward wave · Phase velocity · Group velocity · Signal velocity · Dispersive medium · Poynting vector · Power density · Conductivity · Equivalent permittivity · Propagation constant · Skin effect · Skin depth

Introduction

In Chap. 2, we derive wave equation from Maxwell's equations. The derivation reveals how an EM wave is generated and propagates in space. When considering plane waves, we then attain the solutions of wave equation and these solutions precisely describe behaviors of EM waves mathematically. In this chapter, we further investigate these solutions and explore important properties of EM waves. By doing this, we gradually accumulate fundamental knowledge of EM waves in order to see the whole picture eventually.

   This chapter consists of four sections. Each section introduces at least a critical parameter of EM waves and the associated properties. Firstly, we introduce an important relationship between the electric field and the magnetic field of a plane wave. Secondly, we introduce a "featuring" velocity which can represent the overall speed of a group of EM waves although each wave has a distinctive phase velocity. Thirdly, we introduce a specific vector which effectively represents the power transfer of dynamic EM waves. Finally, we introduce characteristics of EM waves in a conductor including the skin effect. Being familiar with these parameters characterizing an EM wave, we can get a feel for what EM waves may behave in different environments.

3.1 Wave Impedance

In this section, we investigate the relationship between the electric field and the magnetic field of an EM wave. From Maxwell's equations, a simple and useful relationship between E-field and M-field of a uniform plane wave can be extracted. By using this relationship, we can directly derive M-field from the associated E-field, and vice versa.

A   M-Field of Plane Waves

In Chap. 2, we have derived the E-field of a plane wave. Here we will derive the associated M-field of the plane wave. First, from Faraday's law, we have

$$\nabla \times \vec{E} = -\mu \frac{\partial \vec{H}}{\partial t}. \tag{3.1}$$

As we use phasor representations and the differentiation property of phasors to simplify the analysis, Eq. (3.1) can be rewritten as

$$\nabla \times \vec{E} = -j\omega\mu\vec{H}. \tag{3.2}$$

Because phasor representations are widely used in analyzing EM waves, in the following context, we will use $\vec{E}$ and $\vec{H}$ to represent phasors of E-field and M-field, respectively.

Next, we consider a uniform plane wave propagating in $+z$, and the associated E-field only has x-component given by

$$\vec{E} = E_x \cdot \hat{x}, \tag{3.3}$$

where E_x is a phasor and it only depends on z, i.e., $E_x = E_x(z)$.

In Eq. (3.2), $\nabla \times \vec{E}$ can be replaced by

$$\nabla \times \vec{E} = \left(\frac{\partial E_z}{\partial y} - \frac{\partial E_y}{\partial z}\right)\hat{x} + \left(\frac{\partial E_x}{\partial z} - \frac{\partial E_z}{\partial x}\right)\hat{y} + \left(\frac{\partial E_y}{\partial x} - \frac{\partial E_x}{\partial y}\right)\hat{z}. \tag{3.4}$$

Because $E_y = E_z = 0$ and $E_x = E_x(z)$, from Eqs. (3.2) and (3.4), we have

$$\nabla \times \vec{E} = \frac{\partial E_x}{\partial z} \cdot \hat{y} = -j\omega\mu\vec{H}. \tag{3.5}$$

Hence, the associated M-field can be derived from Eq. (3.5) and given by

Fig. 3.1 Illustrating uniform
plane wave

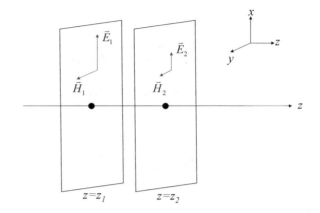

$$\vec{H} = \frac{1}{-j\omega\mu}\frac{\partial E_x}{\partial z}\cdot\hat{y} = H_y\cdot\hat{y}, \tag{3.6}$$

where H_y denotes the y-component of $\vec{H}$ and is given by

$$H_y = \frac{1}{-j\omega\mu}\frac{\partial E_x}{\partial z}. \tag{3.7}$$

From Eq. (3.6), we find that $\vec{H}$ only has y-component and its magnitude depends on $\partial E_x/\partial z$. This plane wave is shown in Fig. 3.1, where the E-field only has x-component, the M-field only has y-component, and the wave propagates along z-direction. Hence, the E-field, M-field, and the propagation direction are mutually perpendicular. As shown in Fig. 3.1, the wave at $z = z_1$ plane and $z = z_2$ plane has corresponding EM field $(\vec{E}_1, \vec{H}_1)$ and $(\vec{E}_2, \vec{H}_2)$, respectively. The direction of $\vec{E}_1$ and $\vec{E}_2$ points along x, while that of $\vec{H}_1$ and $\vec{H}_2$ points along y. In general, the EM fields at $z = z_1$ and $z = z_2$ are different. Hence, $\vec{E}_1 \neq \vec{E}_2$ and $\vec{H}_1 \neq \vec{H}_2$. In the following, we further investigate the relationship between $\vec{E}$ and $\vec{H}$ considering the propagation direction.

B   Forward Wave

First, we consider a forward wave having frequency ω and wave number k. It propagates toward +z and the corresponding E-field is given by

$$E_x(z) = E_a e^{-jkz}, \tag{3.8}$$

where E_a is the phasor of E-field at $z = 0$ (the detailed discussions of forward waves can be referred to Sect. 2.3). From Eq. (3.8), we have

$$\frac{\partial E_x(z)}{\partial z} = -jk \cdot E_a e^{-jkz} = -jk E_x(z). \tag{3.9}$$

Inserting Eq. (3.9) into Eq. (3.7), we have

$$H_y(z) = \frac{-jk E_x(z)}{-j\omega\mu} = \frac{k}{\omega\mu} E_x(z). \tag{3.10}$$

From Eq. (3.10), we find that the M-field, $H_y(z)$, is proportional to the corresponding E-field, $E_x(z)$. The ratio of $E_x(z)$ to $H_y(z)$ can be derived by

$$\eta = \frac{E_x(z)}{H_y(z)} = \frac{\omega\mu}{k}. \tag{3.11}$$

From Eq. (3.11), the ratio of $E_x(z)$ to $H_y(z)$ is independent of z. In other words, the ratio is a constant and does not depend on the position. Besides, Eq. (3.11) tells us that when the ratio η is given, we can derive $E_x(z)$ directly from $H_y(z)$, and vice versa. We call the parameter η **wave impedance** because it is the ratio of E-field to the associated M-field. The concept is similar to electrical impedance which we have in circuitry, where electrical impedance is a ratio of voltage to the associated current in a circuit.

Because the wave number $k = \omega\sqrt{\mu\,\epsilon}$, Eq. (3.11) can be rewritten as

$$\eta = \frac{\omega\mu}{\omega\sqrt{\mu\,\epsilon}} = \sqrt{\frac{\mu}{\epsilon}}. \tag{3.12}$$

Hence, wave impedance only depends on (μ, ϵ). It means that the ratio of $E_x(z)$ to $H_y(z)$ only depends on EM properties of the propagation medium. Hence, the wave impedance is also called the **intrinsic impedance**. In free space, because $\mu = \mu_0$ and $\epsilon = \epsilon_0$, the wave impedance is given by

$$\eta = \eta_0 = \sqrt{\frac{\mu_0}{\epsilon_0}} = \sqrt{\frac{4\pi \times 10^{-7}(H/m)}{\frac{1}{36\pi} \times 10^{-9}(F/m)}} = 120\pi. \tag{3.13}$$

For a non-magnetic dielectric medium, $\mu = \mu_0$ and $\epsilon = \epsilon_r \epsilon_0$, where ϵ_r is the relative permittivity. Hence

$$\eta = \sqrt{\frac{\mu_0}{\epsilon_r \epsilon_0}} = \frac{\eta_0}{n}, \tag{3.14}$$

where $n = \sqrt{\epsilon_r}$ is the refractive index of the medium. Since $n \geq 1$, from Eq. (3.14), the free space has the largest wave impedance. In the remaining context, we use η_0 to stand for the wave impedance of free space.

Now, we compare wave impedance with electrical impedance. Suppose an electrical impedance Z is given by

$$Z = \frac{V}{I},$$

where V denotes the voltage phasor and I denotes the current phasor. There are two major differences between η and Z:

1. Wave impedance η is a positive real number, while electrical impedance Z can be a complex number. For instance, the electrical impedance of an inductor is $Z = j\omega L$.
2. The maximal value of wave impedance is $\eta_0 = 120\pi = 377(\Omega)$. Hence, any wave impedance will not exceed 377 Ω. On the other hand, an electrical impedance Z may approach infinity!

From above, for a general dielectric medium, η is much simpler than Z because it is a real positive number equal to or less than 377 Ω. When an electrical engineer says "The impedance of free space is 377 ohms", which he means is wave impedance, because the electrical impedance of free space approaches infinity!

Example 3.1

Suppose a forward wave has the electric field $\vec{E} = E_x \cdot \hat{x} = 5 \cdot \cos(\omega t - kz + 45°) \cdot \hat{x}$ propagating in a dielectric medium having the permittivity $\epsilon = 4 \ \epsilon_0$, please derive the M-field $\vec{H} = H_y \cdot \hat{y}$.

Solution

First, the phasor of E-field is given by

$$E_x(z) = 5 \cdot e^{j(-kz+45°)}.$$

Next, the refractive index is given by

$$n = \sqrt{\epsilon_r} = \sqrt{4} = 2.$$

Hence, the wave impedance is derived by

$$\eta = \frac{\eta_0}{n} = \frac{377}{2}(\Omega).$$

Applying Eq. (3.11), the phasor of M-field is given by

$$H_y(z) = \frac{E_x(z)}{\eta} = \frac{10}{377} \cdot e^{j(-kz+45°)}.$$

Hence, the corresponding M-field $H_y(z, t)$ can be derived by

$$H_y(z, t) = \mathrm{Re}\{H_y(z) \cdot e^{j\omega t}\} = \frac{10}{377} \cdot \cos(\omega t - kz + 45°).$$

And it can be represented by a vector field

$$\vec{H} = H_y(z, t) \cdot \hat{y} = \frac{10}{377} \cdot \cos(\omega t - kz + 45°) \cdot \hat{y}.$$

∎

From Eq. (3.11), the ratio of E-field phasor to M-field phasor is η. This can be directly extended to the time-dependent EM field. Suppose $E_x(z, t)$ and $H_y(z, t)$ are E-field and the associated M-field, respectively. We have

$$E_x(z, t) = \mathrm{Re}\{E_x(z) \cdot e^{j\omega t}\}, \tag{3.15}$$

$$H_y(z, t) = \mathrm{Re}\{H_y(z) \cdot e^{j\omega t}\}, \tag{3.16}$$

where $E_x(z)$ and $H_y(z)$ are phasors. According to Eq. (3.11), we have $E_x(z) = \eta H_y(z)$. Hence

$$\frac{E_x(z, t)}{H_y(z, t)} = \frac{\mathrm{Re}\{\eta H_y(z) \cdot e^{j\omega t}\}}{\mathrm{Re}\{H_y(z) \cdot e^{j\omega t}\}}. \tag{3.17}$$

Since η is a real positive number, we have

$$\mathrm{Re}\{\eta H_y(z) e^{j\omega t}\} = \eta \cdot \mathrm{Re}\{H_y(z) e^{j\omega t}\}. \tag{3.18}$$

From Eqs. (3.17) and (3.18), we get

$$\frac{E_x(z, t)}{H_y(z, t)} = \eta. \tag{3.19}$$

Hence, if the time-dependent E-field is given by

$$E_x(z, t) = A \cdot \cos(\omega t - kz + \theta), \tag{3.20}$$

the associated time-dependent M-field $H_y(z, t)$ can be directly derived by

$$H_y(z, t) = \frac{A}{\eta} \cdot \cos(\omega t - kz + \theta). \tag{3.21}$$

The simple relationship in Eq. (3.19) can greatly simplify problem-solving as illustrated below.

Example 3.2

Please redo Example 3.1 by utilizing Eq. (3.19).

Solution

From Eq. (3.19), we get

$$H_y(z, t) = \frac{E_x(z, t)}{\eta} = \frac{10}{377} \cdot \cos(\omega t - kz + 45°).$$

Hence

$$\vec{H} = H_y \cdot \hat{y} = \frac{10}{377} \cdot \cos(\omega t - kz + 45°) \cdot \hat{y}.$$

From Example 3.2, once the E-field is known, we can skip phasor approach and directly derive the M-field from Eq. (3.19).

∎

C   Backward Wave

In the previous subsection, we consider a forward wave. Here we consider a wave having an inverse direction: backward wave (The detail discussions of backward waves can be referred to Sect. 2.3). A backward wave propagates toward −z direction and the corresponding E-field is given by

$$\vec{E} = E_x(z) \cdot \hat{x}, \tag{3.22}$$

where $E_x(z)$ is the phasor of E-field, given by

$$E_x(z) = E_b e^{jkz}. \tag{3.23}$$

In Eq. (3.23), E_b is the phasor of E-field at $z = 0$ and k is the wave number. As we take differentiation of Eq. (3.23), we have

$$\frac{\partial E_x(z)}{\partial z} = jk E_b e^{jkz} = jk E_x(z). \tag{3.24}$$

Inserting Eq. (3.24) into Eq. (3.5), we have

$$jkE_x(z) \cdot \hat{y} = -j\omega\mu\vec{H}. \tag{3.25}$$

Hence, the M-field $\vec{H}$ only has y-component and can be represented by

$$\vec{H} = H_y(z) \cdot \hat{y} = -\frac{k}{\omega\mu} E_x(z) \cdot \hat{y}, \tag{3.26}$$

where

$$H_y(z) = -\frac{k}{\omega\mu} E_x(z) = -\frac{E_x(z)}{\eta}. \tag{3.27}$$

Hence, the ratio of $E_x(z)$ to $H_y(z)$ is given by

$$\frac{E_x(z)}{H_y(z)} = -\eta. \tag{3.28}$$

Comparing Eq. (3.28) with Eq. (3.11), we find that the difference between a backward wave and a forward wave is simply a sign.

Following the same logics, suppose $E_x(z, t)$ and $H_y(z, t)$ represent the time-dependent E-field and time-dependent M-field, respectively. Then we have

$$\frac{E_x(z, t)}{H_y(z, t)} = -\eta \tag{3.29}$$

for a backward wave. Hence, if the E-field is given by

$$E_x(z, t) = B \cdot \cos(\omega t + kz + \varphi). \tag{3.30}$$

The associated M-field $H_y(z, t)$ can be directly derived by

$$H_y(z, t) = -\frac{B}{\eta} \cdot \cos(\omega t + kz + \varphi). \tag{3.31}$$

Example 3.3

Suppose the E-field of a backward wave is given by $\vec{E} = E_x \cdot \hat{x} = 5 \cdot \cos(\omega t + kz + 45°) \cdot \hat{x}$. If the permittivity $\epsilon = 4\ \epsilon_0$, please derive the associated M-field $\vec{H} = H_y \cdot \hat{y}$.

Solution

First, the refractive index is given by

$$n = \sqrt{4} = 2.$$

Hence, the wave impedance is given by

$$\eta = \frac{\eta_0}{n} = \frac{377}{2} (\Omega).$$

Then from Eq. (3.29), we derive the time-dependent M-field as

$$H_y(z, t) = -\frac{E_x(z, t)}{\eta} = -\frac{10}{377} \cos(\omega t + kz + 45°).$$

The vector M-field is given by

$$\vec{H} = H_y(z, t) \cdot \hat{y} = -\frac{10}{377} \cos(\omega t + kz + 45°) \cdot \hat{y}.$$

∎

D   Forward Wave and Backward Wave Coexist

Finally, we consider the case when a forward wave and a backward wave coexist. Suppose we have a plane wave consisting of a forward wave and a backward wave. The E-field $\vec{E} = E_x(z) \cdot \hat{x}$ is given by

$$E_x(z) = E_a e^{-jkz} + E_b e^{jkz}, \tag{3.32}$$

where E_a and E_b are E-field phasors of the forward wave and the backward wave, respectively, at $z = 0$. Suppose the associated M-field of $\vec{E}$ is $\vec{H} = H_y(z) \cdot \hat{y}$, and $H_y(z)$ is represented by

$$H_y(z) = H_a e^{-jkz} + H_b e^{jkz}, \tag{3.33}$$

where H_a and H_b are M-field phasors of the forward wave and the backward wave, respectively, at $z = 0$. From the previous results, we have

$$H_a = \frac{E_a}{\eta}, \tag{3.34}$$

$$H_b = -\frac{E_b}{\eta}. \tag{3.35}$$

Hence, $H_y(z)$ can be rewritten as

$$H_y(z) = \frac{E_a}{\eta}e^{-jkz} - \frac{E_b}{\eta}e^{jkz} = \frac{1}{\eta}\left(E_a e^{-jkz} - E_b e^{jkz}\right). \tag{3.36}$$

From Eqs. (3.32) and (3.36), we have

$$\frac{E_x(z)}{H_y(z)} = \eta \cdot \frac{E_a e^{-jkz} + E_b e^{jkz}}{E_a e^{-jkz} - E_b e^{jkz}}. \tag{3.37}$$

Equation (3.37) shows that when a forward wave and a backward wave coexist, the relationship between E-field and M-field becomes much more complicated and can not be simply described by a constant ratio. Actually, it depends not only on η, but also on (E_a, E_b) and (e^{-jkz}, e^{jkz}). In other words, it depends on wave impedance, E-field magnitude, phase, and position.

Example 3.4

Given a plane wave consisting of a forward wave and a backward wave, the E-field is $\vec{E} = E_x \cdot \hat{x}$, where $E_x = E_a e^{-jkz} + E_b e^{jkz}$. Suppose $E_a = 10e^{j\frac{\pi}{3}}$ and $E_b = 7e^{j\frac{2\pi}{5}}$. If the medium has the refractive index $n = 3$ and the wave number $k = \frac{\pi}{4}$ (m^{-1}). Please derive the associated M-field $\vec{H} = H_y \cdot \hat{y}$ at $z = 5$ m.

Solution

First, the phasor of E-field of the forward wave at $z = 5$ m is given by

$$E^+ = E_a e^{-jkz} = (10e^{j\frac{\pi}{3}}) \cdot e^{-j(\frac{\pi}{4}\times 5)} = 10e^{-j\frac{11}{12}\pi}.$$

Besides, the phasor of E-field of the backward wave at $z = 5$ m is given by

$$E^- = E_b e^{jkz} = (7e^{j\frac{2\pi}{5}}) \cdot e^{j(\frac{\pi}{4}\times 5)} = 7e^{j\frac{33}{20}\pi}.$$

Next, the wave impedance of the medium is derived by

$$\eta = \frac{\eta_0}{n} = \frac{377}{3}.$$

Therefore, the associated M-fields of the forward wave and the backward wave at $z = 5$ m are given by

$$H^+ = \frac{E^+}{\eta} = \frac{30}{377} \cdot e^{-j\frac{11}{12}\pi},$$

$$H^- = -\frac{E^-}{\eta} = -\frac{21}{377} \cdot e^{j\frac{33}{20}\pi},$$

respectively. Hence, the M-field phasor at $z = 5$ m is given by

$$H_y = H^+ + H^- = \frac{30}{377} \cdot e^{-j\frac{11}{12}\pi} - \frac{21}{377} \cdot e^{j\frac{33}{20}\pi}.$$

Finally, the time-dependent M-field is given by

$$\vec{H}(z,t) = \text{Re}\{H_y \cdot e^{j\omega t}\} \cdot \hat{y} = [\frac{30}{377} \cdot \cos(\omega t - \frac{11}{12}\pi) - \frac{21}{377} \cdot \cos(\omega t + \frac{33}{20}\pi)] \cdot \hat{y}.$$

∎

From Example 3.4, we learn that when forward waves and backward waves coexist, we shall deal with each individually using the property of the wave impedance, and finally linearly combine them to attain the overall result. This approach is effective in dealing with related problems.

Summary

We summarize this section as follows:

Case 1: *Forward wave* $\Rightarrow \frac{E_x}{H_y} = \eta$,

Case 2: *Backward wave* $\Rightarrow \frac{E_x}{H_y} = -\eta$,

Case 3: *Forward wave* + *Backward wave* $\Rightarrow \frac{E_x}{H_y} \neq constant$.

When a plane wave consists of only forward waves or backward waves, the relationship between E-field and M-field is quite simple. The wave impedance characterizes the relationship. The M-field can be directly derived from the associated E-field, and vice versa. However, when both forward waves and backward waves coexist, the problem becomes much more complicated as illustrated in Example 3.4.

3.2 Group Velocity

The velocity is a physical quantity we can observe everywhere in our daily life. For example, when a person walks a distance of Δz during a time interval Δt, the velocity is given by

$$v = \frac{\Delta z}{\Delta t}. \tag{3.38}$$

Equation (3.38) defines the velocity of a moving object by the rate of change of its position. This definition is simple and consistent with our intuition.

Next, let us consider a more complicated case:

Suppose we do not have only one moving object, but many. Then how do we define the velocity of these moving objects?

For example, we can easily define the walking velocity of a person by Eq. (3.38). But what if we have a group of people walking individually and each has a distinct velocity? How do we define the velocity of the whole group?

The above question occurs not only in our daily life, but also in EM waves. This is what we are going to study in this section.

A Phase Velocity

First, we consider an EM wave consisting of only a sinusoid. Suppose a plane wave has the frequency ω and propagates toward $+z$. The E-field is given by

$$\vec{E} = E_x \cdot \hat{x} = A \cos(\omega t - kz) \cdot \hat{x}, \tag{3.39}$$

where A is the amplitude and k is the wave number given by

$$k = \frac{n\omega}{c}, \tag{3.40}$$

where n is the refractive index and c is the speed of light.

It is easy to define the velocity of the EM wave. Similar to how we define the walking velocity of a person in Eq. (3.38), we can rewrite Eq. (3.39) as

$$E_x = A \cos \Omega, \tag{3.41}$$

where Ω is the phase given by

$$\Omega = \omega t - kz. \tag{3.42}$$

Hence, Ω depends on time t and position z.

In Fig. 3.2, when $t = t_1$ and $z = z_1$, the associated E-field at the point P of the EM wave is given by

$$E_x = \cos(\omega t_1 - kz_1) = \cos \Omega_P, \tag{3.43}$$

where Ω_P is the phase of point P and

$$\Omega_P = \omega t_1 - kz_1. \tag{3.44}$$

In Fig. 3.3, when $t = t_1 + \Delta t$, the point P propagates from $z = z_1$ to $z = z_2 = z_1 + \Delta z$. As derived in Sect. 2.4, the velocity of the point P is given by

$$v = \frac{\Delta z}{\Delta t} = \frac{\omega}{k}. \tag{3.45}$$

Fig. 3.2 Explaining phase
velocity of a sinusoidal wave
(I)

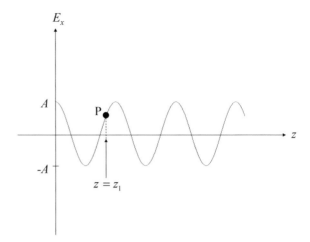

Fig. 3.3 Explaining phase
velocity of a sinusoidal wave
(II)

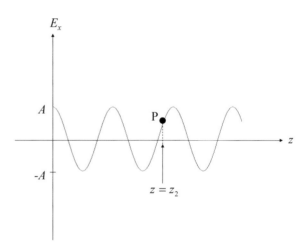

Since $k = n\omega/c$, we have

$$v = v_p = \frac{c}{n}. \tag{3.46}$$

From Eq. (3.46), the velocity depends on the refractive index n.

In above, for a single-frequency EM wave, we can derive the velocity of EM wave by Eq. (3.46). We call it the **phase velocity** and denote it by v_p. Conventionally, we use v_p describing the velocity of a single-frequency EM wave.

B Group Velocity

For an EM wave consisting of a single sinusoid, its velocity can be simply defined by the phase velocity v_p. Unfortunately, an EM wave always consists of multiple sinusoids in practice. Besides, even if we have the best oscillator in the world, we cannot generate an EM wave purely consisting of a single-frequency component. Therefore, we shall treat an EM wave as a combination of sinusoids and each sinusoid has a distinct frequency.

Next, suppose we have an EM wave consisting of a group of sinusoids having different frequencies, and we want to derive the velocity of this wave. Considering the simplest case so that each sinusoid has the same phase velocity, then the velocity of the whole group shall be equal to the phase velocity. It can be seen in the example of a walking group. If everyone walks at the same speed, the velocity of the whole group shall be equal to the speed. However, it is not usually the case in real world.

From Eq. (3.46), the phase velocity depends on the refractive index n of the medium. Generally, n is a function of frequency, i.e., $n = n(\omega)$. The physical meaning of refractive index n can be referred in Appendix C. For a plane wave consisting of multiple frequency components, the associated refractive index $n(\omega)$ is thus different. Hence, different sinusoids shall have different phase velocities. Here, we encounter a rather complicated question:

Q: How do we define the velocity of a group of sinusoids and each sinusoid has a distinct phase velocity?

It is similar to the example of a walking group: how do we define the velocity of the whole group when each one has a distinct speed?

First, let us consider a simple case. Suppose we have a plane wave consisting of two sinusoids. Each one has a distinct frequency and the associated phase velocity. The wave travels toward $+z$ and the associated E-field is represented by

$$\vec{E} = E_x \cdot \hat{x} = [\cos(\omega_1 t - k_1 z) + \cos(\omega_2 t - k_2 z)] \cdot \hat{x}, \tag{3.47}$$

where $\vec{E}$ consists of two sinusoids with different frequencies ω_1 and ω_2. The corresponding wave numbers are given by

$$k_1 = \frac{n_1 \omega_1}{c}, \tag{3.48}$$

$$k_2 = \frac{n_2 \omega_2}{c}, \tag{3.49}$$

where $n_1 = n(\omega_1)$, $n_2 = n(\omega_2)$, and $n_1 \neq n_2$. The respective phase velocities are given by

$$v_{p1} = \frac{c}{n_1}, \tag{3.50}$$

$$v_{p2} = \frac{c}{n_2}. \tag{3.51}$$

Because $n_1 \neq n_2$, we have $v_{p1} \neq v_{p2}$.

Now, we face a problem: how do we define the velocity of the EM wave having two frequency components? Intuitively, we may take the average of the phase velocities as the answer, which is given by

$$v = \frac{v_{p1} + v_{p2}}{2}. \tag{3.52}$$

Unfortunately, this answer lacks of physical insight, and the answer to be derived is much more complicated than Eq. (3.52).

Suppose $\omega_2 = \omega_1 + \Delta\omega$ and $k_2 = k_1 + \Delta k$, where $\Delta\omega = \omega_2 - \omega_1$ and $\Delta k = k_2 - k_1$. Using the trigonometric formula $\cos A + \cos B = 2 \cos \frac{A+B}{2} \cos \frac{A-B}{2}$, the E-field in Eq. (3.47) can be rewritten as

$$E_x = \cos(\omega_1 t - k_1 z) + \cos[(\omega_1 + \Delta\omega)t - (k_1 + \Delta k)z]$$

$$= 2 \cos\left[\left(\omega_1 + \frac{\Delta\omega}{2}\right)t - \left(k_1 + \frac{\Delta k}{2}\right)z\right] \cdot \cos\left(\frac{\Delta\omega}{2}t - \frac{\Delta k}{2}z\right). \tag{3.53}$$

Assuming $\omega_1 >> \Delta\omega$ and $k_1 >> \Delta k$, we can approximate Eq. (3.53) by

$$E_x \approx 2 \cdot \underbrace{\cos(\omega_1 t - k_1 z)}_{\text{high - frequency}} \cdot \underbrace{\cos\left(\frac{\Delta\omega}{2}t - \frac{\Delta k}{2}z\right)}_{\text{low - frequency}}. \tag{3.54}$$

As denoted in Eq. (3.54), the first term is the high-frequency component having the frequency ω_1, and the second term is the low-frequency component having the frequency $\Delta\omega/2$.

In order to investigate the velocity of this EM wave, we rewrite Eq. (3.54) as

$$E_x = S(t, z) \cdot \cos(\omega_1 t - k_1 z), \tag{3.55}$$

where

$$S(t, z) = 2 \cdot \cos\left(\frac{\Delta\omega}{2}t - \frac{\Delta k}{2}z\right). \tag{3.56}$$

In Eq. (3.55), we have a high-frequency component $\cos(\omega_1 t - k_1 z)$ whose amplitude is enveloped by the low-frequency component $S(t, z)$ as shown in Fig. 3.4. For a fixed time t, we draw a dashed line representing the high-frequency component $\cos(\omega_1 t - k_1 z)$ and a solid line representing the low-frequency component $S(t, z)$. From Fig. 3.4, it can be seen that the "envelope" of this waveform is determined by $S(t, z)$, instead of $\cos(\omega_1 t - k_1 z)$.

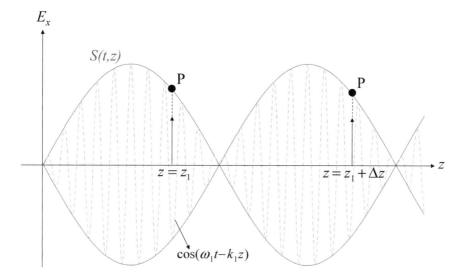

Fig. 3.4 Explaining group velocity

In most applications, what we really care is the velocity of the low-frequency component $S(t, z)$, but not the high-frequency component $\cos(\omega_1 t - k_1 z)$. For example, in a radio system, we usually use a low-frequency audio signal to modulate a high-frequency sinusoidal carrier. Then we can transmit the modulated waveform through a proper antenna. In this case, what we really care is the velocity of the low-frequency audio signal, instead of the high-frequency carrier, since the desired information is embedded in the audio signal.

We also can appreciate this idea from another example in nature. Imagine that you are a hungry shark wandering in the ocean and you see a large group of fishes as shown in Fig. 3.5. Each fish swims at a different velocities in the group. In this case, you do not really care the velocity of an individual fish. What you really care is the velocity of the whole fish group, i.e., the envelope, because it determines whether you can get something to eat or not. In this example, the envelope is like the waveform of $S(t, z)$ in Fig. 3.4, and the velocity of $S(t, z)$ represents the velocity of the whole group.

Next, similar to the derivation of the phase velocity of a sinusoid, we can derive the velocity of $S(t, z)$. From Eq. (3.56), when $t = t_1$ and $z = z_1$, $S(t, z)$ is given by

$$S(t_1, z_1) = 2 \cos\left(\frac{\Delta\omega}{2} t_1 - \frac{\Delta k}{2} z_1 \right) = 2 \cos \Omega_P, \qquad (3.57)$$

where Ω_P can be treated as the phase of $S(t, z)$ and given by

Fig. 3.5 Example of group velocity in nature

$$\Omega_P = \frac{\Delta\omega}{2}t_1 - \frac{\Delta k}{2}z_1. \tag{3.58}$$

It can be illustrated as the point P in Fig. 3.4. Now, suppose when $t = t_1 + \Delta t$, the point P travels to the same phase at location $z = z_1 + \Delta z$. Then we have Ω_P given by

$$\Omega_P = \frac{\Delta\omega}{2}(t_1 + \Delta t) - \frac{\Delta k}{2}(z_1 + \Delta z). \tag{3.59}$$

From Eqs. (3.58) and (3.59), we obtain

$$\Delta\omega \cdot \Delta t - \Delta k \cdot \Delta z = 0. \tag{3.60}$$

Hence, the velocity of $S(t, z)$ is given by

$$v = \frac{\Delta z}{\Delta t} = \frac{\Delta\omega}{\Delta k}. \tag{3.61}$$

From above, when an EM wave consists of two sinusoids, the velocity of the corresponding $S(t, z)$ is $\Delta\omega/\Delta k$. As $\Delta\omega \to 0$ and $\Delta k \to 0$, Eq. (3.61) can be rewritten as

$$v = \frac{d\omega}{dk}. \tag{3.62}$$

In the previous case of an EM wave consisting of two sinusoids, we derive the velocity of $S(t, z)$. Actually, we can extend the result in Eq. (3.62) and derive the velocity of an EM wave consisting of multiple sinusoids. The velocity is given by

$$v = v_g = \frac{d\omega}{dk}\Big|_{\omega=\omega_c}, \tag{3.63}$$

where ω_c is the **center frequency** of the composed sinusoids. The result v_g in Eq. (3.63) is called **group velocity,** which represents the velocity of an EM wave consisting of a group of sinusoids. In fact, this velocity represents the speed of the desired signal bearing information and hence also called **signal velocity**.

In above, the velocity of an EM wave falls into two categories:

1. For an EM wave consisting of a single sinusoid, the speed is called phase velocity and given by

$$v_p = \frac{\omega}{k}.$$

2. For an EM wave consisting of multiple sinusoids, the associated speed is called group velocity and given by

$$v_g = \frac{d\omega}{dk}\Big|_{\omega=\omega_c},$$

where ω_c is the center frequency.

In most cases, these two categories provide what we care about the velocity of an EM wave.

C Dispersive Medium

In general, the wave number k is a function of frequency ω. Hence, the group velocity in Eq. (3.63) is usually rewritten as

$$v_g = \frac{d\omega}{dk} = \frac{1}{dk/d\omega}. \tag{3.64}$$

Because $k = \frac{n\omega}{c}$, we have

$$\frac{dk}{d\omega} = \frac{1}{c} \cdot (n + \omega \cdot \frac{dn}{d\omega}). \tag{3.65}$$

Hence, Eq. (3.64) can be rewritten as

$$v_g = \frac{c}{n + \omega \cdot \frac{dn}{d\omega}}. \tag{3.66}$$

Comparing Eq. (3.46) with Eq. (3.66), we find the difference between phase velocity and group velocity is the term $\omega \frac{dn}{d\omega}$ in denominator. If $\frac{dn}{d\omega} = 0$, then $v_g = v_p$; if $\frac{dn}{d\omega} \neq 0$, then $v_g \neq v_p$.

If the refractive index n of a transmission medium does not depend on the frequency, we call the medium a **non-dispersive medium**. For a non-dispersive medium, because $\frac{dn}{d\omega} = 0$, we have

$$v_g = \frac{c}{n} = v_p. \tag{3.67}$$

Hence, in this case, the group velocity is equal to the phase velocity. This result is reasonable because as the refractive index n is independent of the frequency, different sinusoids will have an identical phase velocity. Thus, the group velocity is equal to each individual phase velocity, i.e., $v_g = v_p$.

When the refractive index of a transmission medium depends on the frequency, we call the medium a **dispersive medium**. When a group of sinusoids propagate in this medium, because each sinusoid has its distinct phase velocity, the waveform of the whole group will "disperse". This is why we call the medium a dispersive medium. The dispersion phenomenon of an EM wave can be illustrated by the following example. Suppose we have 100 runners forming a 10×10 square running at a stadium and each one has a distinct speed. As time goes on, the "square" shape will disperse gradually. This is what happens to the propagation of an EM wave in a dispersive medium.

Finally, when a group of sinusoids propagate in a dispersive medium, we are not interested in the individual velocity of each sinusoid. What we really care is "the velocity of the whole group", which is the group velocity. This group velocity represents the speed of the whole EM wave and plays an important role in lots of applications.

Example 3.5

Suppose the refractive index n is a function of the frequency ω and given by $n(\omega) = n_0 + \alpha(\omega - \omega_0)$, where n_0, α, and ω_0 are given. Please derive the associated phase velocity and group velocity when $\omega = \omega_m$.

Solution

First, when $\omega = \omega_m$, the corresponding refractive index is given by

$$n(\omega_m) = n_0 + \alpha(\omega_m - \omega_0).$$

From Eq. (3.46), the phase velocity is given by

$$v_p = \frac{c}{n(\omega_m)} = \frac{c}{n_0 + \alpha(\omega_m - \omega_0)}.$$

Next, because

$$\frac{dn}{d\omega} = \alpha.$$

From Eq. (3.66), the group velocity is given by

$$v_g = \left(\frac{c}{n + \omega \cdot \frac{dn}{d\omega}}\right)_{\omega=\omega_m} = \frac{c}{n_0 + 2\alpha\omega_m - \alpha\omega_0}.$$

∎

Example 3.6

Suppose the refractive index n is a function of frequency f and given by $n(f) = 2 + (0.01) \cdot \left(1 + \frac{f}{f_0}\right)$, where $f_0 = 300\,\text{MHz}$. Please derive the corresponding phase velocity and group velocity when $f = 500\,\text{MHz}$.

Solution

First, when $f = 500\,\text{MHz}$, we have

$$n = 2 + (0.01) \cdot \left(1 + \frac{500}{300}\right) = 2.027.$$

From Eq. (3.46), the phase velocity is given by

$$v_p = \frac{c}{n} = \frac{3 \times 10^8}{2.027} = 1.48 \times 10^8\,(\text{m/s}).$$

Next, from Eq. (3.66), the group velocity is given by

$$v_g = \frac{c}{n + \omega\frac{dn}{d\omega}}.$$

Because $\omega = 2\pi f$, we obtain

$$\omega \cdot \frac{dn}{d\omega} = 2\pi f \cdot \frac{dn}{2\pi df} = f \cdot \frac{dn}{df} = f \cdot \frac{0.01}{f_0}.$$

Thus, when $f = 500\,\text{MHz}$, we have

$$n + f\frac{dn}{df} = 2.027 + (500) \cdot \frac{0.01}{300} = 2.044.$$

Finally, from Eq. (3.66), we get

$$v_g = \frac{c}{n + f \frac{dn}{df}} = \frac{3 \times 10^8}{2.044} = 1.47 \times 10^8 (\text{m/s}).$$

∎

3.3 Poynting Vector

The sunlight in our daily life is actually an EM wave radiated from the sun and the frequency spectrum of visible light may go up to 10^{14}Hz. The radiation not only warms up the earth, but also gives the energy to every single life. Similar to the sunlight, all the EM waves also radiate energy. In fact, we can regard EM radiation as energy transfer in the form of propagating waves.

In this section, we will focus on the energy transfer of EM waves. We will introduce a useful quantity called **Poynting vector**. This vector is named after the British scientist J. H. Poynting who first derived it. Poynting vector precisely describes the relationship between dynamic EM fields and the associated energy flux. It also indicates the direction of the energy transfer. Hence, Poynting vector is a very important quantity and readers may need to put more effort to understand the associated principles.

A   Dot Product and Cross Product

In order to understand the principles of Poynting vector, we need to be familiar with two vector operators: **dot product** and **cross product**.

Let $\vec{A}$ and $\vec{B}$ be two vectors. The associated dot product is defined as

$$\vec{A} \cdot \vec{B} = |\vec{A}| \cdot |\vec{B}| \cos \theta_{AB}, \tag{3.68}$$

where θ_{AB} is the angle between $\vec{A}$ and $\vec{B}$, and $0 \le \theta_{AB} \le \pi$.

Suppose $\vec{A} = A_x \cdot \hat{x} + A_y \cdot \hat{y} + A_z \cdot \hat{z}$ and $\vec{B} = B_x \cdot \hat{x} + B_y \cdot \hat{y} + B_z \cdot \hat{z}$. From Eq. (3.68), we can show that

$$\vec{A} \cdot \vec{B} = A_x B_x + A_y B_y + A_z B_z. \tag{3.69}$$

Thus, the dot product $\vec{A} \cdot \vec{B}$ is a scalar, which is the sum of the products of the three axial components.

On the other hand, the cross product of $\vec{A}$ and $\vec{B}$ is defined as

Fig. 3.6 Schematic plot of cross product

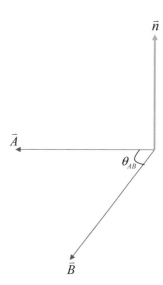

$$\vec{A} \times \vec{B} = (|\vec{A}| \cdot |\vec{B}| \sin \theta_{AB}) \cdot \hat{n}, \tag{3.70}$$

where $\hat{n}$ is a unit vector, i.e., $|\hat{n}| = 1$. The direction of $\hat{n}$ is determined by the right-hand rule as shown in Fig. 3.6. First, we put our right hand on the plane where the vectors $\vec{A}$ and $\vec{B}$ are. Next, we cross our four fingers from $\vec{A}$ to $\vec{B}$. Then the thumb will naturally point to the direction of $\hat{n}$. From above, the cross product $\vec{A} \times \vec{B}$ is a vector and it is perpendicular to both $\vec{A}$ and $\vec{B}$.

From Eq. (3.70), we can show that

$$\vec{A} \times \vec{B} = (A_y B_z - A_z B_y) \cdot \hat{x} + (A_z B_x - A_x B_z) \cdot \hat{y} + (A_x B_y - A_y B_x) \cdot \hat{z}. \tag{3.71}$$

Comparing Eq. (3.71) with Eq. (3.69), it can be found that the computation of cross product is more complicated than that of the dot product.

Finally, it can be proved that the following equality always holds:

$$\nabla \cdot \left(\vec{A} \times \vec{B} \right) = \vec{B} \cdot \left(\nabla \times \vec{A} \right) - \vec{A} \cdot \left(\nabla \times \vec{B} \right), \tag{3.72}$$

where

$\nabla \cdot \left(\vec{A} \times \vec{B} \right)$ is the divergence of $\vec{A} \times \vec{B}$,

$\vec{B} \cdot (\nabla \times \vec{A})$ is the dot product of $\vec{B}$ and $\nabla \times \vec{A}$,
$\vec{A} \cdot (\nabla \times \vec{B})$ is the dot product of $\vec{A}$ and $\nabla \times \vec{B}$.

Equation (3.72) will be used to introduce Poynting vector.

B   Poynting Vector

Because most EM waves propagate in dielectric media, we only consider dielectric media here. First, suppose the current density $\vec{J} = 0$ in a dielectric medium. From Faraday's law and Ampere's law, we have

$$\nabla \times \vec{E} = -\mu \frac{\partial \vec{H}}{\partial t}, \tag{3.73}$$

$$\nabla \times \vec{H} = \epsilon \frac{\partial \vec{E}}{\partial t}. \tag{3.74}$$

These two equations signify the interplay between E-field and M-field. Moreover, from Eq. (3.72), we have the following equality:

$$\nabla \cdot \left(\vec{E} \times \vec{H} \right) = \vec{H} \cdot \left(\nabla \times \vec{E} \right) - \vec{E} \cdot \left(\nabla \times \vec{H} \right). \tag{3.75}$$

Inserting Eq. (3.73) and Eq. (3.74) into Eq. (3.75), we obtain

$$\nabla \cdot (\vec{E} \times \vec{H}) = -\mu (\vec{H} \cdot \frac{\partial \vec{H}}{\partial t}) - \epsilon (\vec{E} \cdot \frac{\partial \vec{E}}{\partial t}). \tag{3.76}$$

For a vector $\vec{A}$, it can be proved that

$$\frac{\partial |\vec{A}|^2}{\partial t} = \frac{\partial (\vec{A} \cdot \vec{A})}{\partial t} = \frac{\partial \vec{A}}{\partial t} \cdot \vec{A} + \vec{A} \cdot \frac{\partial \vec{A}}{\partial t} = 2\vec{A} \cdot \frac{\partial \vec{A}}{\partial t}. \tag{3.77}$$

Using Eq. (3.77), we can rewrite Eq. (3.76) as

$$\nabla \cdot (\vec{E} \times \vec{H}) = -\mu \cdot \frac{1}{2} \frac{\partial |\vec{H}|^2}{\partial t} - \epsilon \cdot \frac{1}{2} \frac{\partial |\vec{E}|^2}{\partial t} = -\frac{\partial}{\partial t} \left(\frac{1}{2} \mu |\vec{H}|^2 + \frac{1}{2} \epsilon |\vec{E}|^2 \right). \tag{3.78}$$

Equation (3.78) is directly derived from Maxwell's equations, and the associated physical meaning will reveal an important principle of EM waves.

First, from static electromagnetics, the first term $\frac{1}{2}\mu |\vec{H}|^2$ at the right side of Eq. (3.78) denotes the stored energy density in the magnetic field, and the second term $\frac{1}{2} \in |\vec{E}|^2$ denotes the stored energy density in the electric field.[1] The unit is Joule/m^3, which means the energy per unit volume. Hence, Eq. (3.78) is related to the energy of EM fields.

In order to get more insight of Eq. (3.78), we consider the variation of the energy stored in an arbitrary volume V. We do volume integration at both sides of Eq. (3.78) over V and obtain

$$\int_V \nabla \cdot (\vec{E} \times \vec{H}) dv = -\frac{\partial}{\partial t} \int_V \left(\frac{1}{2} \mu |\vec{H}|^2 + \frac{1}{2} \in |\vec{E}|^2 \right) dv. \qquad (3.79)$$

From divergence theorem, we have

$$\int_V \nabla \cdot \left(\vec{E} \times \vec{H} \right) dv = \oint_S \left(\vec{E} \times \vec{H} \right) \cdot d\vec{s}, \qquad (3.80)$$

where S is the surface of V. Thus, Eq. (3.79) becomes

$$\oint_S \left(\vec{E} \times \vec{H} \right) \cdot d\vec{s} = -\frac{\partial}{\partial t} \int_V \left(\frac{1}{2} \mu |\vec{H}|^2 + \frac{1}{2} \in |\vec{E}|^2 \right) dv. \qquad (3.81)$$

The equality in Eq. (3.81) holds for any volume V.

In order to explore the physical meaning of Eq. (3.81), we define a parameter M as

$$M = \int_V \left(\frac{1}{2} \mu |\vec{H}|^2 + \frac{1}{2} \in |\vec{E}|^2 \right) dv. \qquad (3.82)$$

Note that in Eq. (3.82), the unit of $\frac{1}{2} \mu |\vec{H}|^2$ and $\frac{1}{2} \in |\vec{E}|^2$ is Joule/m^3, the unit of dv is m^3, and hence the unit of M is Joule. Because $\frac{1}{2} \mu |\vec{H}|^2$ and $\frac{1}{2} \in |\vec{E}|^2$ are energy density stored in M-field and E-field, respectively, the parameter M is actually the total energy stored in the volume V.

Furthermore, we define a vector given by

$$\vec{P} = \vec{E} \times \vec{H}, \qquad (3.83)$$

where $\vec{P}$ is called **Poynting vector**, which is the cross product of $\vec{E}$ and $\vec{H}$. As we insert Eq. (3.82) and Eq. (3.83) into Eq. (3.81), we have

$$\oint_S \vec{P} \cdot d\vec{s} = -\frac{\partial M}{\partial t}, \qquad (3.84)$$

---

[1] Hugh Young and Roger Freedman, "University Physics and Modern Physics," 13th edition, Pearson, 2012.

where M is the total energy of EM fields stored in a volume V. The right side of Eq. (3.84) denotes the time variation of M. The left side denotes the surface integral of $\vec{P}$ over S, where S is the surface of V. Hence, Eq. (3.84) describes the relationship between the variation of stored energy in V with respect to the Poynting vector $\vec{P}$ over the surface S.

The physical meaning of Eq. (3.84) can be illustrated with Fig. 3.7. Suppose at the time $t = t_0$, the energy stored in a volume V is M_0; when $t = t_0 + \Delta t$, the energy stored in V becomes $M_0 + \Delta M$. It means that during Δt, the energy ΔM flows into the volume V. Hence, the average power flowing into V during Δt is $\Delta M / \Delta t$. When $\Delta t \to 0$, the instant power flowing into V, denoted as P_{in}, can be represented by

$$P_{in} = \frac{\partial M}{\partial t}. \tag{3.85}$$

Thus, the outward power leaving the volume V can be defined as

$$P_{out} = -P_{in} = -\frac{\partial M}{\partial t}. \tag{3.86}$$

Hence, Eq. (3.84) can be interpreted as

$$\oint_S \vec{P} \cdot d\vec{s} = P_{out} \equiv The\ outward\ power\ leaving\ the\ volume\ V. \tag{3.87}$$

Fig. 3.7 Illustrating Poynting vector

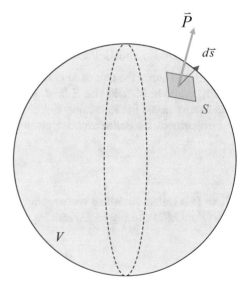

As shown in Fig. 3.7, the power leaving the volume V must transfer through its enclosed surface S, and thus $\vec{P} \cdot d\vec{s}$ represents the outward power flowing through a small area ds. The sum $\oint_S \vec{P} \cdot d\vec{s}$ represents the total power leaving V. Hence, the physical meaning of Poynting vector $\vec{P}$ can be interpreted as follows:

1. The Poynting vector represents the power flow of an EM wave. The direction of $\vec{P}$ is actually the direction of power flow of EM waves. In other words, it indicates the direction of energy transfer of an EM wave.

2. Because the unit of $\vec{E}$ is V/m and the unit of $\vec{H}$ is A/m, the unit of $\vec{P} = \vec{E} \times \vec{H}$ is $V \cdot A/m^2$. Since $V \cdot A = W$ (voltage $\times$ current $=$ power), the unit of $\vec{P}$ can be rewritten as W/m^2, which represents the power per unit area, or equivalently, **power density**. Hence, the magnitude $|\vec{P}|$ of Poynting vector represents the power density of an EM wave.

From the above illustration, $\vec{P}$ not only tells us the direction of energy transfer of EM waves, but also indicates its power density. In a word, it gives all the information we need concerning the power flow of EM waves.

Moreover, because the direction of energy transfer is the propagation direction of an EM wave, we get a critical idea:

The direction of $\vec{P}$ is the propagation direction of an EM wave.

This is shown in Fig. 3.8, where the Poynting vector $\vec{P} = \vec{E} \times \vec{H}$ and thus $\vec{P}$ is perpendicular to both $\vec{E}$ and $\vec{H}$. It means the propagation direction of an EM wave is perpendicular to the associated E-field and M-field. Note that in Eq. (3.87), the Poynting vector $\vec{P}$ denotes the power density (W/m^2), $\vec{P} \cdot d\vec{s}$ denotes the power (W), and P_{out} denotes power (W) too. Readers should pay attention to this difference, which is quite important in solving related problems.

In above, the Poynting vector $\vec{P}$ not only tells us the power density of an EM wave, but also indicates the propagation direction. Hence, it is an important and useful quantity in many applications. For example, suppose we have $\vec{E} = E_0 \cdot \hat{x}$ and $\vec{H} = H_0 \cdot \hat{y}$. Then the Poynting vector is given by

Fig. 3.8 Explaining direction of Poynting vector

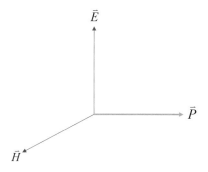

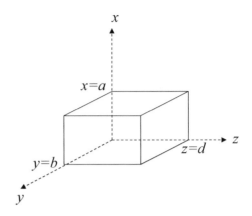

$$\vec{P} = \vec{E} \times \vec{H} = (E_0 H_0) \cdot (\hat{x} \times \hat{y}) = (E_0 H_0) \cdot \hat{z}. \qquad (3.88)$$

Equation (3.88) tells us that the associated EM wave propagates toward $+z$ and the power density is given by $|\vec{P}| = |E_0 H_0|$. These information is very useful in dealing with problems involving power flow of EM waves.

Example 3.7

As shown in Fig. 3.9, consider a rectangular box enclosed by $0 \le x \le a$, $0 \le y \le b$, and $0 \le z \le d$. Suppose a plane wave propagates toward $+z$ and its E-field and M-field are given by

$$\vec{E} = E_0 \cos(\omega t - kz) \cdot \hat{x},$$
$$\vec{H} = \frac{E_0}{\eta} \cos(\omega t - kz) \cdot \hat{y},$$

respectively, where k is the wave number and η is the wave impedance. Please derive

(a) the associated Poynting vector.
(b) the power flowing into the box at $z = 0$.
(c) the power flowing into the box at $z = d$.
(d) the power flowing into this box at time instant t.

Solution

(a) According to Eq. (3.83), we have

$$\vec{P} = \vec{E} \times \vec{H} = \frac{E_0^2}{\eta} \cos^2(\omega t - kz) \cdot (\hat{x} \times \hat{y})$$

$$= \frac{E_0^2}{\eta} \cos^2(\omega t - kz) \cdot \hat{z}.$$

Hence, the direction of $\vec{P}$ is $\hat{z}$. This is exactly the propagation direction of EM wave. In addition, $\vec{P}$ is perpendicular to $\vec{E}$ and $\vec{H}$.

(b) From the result in (a), we can derive the Poynting vector at $z = 0$ given by

$$\vec{P}_0 = \left(\frac{E_0^2}{\eta} \cos^2 \omega t \right) \cdot \hat{z}.$$

Since the direction of $\vec{P}_0$ is $+z$, the direction of power flow is $+z$ too. From Fig. 3.9, at $z = 0$, because $\vec{P}_0$ is perpendicular to xy-plane and all the points on the xy-plane have identical power density $|\vec{P}_0|$, the power flowing into the box with the area ab is given by

$$P_A = ab \cdot |\vec{P}_0| = ab \cdot \frac{E_0^2}{\eta} \cos^2 \omega t.$$

(c) From the result in (a), we can derive the Poynting vector at $z = d$ given by

$$\vec{P}_1 = \frac{E_0^2}{\eta} \cos^2(\omega t - kd) \cdot \hat{z}.$$

From Fig. 3.9, at $z = d$, the direction of $\vec{P}_1$ $(+z)$ is leaving the box and $\vec{P}_1$ is perpendicular to the plane of $z = d$. Besides, all the points on the plane have identical power density $|\vec{P}_1|$. Thus, the power flowing into the box at $z = d$ with the area ab is given by

$$P_B = -ab \cdot |\vec{P}_1| = -ab \cdot \frac{E_0^2}{\eta} \cos^2(\omega t - kd).$$

Note that the negative sign of P_B indicates that the power is leaving the box, instead of flowing into the box.

(d) The total power flowing into the box at time instant t is the sum of power flow at $z = 0$ and the power flow at $z = d$. That is,

$$P_{total} = P_A + P_B = ab \cdot \frac{E_0^2}{\eta} \cdot [\cos^2 \omega t - \cos^2(\omega t - kd)].$$

∎

C   Instantaneous Power Density and Average Power Density

In Chap. 2, we learned that EM fields can be represented by phasors. In the following, we will derive the average power density using the concept of Poynting vector and phasors. First, suppose we have a forward plane wave traveling toward $+z$ and the E-field is $\vec{E} = E_x(z) \cdot \hat{x}$, where $E_x(z)$ is the associated phasor given by

$$E_x(z) = E_a \cdot e^{-jkz}. \tag{3.89}$$

In Eq. (3.89), E_a is the E-field phasor at $z = 0$ and k is the wave number. On the other hand, the M-field is $\vec{H} = H_y(z) \cdot \hat{y}$ and

$$H_y(z) = \frac{E_x(z)}{\eta} = \frac{E_a}{\eta} \cdot e^{-jkz}, \tag{3.90}$$

where η is the wave impedance of the medium.

Let $E_a = Ae^{j\theta}$. From Eqs. (3.89) and (3.90), the instantaneous E-field and M-field are given by

$$E_x(z, t) = \mathrm{Re}\{E_x(z) \cdot e^{j\omega t}\} = A\cos(\omega t - kz + \theta), \tag{3.91}$$

$$H_y(z, t) = \mathrm{Re}\{H_y(z) \cdot e^{j\omega t}\} = \frac{A}{\eta}\cos(\omega t - kz + \theta), \tag{3.92}$$

respectively. Hence, at $z = z_0$, the magnitude of the Poynting vector is given by

$$|\vec{P}(z_0, t)| = |E_x(z_0, t) \cdot H_y(z_0, t)| = \frac{A^2}{\eta} \cdot \cos^2(\omega t - kz_0 + \theta). \tag{3.93}$$

In Eq. (3.93), $|\vec{P}(z_0, t)|$ represents the **instantaneous power density** at time instant t. Because the power density changes with time, $|\vec{P}(z_0, t)|$ is a function of t. Next, the **average power density** at $z = z_0$ is defined as

$$|\vec{P}(z_0)|_{avg} = \frac{1}{T} \int_0^T |\vec{P}(z_0, t)| dt, \tag{3.94}$$

where $T = 2\pi/\omega$ is the period of the EM wave. During a sinusoidal period, the average of $\cos^2(\omega t - kz_0 + \theta)$ is equal to $1/2$. That is,

$$\frac{1}{T} \int_0^T \cos^2(\omega t - kz_0 + \theta) dt = \frac{1}{2}. \tag{3.95}$$

Inserting Eq. (3.93) into Eq. (3.94) and using Eq. (3.95), we have

$$|\vec{P}(z_0)|_{avg} = \frac{1}{2} \cdot \frac{A^2}{\eta}. \tag{3.96}$$

From Eq. (3.96), we find that the average power density only depends on the magnitude A and the wave impedance η. It does not depend on the frequency ω or wave number k.

In addition, the phasor of E-field at $z = z_0$ is given by

$$E_x(z_0) = E_a \cdot e^{-jkz_0} = A e^{j\theta} \cdot e^{-jkz_0}. \tag{3.97}$$

Hence

$$|E_x(z_0)| = A. \tag{3.98}$$

Using Eq. (3.98), we can rewrite Eq. (3.96) as

$$|\vec{P}(z_0)|_{avg} = \frac{|E_x(z_0)|^2}{2\eta}. \tag{3.99}$$

In the above context, we use a forward plane wave as an example to explain the relationship between the E-field phasor $E_x(z_0)$ and the average power density $|\vec{P}(z_0)|_{avg}$. This relationship can be generalized to an arbitrary plane wave. For example, when we have a plane wave with E-field phasor E at some point and the wave impedance of the medium is η, the average power density can be immediately derived by

$$|\vec{P}|_{avg} = \frac{|E|^2}{2\eta}. \tag{3.100}$$

Example 3.8

Continued from Example 3.7 as shown in Fig. 3.9, please derive

(a) the average power flowing into the box at $z = 0$,
(b) the average power flowing into the box at $z = d$,
(c) the total power flowing into the box.

Solution

(a) From the result of Example 3.7 (b) and because the average of $\cos^2 \omega t$ is equal to $1/2$, the average power density at $z = 0$ is given by

$$|\vec{P}(0)|_{avg} = \frac{E_0^2}{2\eta}.$$

Since the wave is entering the box, the average power flowing into the box is given by

$$P_A = ab \cdot \left|\vec{P}(0)\right|_{avg} = ab \cdot \frac{E_0^2}{2\eta}.$$

(b) From the result of Example 3.7 (c) and because the average of $\cos^2(\omega t - kd)$ is 1/2, the average power density at $z = d$ is given by

$$|\vec{P}(d)|_{avg} = \frac{E_0^2}{2\eta}.$$

Since the wave is leaving the box at $z = d$, the average power flowing into the box is given by

$$P_B = -ab \cdot \left|\vec{P}(d)\right|_{avg} = -ab \cdot \frac{E_0^2}{2\eta}.$$

Note that the negative sign indicates that the power is leaving the box.

(c) As we utilize the results in (a) and (b), the total average power flowing into the box is given by

$$P_{total} = P_A + P_B = 0.$$

The result is consistent with our intuition because the EM wave propagates through the box toward +z. Hence, the average power flowing into the box at $z = 0$ will be equal to the average power leaving the box at $z = d$. The total average power flowing into the box is zero!

■

Example 3.9

Suppose we have an EM wave propagating in a dielectric medium with the refractive index $n = 4$. For a specific point, we have the E-field phasor $E = 10e^{j\frac{\pi}{3}}$. Please derive the average power density at the point.

Solution

First, the wave impedance of this medium can be derived by

$$\eta = \frac{\eta_0}{n} = \frac{377}{4}.$$

Then from Eq. (3.100), the average power density can be readily derived by

$$|\vec{P}|_{avg} = \frac{|E|^2}{2\eta} = \frac{10^2}{2 \cdot \frac{377}{4}} = \frac{200}{377}(\text{W/m}^2).$$

∎

3.4 Characteristics of EM Waves in Conductors

In the previous sections, we focus on EM wave propagation in a dielectric medium. A dielectric medium is regarded as an electrical insulator that can be polarized by an applied electric field. In this section, we will investigate EM wave propagation in a conductor. We will find that a conductor is not a good medium for wave propagation because an EM wave vanishes shortly after it enters a conductor.

A conductor contains a lot of free electrons. Because free electrons can move freely, a conductor has a good electrical conductivity. On the other hand, an insulator contains mainly bound electrons and very few free electrons. Therefore, when an EM wave enters a dielectric medium, the incoming electric field usually cannot move the bound electrons, but make them slightly shift from their average equilibrium positions and cause dielectric polarization. In this case, very little energy is consumed. However, when an EM wave enters a conductor, the EM energy is absorbed by lots of free electrons and makes them move. The movement consumes energy significantly and hence the EM wave vanishes quickly in a very short distance. In the following, we start from Maxwell's equations and obtain the wave equation for a conductor. Using the wave equation, we can explore the behavior of an EM wave in a conductor.

A Wave Equation for a Conductor

First, from Faraday's law and Ampere's law, we have

$$\nabla \times \vec{E} = -j\omega\mu\vec{H}, \tag{3.101}$$

$$\nabla \times \vec{H} = \vec{J} + j\omega \in \vec{E}, \tag{3.102}$$

Fig. 3.10 A rectangular medium

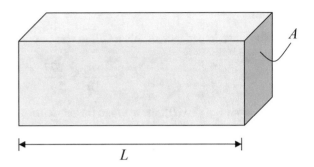

$$L$$

where $\vec{E}$ is the phasor of the electric field, $\vec{H}$ is the phasor of the magnetic field, and $\vec{J}$ is the phasor of the current density.

Because the magnitude of the induced current density $\vec{J}$ is proportional to the applied electric field $\vec{E}$, and both $\vec{E}$ and $\vec{J}$ have the same direction, we formulate their relationship as follows:

$$\vec{J} = \sigma \vec{E}, \tag{3.103}$$

where σ is the **conductivity** of the medium. For a conductor, the conductivity is very large. On the other hand, the conductivity of an insulator approximates zero.

The conductivity is closely related to the resistance. An example is shown in Fig. 3.10, where we have a rectangular medium having a cross-sectional area A and length L. Suppose the conductivity of this medium is σ and the current flows along the direction of L. From fundamental principles of electricity, the resistance of this medium is given by

$$R = \frac{1}{\sigma} \cdot \frac{L}{A}. \tag{3.104}$$

In Eq. (3.104), the resistance R is proportional to the length L and inversely proportional to the cross-sectional area A. Because the unit of the resistance is ohm (Ω), the unit of L is m (meter), and the unit of A is m^2, the unit of σ is $1/(\Omega \cdot \text{m})$. Traditionally, the unit of σ is expressed by S/m, where S denotes Simon and $1S = 1\Omega^{-1}$. The conductivity of a metal is usually very large. For example, the conductivity of copper is $5.8 \times 10^7 \, (\text{S}/\text{m})$ and the conductivity of silver is $6.17 \times 10^7 \, (\text{S/m})$. Both of them are very good conductors.

Next, suppose that we have a conductor having conductivity σ. Inserting Eq. (3.103) into Eq. (3.102), we have

$$\nabla \times \vec{H} = (\sigma + j\omega \, \epsilon) \cdot \vec{E} = j\omega \, \epsilon_{eq} \cdot \vec{E}, \tag{3.105}$$

where

$$\epsilon_{eq} = \epsilon + \frac{\sigma}{j\omega}. \tag{3.106}$$

The parameter ϵ_{eq} is called **equivalent permittivity**. Obviously, the equivalent permittivity includes the effect of the conductivity.

Readers may already find that Eq. (3.105) has the same mathematical form as that of Ampere's law for an insulator (dielectric medium). The only difference is the replacement of ϵ by ϵ_{eq}. Hence, following the same procedure as we derived the wave equation for a dielectric medium in Sect. 2.1, we get the wave equation for a conductor as follows:

$$\nabla^2 \vec{E} + k_{eq}^2 \vec{E} = 0, \tag{3.107}$$

where

$$k_{eq} = \omega \sqrt{\mu \, \epsilon_{eq}}. \tag{3.108}$$

Obviously, Eq. (3.107) shares the same mathematical form with the wave equation of a dielectric medium. The only difference is the replacement of k by k_{eq}.

B Solutions of Wave Equations

We consider a uniform plane wave propagating along z-direction with frequency ω and the electric field is $\vec{E} = E_x \cdot \hat{x}$, where E_x is a function of z only. Because Eq. (3.107) has identical mathematical form as what we have for a dielectric medium, we also have two solutions, **forward wave** and **backward wave**, given by

$$E_x(z) = E_a e^{-jk_{eq}z}, \quad \text{(forward wave)} \tag{3.109}$$

$$E_x(z) = E_b e^{jk_{eq}z}, \quad \text{(backward wave)} \tag{3.110}$$

where E_a and E_b denote the phasors of the forward wave and the backward wave at $z = 0$, respectively. The forward wave travels toward $+z$ and the backward wave travels toward $-z$.

In order to simplify Eqs. (3.109) and (3.110), we define a new parameter

$$\gamma = jk_{eq}. \tag{3.111}$$

Because γ is closely related to the propagation behavior of an EM wave, it is called **propagation constant**. From Eqs. (3.106) and (3.108), we have

$$\gamma = j\omega\sqrt{\mu \,\epsilon_{eq}} = j\omega\sqrt{\mu \,\epsilon \left(1 + \frac{\sigma}{j\omega \,\epsilon}\right)}. \tag{3.112}$$

Because the conductivity of a conductor is usually very large, we assume $\sigma \gg \omega \,\epsilon$. Then $1 + \frac{\sigma}{j\omega\epsilon} \approx \frac{\sigma}{j\omega\epsilon}$ and γ can be further simplified as

$$\gamma = j\omega\sqrt{\mu \,\epsilon \left(\frac{\sigma}{j\omega \,\epsilon}\right)} = \sqrt{j\omega\mu\sigma}. \tag{3.113}$$

Because $j = \sqrt{-1} = e^{j90°}$, we have

$$\sqrt{j} = j^{\frac{1}{2}} = e^{j45°} = \frac{1}{\sqrt{2}} + j\frac{1}{\sqrt{2}}. \tag{3.114}$$

Thus

$$\gamma = \sqrt{\frac{\omega\mu\sigma}{2}} + j\sqrt{\frac{\omega\mu\sigma}{2}} = \alpha + j\beta, \tag{3.115}$$

where

$$\alpha = \beta = \sqrt{\frac{\omega\mu\sigma}{2}} = \sqrt{\pi f \mu\sigma}. \tag{3.116}$$

In Eq. (3.115), α is the real part of γ and β is the imaginary part. They have the same value and both increase as the frequency f increases. Besides, they are proportional to $\sqrt{\sigma}$. On the other hand, α and β do not depend on permittivity ϵ. It implies that the propagation behavior of an EM wave in a conductor is determined by the conductivity, instead of the permittivity ϵ. More details of propagation constant regarding conductivity σ and permittivity ϵ are provided in Appendix D.

In the following, we explain the physical meaning of α and β.

- Forward wave

Because $\gamma = jk_{eq}$, from Eq. (3.109), we get the E-field phasor of forward wave given by

$$E_x(z) = E_a e^{-\gamma z} = E_a \cdot e^{-\alpha z} \cdot e^{-j\beta z}. \tag{3.117}$$

Suppose $E_a = Ae^{j\theta}$. The time-varying E-field is represented by

$$E_x(z, t) = \text{Re}\{E_x(z) \cdot e^{j\omega t}\} = Ae^{-\alpha z} \cdot \cos(\omega t - \beta z + \theta). \tag{3.118}$$

Equation (3.118) reveals the physical meanings of α and β. When a forward wave travels toward $+z$, its magnitude decays with $e^{-\alpha z}$. The greater the α, the more the decay. Hence, α is called **attenuation constant**. In addition, β determines the varying rate of phase with respect to z. The greater the β, the greater the varying rate of the phase along z. Hence, β is called **phase constant**. Because $\alpha = \beta = \sqrt{\pi f \mu \sigma}$, the decay and the varying rate of phase will increase when the frequency increases.

- Backward wave

From Eq. (3.110), the E-field phasor of backward wave is given by

$$E_x(z) = E_b e^{\gamma z} = E_b \cdot e^{\alpha z} \cdot e^{j\beta z}. \tag{3.119}$$

Suppose $E_b = B e^{j\varphi}$. The time-varying E-field is represented by

$$E_x(z, t) = \mathrm{Re}\{E_x(z) \cdot e^{j\omega t}\} = B e^{\alpha z} \cdot \cos(\omega t + \beta z + \varphi). \tag{3.120}$$

When a backward wave travels toward $-z$, its magnitude decays with $e^{\alpha z}$. The greater the α, the more the decay. In addition, β determines the varying rate of phase. This result is similar to the case of a forward wave.

Example 3.10

A copper has the conductivity $\sigma = 5.8 \times 10^7 \, (S/m)$ and $\mu = \mu_0 = 4\pi \times 10^{-7} (H/m)$. Suppose an EM wave has the frequency $f = 1 \, \mathrm{MHz}$.

(a) Please derive α and β.
(b) Suppose the associated forward wave has E-field phasor $E_a = 10 e^{j\frac{\pi}{3}}$ at $z = 0$. Please derive the E-field phasor at $z = 1 \, \mathrm{mm}$.
(c) Suppose the associated backward wave has the E-field phasor $E_b = 5 e^{j\frac{2\pi}{3}}$ at $z = 0$. Please derive the E-field phasor at $z = -2 \, \mathrm{mm}$.

***

Solution

(a) From Eq. (3.116), we have

$$\alpha = \beta = \sqrt{\pi f \mu \sigma} = \sqrt{\pi \cdot 10^6 \cdot (4\pi \times 10^{-7}) \cdot (5.8 \times 10^7)} = 1.5 \times 10^4 \, (m^{-1}).$$

Note that the unit of α and β is m^{-1}.

(b) When $z = 1 \mathrm{mm} = 10^{-3} \mathrm{m}$, we have

$$\alpha z = \beta z = (1.5 \times 10^4) \cdot (10^{-3}) = 15.$$

From Eq. (3.117), we obtain

$$E(z) = E_a e^{-\alpha z} e^{-j\beta z} = (10 e^{j\frac{\pi}{3}}) \cdot e^{-15} \cdot e^{-j15} = (3.06 \times 10^{-6}) \cdot e^{j(\frac{\pi}{3} - 15)}.$$

(c)   When $z = -2$mm, we have

$$\alpha z = \beta z = (1.5 \times 10^4) \cdot (-2 \times 10^{-3}) = -30.$$

From Eq. (3.119), we have

$$E(z) = E_b e^{\alpha z} e^{j\beta z} = (5 e^{j\frac{2\pi}{3}}) \cdot e^{-30} \cdot e^{-j30} = (4.68 \times 10^{-13}) \cdot e^{j(\frac{2\pi}{3} - 30)}.$$

■

From Example 3.10, when an EM wave enters a conductor, it decays significantly even in a very short traveling distance.

C   Phase Velocity

From Eq. (3.115), the propagation constant for a conductor is $\gamma = \alpha + j\beta$, which is different from the term jk we get for a dielectric medium in Sect. 2.3. It brings two major differences. The first difference is induced by the imaginary part β.
   Suppose a forward wave travels along $+z$ and the E-field is given by

$$E_x(z, t) = A e^{-\alpha z} \cdot \cos(\omega t - \beta z + \theta). \qquad (3.121)$$

We can rewrite Eq. (3.121) as

$$E_x(z, t) = A e^{-\alpha z} \cdot \cos \Omega, \qquad (3.122)$$

where Ω is representing the phase given by

$$\Omega = \omega t - \beta z + \theta. \qquad (3.123)$$

Using Eq. (3.123) and the similar derivation in Sect. 3.2, we can obtain the associated phase velocity given by

$$v_p = \frac{\omega}{\beta}. \qquad (3.124)$$

Inserting Eq. (3.116) into Eq. (3.124), we get

$$v_p = \sqrt{\frac{4\pi f}{\mu \sigma}}. \qquad (3.125)$$

Equation (3.125) shows that the phase velocity of a conductor depends on the frequency. On the other hand, the phase velocity of a dielectric medium is $v_p = \frac{c}{n}$, which is determined by the refractive index and is independent of the frequency. Besides, the phase velocity of a conductor is usually much lower than that of a dielectric medium. This can be seen in the following example.

Example 3.11

Suppose a copper has conductivity $\sigma = 5.8 \times 10^7 (S/m)$ and $\mu = \mu_0 = 4\pi \times 10^{-7} (H/m)$. Please derive the phase velocity of an EM wave in a copper when the frequency is $f = 1\,MHz$.

Solution

From Eq. (3.125), we have

$$v_p = \sqrt{\frac{4\pi f}{\mu \sigma}} = \sqrt{\frac{4\pi \cdot (1 \times 10^6)}{(4\pi \times 10^{-7}) \cdot (5.8 \times 10^7)}} = 415\,(m/s).$$

From this example, we find that the phase velocity of a conductor is much lower than that of a dielectric medium. For example, the air has refractive index n approximately 1 and the associated phase velocity is about the speed of light, which is $3 \times 10^8 (m/s)$!

∎

D Skin Effect

The second difference between a conductor and that of a dielectric medium comes from the attenuation constant α. It induces an important effect called **skin effect**.

Suppose we have a forward wave traveling along $+z$. From Eq. (3.121), the E-field decays with $e^{-\alpha z}$. We define a new parameter

$$\delta = \frac{1}{\alpha} = \frac{1}{\sqrt{\pi f \mu \sigma}}. \tag{3.126}$$

Because the unit of α is $1/m$, the unit of δ is m. The physical meaning of δ is simple and straightforward. When an EM wave travels a distance of δ in a conductor, the E-field will decay by a factor of $e^{-\alpha\delta} = e^{-1} = 0.368$. When traveling a distance of 2δ, the E-field will decay $e^{-2} = 0.135$. When traveling a distance of $q\delta$, the

E-field will decay e^{-q}. Hence, the smaller the δ, the greater the decay rate for a given distance.

Example 3.12

Suppose a copper has the conductivity $\sigma = 5.8 \times 10^7 \, (\text{S/m})$ and $\mu = \mu_0 = 4\pi \times 10^{-7} \, (\text{H/m})$. Please derive the corresponding δ when the frequency $f = 9 \, \text{MHz}$.

Solution

From Eq. (3.126), we have

$$\delta = \frac{1}{\sqrt{\pi f \mu \sigma}} = \frac{1}{\sqrt{\pi \cdot (9 \times 10^6) \cdot (4\pi \times 10^{-7}) \cdot (5.8 \times 10^7)}}$$
$$= 2.2 \times 10^{-5} \, (\text{m}) = 0.022 \, (\text{mm}).$$

■

In Example 3.12, since δ is quite small, an EM wave decays significantly in a very short distance in a conductor. From Example 3.12 of a copper, if the frequency is $f = 9 \, \text{MHz}$, the corresponding δ is 0.022 mm. Hence, when the EM wave enters the surface of a copper, at a depth of 0.022 mm, the E-field decays by a factor of $e^{-1} = 0.368$. Furthermore, at a depth of 0.044 mm, the E-field decays $e^{-2} = 0.135$. At a depth of 1 mm, the E-field decays $e^{-45} = 2.9 \times 10^{-20}$ (1 mm is about 45δ)!

From the previous example, we learn that when an EM wave enters a conductor, the energy vanishes quickly in a very short distance. This phenomenon is called **skin effect**. In practice, when an EM wave enters a conductor, we still can detect the wave at a depth of δ. However, when the depth is larger than several δ, the energy is so small that we can neglect it. Because δ is usually very small, the EM field is considered as **"concentrated" at the skin (surface) of a "conductor"** and thus we name the phenomenon skin effect and δ is called **skin depth**.

By applying skin effect, we can use a thin plate of metal to shield a circuit because EM waves cannot penetrate the plate. It can prevent the circuit being interfered by undesired environmental EM waves. In addition, because skin depth δ decreases when the frequency increases, skin effect becomes more significant for a high-frequency EM wave.

Finally, it is worth to mention a concept which will be used in the remaining context. When dealing with EM waves in a conductor, because of skin effect, we usually assume that

All the prominent behaviors of an EM wave only occur at a depth of δ under the surface of a conductor.

As an example, a copper block with a thickness of 10 cm is illustrated in Fig. 3.11. When an EM wave having the frequency $f = 9 \, \text{MHz}$ enters the copper block, the skin depth δ is only 0.022 mm as derived in Example 3.12. Therefore, **we may**

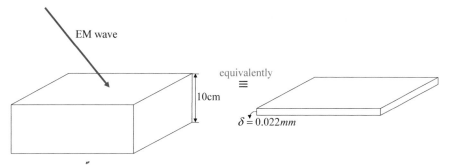

Fig. 3.11 Illustrating skin effect

consider the copper block as a metal plate having a thickness of 0.022 mm. This is because all the prominent behaviors of the EM wave concentrate within the depth of δ. The concept is very useful, which greatly simplify problems when dealing with EM waves involving conductors.

Summary

We learn important parameters of EM waves in this chapter, which includes four sections:

3.1: Wave impedance
   We learn the principles of wave impedance. Once the wave impedance is given, the M-field can be directly derived from the associated E-field, and vice versa.
3.2: Group velocity
   We learn the physical meaning of group velocity and introduce dispersive media.
3.3: Poynting vector
   We learn the principle of energy transfer and power flow of EM waves. The physical meaning of Poynting vector follows.
3.4: Characteristics of EM waves in conductors
   We learn the behavior of EM waves in conductors. The skin effect and the skin depth are introduced.

   From Chap. 1 through this chapter, we learn principles and behaviors of an EM wave in a single medium. In the next chapter, we will learn the behaviors of an EM wave when it crosses the interface between two different media.

Exercises

1. Suppose $\overrightarrow{H} = H_0 e^{-jkz} \cdot \hat{y}$ in a medium with permittivity $\in$ and permeability μ_0. Please derive the wave impedance of this dielectric medium. (Hint: $\nabla \times \vec{H} = j\omega \in \vec{E}$)

2. A forward wave propagating along +z-direction has $\vec{E} = E_0 \cos(\omega t - kz) \cdot \hat{x}$ and $E_0 = 10 \, (\text{V/m})$. Suppose $\vec{H} = H_y \cdot \hat{y}$. Please derive H_y of the following medium:

   (a) water ($\epsilon_r = 80$).
   (b) dry soil ($\epsilon_r = 3.5$).
   (c) Mica ($\epsilon_r = 6$).

   (Hint: Example 3.2)

3. A forward wave propagates along +z-direction in free space. The associated E-field is $\vec{E} = E_0 \cos(\omega t - kz + \pi/6) \cdot \hat{x}$. Suppose $\omega = 2\pi \times 10^7 (\text{rad/s})$. Please derive the E-field and M-field at $z = 10$ m.

4. A backward wave propagating along $-z$-direction has $\vec{E} = E_0 \cos(\omega t + kz) \cdot \hat{x}$ and $E_0 = 5 \, (\text{V/m})$. Suppose $\vec{H} = H_y \cdot \hat{y}$. Please derive H_y of the following material:

   (a) water ($\epsilon_r = 80$),
   (b) dry soil ($\epsilon_r = 3.5$),
   (c) Mica ($\epsilon_r = 6$).

   (Hint: Example 3.3)

5. A backward wave propagates along $-z$-direction in free space. The associated E-field is $\vec{E} = E_0 \cos(\omega t + kz - \pi/3) \cdot \hat{x}$. Suppose $\omega = 2\pi \times 10^8 (\text{rad/s})$. Please derive the E-field and M-field at $z = 20$ m.

6. An EM wave consisting of a forward wave and a backward wave is propagating in free space. The associated E-field is given by $\vec{E} = E_x \cdot \hat{x} = [4\cos(\omega t - kz) + 10\cos(\omega t + kz)] \cdot \hat{x}$.
   Suppose $\omega = 2\pi \times 10^6 (\text{rad/sec})$.

   (a) Please derive the E-field and M-field at $z = 100$ m.
   (b) Calculate the ratio of E_x/H_y at $z = 100$ m.

7. An EM wave consisting of a forward wave and a backward wave is propagating in a dielectric medium with the refractive index $n = 4$. The associated E-field is given by $\vec{E} = E_x \cdot \hat{x} = [5\cos(\omega t - kz + \pi/4) + 8\cos(\omega t + kz - \pi/3)] \cdot \hat{x}$.
   Suppose $f = 100 \, \text{MHz}$.

   (a) Please derive the E-field and M-field at $z = 10$ m when $t = 0$.
   (b) Calculate the ratio of E_x/H_y at $z = 10$ m when $t = 0$.
   (c) Please derive E-field and M-field at $z = 20$ m when $t = 1$ ns.
   (d) Calculate the ratio of E_x/H_y at $z = 20$ m when $t = 1$ ns.

   (Hint: Example 3.4)

8. The refractive index of a dispersive medium is a function of frequency and given by

$$n(f) = n_0 + \alpha \cdot (1 + \frac{f}{f_0}),$$

where $n_0 = 3$, $\alpha = 0.01$, and $f_0 = 100\,\mathrm{MHz}$. Please derive the associated phase velocity and group velocity at (a) $f = 70\,\mathrm{MHz}$, (b) $f = 150\,\mathrm{MHz}$.
(Hint: Example 3.6)

9. A wideband signal is transmitted in a dielectric medium from location A to location B, where the distance between A and B is $L = 20\,\mathrm{km}$. Suppose the signal spectrum spreads from 80 to 120 MHz, i.e., the lowest frequency component of the signal is $f_1 = 80$ and the highest one is $f_2 = 120\,\mathrm{MHz}$. The central frequency of the signal is $f_C = 100\,\mathrm{MHz}$. Suppose the dielectric medium has the same refractive index as that in Exercise 8.

(a) Please calculate the group velocity of this signal.
(b) Calculate the time delay between the signal components at f_1 and f_2.
(c) Suppose the signal is a rectangular pulse whose pulse width at location A is $T = 1\,\mu\mathrm{s}$. Please derive the pulse width at location B.

10. The wave number of an ionized medium is given by

$$k = \frac{\omega}{c}\sqrt{1 - (\frac{\omega_p}{\omega})^2}.$$

Please derive the phase velocity and the group velocity at $\omega = 2\omega_p$.
(Hint: Example 3.5)

11. For a vector field $\vec{A} = A_x \cdot \hat{x} + A_y \cdot \hat{y} + A_z \cdot \hat{z}$, prove that $\frac{\partial |\vec{A}|^2}{\partial t} = \frac{\partial (\vec{A} \cdot \vec{A})}{\partial t} = 2\vec{A} \cdot \frac{\partial \vec{A}}{\partial t}$.

12. For vector fields $\vec{A} = A_x \cdot \hat{x} + A_y \cdot \hat{y} + A_z \cdot \hat{z}$ and $\vec{B} = B_x \cdot \hat{x} + B_y \cdot \hat{y} + B_z \cdot \hat{z}$,
prove that $\nabla \cdot \left(\vec{A} \times \vec{B} \right) = \vec{B} \cdot \left(\nabla \times \vec{A} \right) - \vec{A} \cdot \left(\nabla \times \vec{B} \right)$.

13. (a) Derive the formula of Poynting vector by your way.
(b) Provide the physical meanings of the direction and magnitude of Poynting vector.
(Hint: Refer to Sect. 3.3)

14. A forward wave propagates along $+z$-direction in free space, where $\vec{E} = E_x \cdot \hat{x} = E_a e^{-ikz} \cdot \hat{x}$ and $E_a = 20(V/m)$.

(a) Assume $\vec{H} = H_y \cdot \hat{y}$. Please derive H_y.
(b) Derive the Poynting vector of this wave.
(c) Suppose the effective area of this plane wave is 100 m². Please calculate the average power flow of this wave in $+z$-direction.

15. A backward wave propagates along $-z$-direction in free space, where $\vec{E} = E_x \cdot \hat{x} = E_b e^{ikz} \cdot \hat{x}$ and $E_b = 10\,(V/m)$.

(a) Assume $\vec{H} = H_y \cdot \hat{y}$. Please derive H_y.
(b) Derive the Poynting vector of this wave.
(c) Suppose the effective area of this plane wave is 50 m². Please calculate the maximum power flow of this wave in $-z$-direction.

16. An EM wave consisting of a forward wave and a backward wave is propagating in free space, where $\vec{E} = (E_a e^{-jkz} + E_b e^{jkz}) \cdot \hat{x}$. Suppose $E_a = 8 \, (\text{V/m})$ and $E_b = 6 \, (\text{V/m})$.

   (a) Please derive the average power densities of forward wave and backward wave, respectively.
   (b) Suppose the effective area of the plane wave is 100 m². Please calculate the net power flow in +z-direction.

17. The associated Poynting vector of a uniform plane wave is $\vec{P} = 20 \cdot \hat{z} \, (\text{W/m}^2)$ in a dielectric medium. Let $\vec{E} = 30\hat{x} + 40\hat{y} \, (\text{V/m})$. Please derive

   (a) the magnetic field $\vec{H}$.
   (b) the wave impedance of the medium.
   (c) the refractive index of the medium.

18. The associated Poynting vector of a uniform plane wave is $\vec{P} = 10\hat{x} + 20\hat{y} + 75\hat{z} \, (\text{W/m}^2)$ in a dielectric medium. Let $\vec{E} = 30\hat{x} - 15\hat{y} \, (\text{V/m})$. Please derive

   (a) the magnetic field $\vec{H}$.
   (b) the wave impedance of the medium.
   (c) the refractive index of the medium.

19. Copper has $\sigma = 5.8 \times 10^7 (\text{S/m})$ and $\epsilon_r = 12$. At the following frequencies (a) $f = 1$ kHz, (b) $f = 1$ MHz, (c) $f = 1$ GHz, (d) $f = 1$ THz (10^{12} Hz), show whether it is a good conductor or not.
       (Refer to Sect. 3.4)

20. Seawater has $\sigma = 4 \, (\text{S/m})$ and $\epsilon_r = 72$. At the following frequencies
       (a) $f = 1$ kHz, (b) $f = 1$ MHz, (c) $f = 1$ GHz, show whether it is a good conductor or not.

21. Copper has $\sigma = 5.8 \times 10^7 (\text{S/m})$ and $\epsilon_r = 12$. Please calculate the associated phase velocity at the following frequencies: (a) $f = 1$ Hz, (b) $f = 1$ kHz, (c) $f = 1$ MHz.

22. Copper has $\sigma = 5.8 \times 10^7 (\text{S/m})$. In the following three cases: (1) $f = 1$ Hz, (2) $f = 100$ kHz, (3) $f = 1$ GHz, please calculate

   (a) the associated attenuation constant.
   (b) the associated skin depth.
   (c) Suppose the input power of EM wave is P_{in}. Please calculate its power after transmitting 1 mm in copper.

       (Hint: Example 3.10 and Example 3.12)

23. Seawater has $\sigma = 4 \, (\text{S/m})$. In the following three cases: (1) $f = 1$ Hz, (2) $f = 100$ kHz, (3) $f = 1$ GHz, please calculate

   (a) the associated attenuation constant.
   (b) the associated skin depth.
   (c) Suppose the input power of EM wave is P_{in}. Please calculate its power after transmitting 1 mm in seawater.

Chapter 4
Interface Behavior of EM Waves

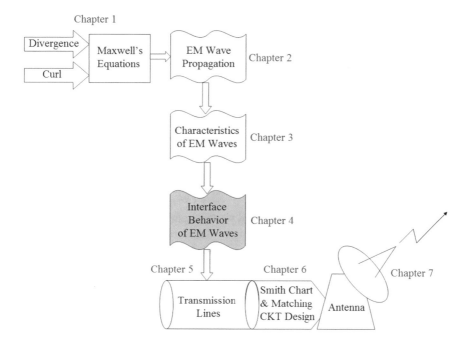

© Springer Nature Switzerland AG 2020

M.-S. Kao and C.-F. Chang, *Understanding Electromagnetic Waves*,

https://doi.org/10.1007/978-3-030-45708-2_4

Abstract In practical applications, EM waves may propagate from air to a building or from air to water. Hence during propagation, they may encounter the interface between two different mediums. At the interface, lots of interesting phenomena occur and we can utilize these phenomena in different applications. For example, electrical engineers use the interface between air and conductor to deliver signals to a far destination. Hence it is worthwhile to study the interface behavior of EM waves. Unfortunately, the study of interface is generally not an easy task, so we break the study into three steps:

> Step 1. Using Maxwell's equations to obtain the boundary conditions which EM fields must satisfy at interface.
> Step 2. Any EM wave can be decomposed into two components: the parallel polarized wave and the perpendicular polarized wave. We study each one's behavior at a dielectric interface using boundary conditions, and then we understand the behavior of any EM wave at the interface.
> Step 3. We proceed to learn the behaviors of EM waves at the conductor interface.

In this chapter, we focus on important physics and phenomena, instead of rigorous mathematics. In addition, many examples and illustrations are provided to help readers understand why and how EM waves change their behavior at an interface. The understanding forms a good background when utilizing EM waves in lots of applications.

Keywords Boundary conditions · Dielectric interface · Conductor interface · Surface charge density · Surface current density · Plane of incidence · Parallel polarized wave · Perpendicular polarized wave · Reflection · Law of reflection · Snell's law · Total reflection · Critical angle · Reflection coefficient · Transmission coefficient · Brewster angle · Polarizer · Surface resistance

Introduction

When living creatures come across an interface between two different environments, some changes may occur and they may have different responses. For example, when a group of fishes swim to the interface between a river and sea, they can feel the variation of the environment. Some fishes may keep going, some may swim around, and others may go back. In fact, not only living creatures, EM waves also change their behaviors when they encounter the interface between two media.

In a dielectric medium, EM waves propagate at a high speed with very limited energy loss. However, in a conductor, EM waves slow down and the energy loses quickly in a short distance. In many situations and applications, EM waves may propagate from a medium to another. For example in wireless communications, they may propagate from air to water. The question we need to ask is "what change may occur across the interface and how EM waves respond to these changes?" This is what we are going to find out here.

In this chapter, we explore behaviors of EM waves across an interface. Their behaviors, especially the changes between two media, are based on the **boundary conditions** derived from Maxwell's equations. By learning boundary conditions, we can understand and predict behaviors of EM waves across an interface.

4.1 Boundary Conditions of Electric Fields

An EM wave consists of dynamic electric fields and magnetic fields. In the first section, we discuss two cases of the boundary conditions of electric fields: at an interface between two dielectric media, and at an interface between a dielectric medium and a conductor. The corresponding boundary conditions of magnetic fields will be discussed in the next section.

A Interface between Two Dielectric Media

Suppose medium 1 and medium 2 are dielectric media; their interface is shown in Fig. 4.1. The associated permittivity and permeability are (ϵ_1, μ_1) for medium 1 and (ϵ_2, μ_2) for medium 2, where $\epsilon_1 \neq \epsilon_2$ and we assume $\mu_1 = \mu_2 = \mu_0$. In the following, we show the boundary conditions that an electric field must satisfy at this interface.

First, Maxwell's equations still hold at interfaces. For the boundary conditions of electric field, we need **Faraday's Law** and **Gauss's Law**. Suppose the charge density is zero in these two media, i.e., $\rho = 0$. Then we have

$$\nabla \times \vec{E} = -\frac{\partial \vec{B}}{\partial t} \tag{4.1}$$

$$\nabla \cdot \vec{D} = 0 \tag{4.2}$$

where Eq. (4.1) is Faraday's Law and Eq. (4.2) is Gauss's Law.

For an arbitrary electric field $\vec{E}$ at the interface between medium 1 and medium 2 as shown in Fig. 4.2, it can be decomposed as a **tangential component** $\vec{E}_t$ and a **normal component** $\vec{E}_n$. Hence we have

Medium 1: (ϵ_1, μ_1)
_____ Interface
Medium 2: (ϵ_2, μ_2)

Fig. 4.1 Interface between two media

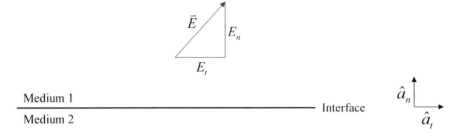

Fig. 4.2 Decomposition of electric field at interface (I)

$$\vec{E} = \vec{E}_t + \vec{E}_n = E_t \hat{a}_t + E_n \hat{a}_n, \tag{4.3}$$

where $\hat{a}_t$ is the unit vector parallel to the interface, $\hat{a}_n$ is the unit vector perpendicular to the interface, and E_t and E_n are their respective magnitudes. For example, if the interface is the xy-plane, then $\hat{a}_t$ is a vector parallel to the xy-plane, and $\hat{a}_n$ is a vector perpendicular to the xy-plane, i.e., $\hat{a}_n = \hat{z}$.

Furthermore, the electric field $\vec{E}_1$ in medium 1 and the electric field $\vec{E}_2$ in medium 2 are shown in Fig. 4.3. They can be decomposed as

$$\vec{E}_1 = \vec{E}_{t1} + \vec{E}_{n1} = E_{t1} \hat{a}_t + E_{n1} \hat{a}_n, \tag{4.4}$$

$$\vec{E}_2 = \vec{E}_{t2} + \vec{E}_{n2} = E_{t2} \hat{a}_t + E_{n2} \hat{a}_n, \tag{4.5}$$

where E_{t1} and E_{t2} are tangential components to the interface, while E_{n1} and E_{n2} are normal components to the interface. Interestingly, E_{t1} and E_{t2} have a universal

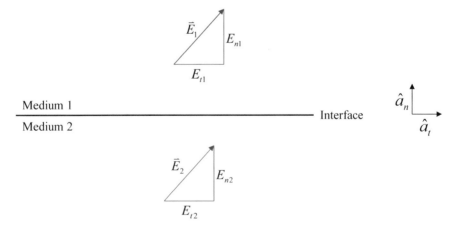

Fig. 4.3 Decomposition of electric field at interface (II)

relation for different dielectric media, and so do E_{n1} and E_{n2}. These two relations are called the **boundary conditions of electric fields**.

First, we derive the relation between E_{t1} and E_{t2} based on Faraday's Law. Suppose S is an arbitrary surface and we do surface integration at both sides of Eq. (4.1). We obtain

$$\int_S (\nabla \times \vec{E}) \cdot d\vec{s} = - \int_S \frac{\partial \vec{B}}{\partial t} \cdot d\vec{s}. \tag{4.6}$$

Utilizing Stokes' theorem, we can convert the left side into a line integral given by

$$\int_S (\nabla \times \vec{E}) \cdot d\vec{s} = \oint_C \vec{E} \cdot d\vec{l}, \tag{4.7}$$

where C is the contour surrounding S as shown in Fig. 4.4. Equation (4.7) tells us that the surface integral of $\nabla \times \vec{E}$ over S is equal to the line integral of $\vec{E}$ along C.

From Eqs. (4.6) and (4.7), we get

$$\oint_C \vec{E} \cdot d\vec{l} = - \int_S \frac{\partial \vec{B}}{\partial t} \cdot d\vec{s}. \tag{4.8}$$

Equation (4.8) means that the line integral of $\vec{E}$ along the surrounding contour of an arbitrary surface S is equal to the surface integral of $-\partial \vec{B}/\partial t$ over S.

Fig. 4.4 An arbitrary surface and its surrounding contour

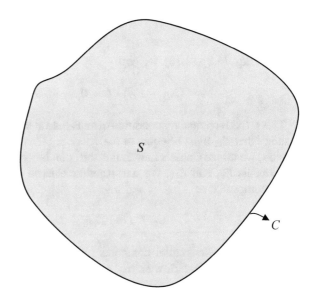

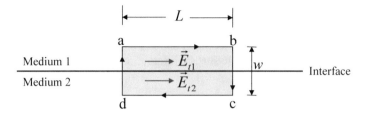

Fig. 4.5 Boundary condition of tangential component of electric field

Next, we apply Eq. (4.8) at the boundary between medium 1 and medium 2. Let S be a rectangular surface with the length L and width w across the interface as shown in Fig. 4.5. Suppose L is very short so that E_{t1} and E_{t2} remain constants along L. Then we let w approach zero, i.e., $w \to 0$, so that the path ab and the path cd of the contour of S approach the interface. In Fig. 4.5, when $w \to 0$, the contribution of line integral along the path bc and the path da vanishes. Therefore, the left side of Eq. (4.8) becomes

$$\oint_C \vec{E} \cdot d\vec{l} = E_{t1} \cdot L - E_{t2} \cdot L = (E_{t1} - E_{t2}) \cdot L. \tag{4.9}$$

On the other hand, when $w \to 0$, the area of S approaches zero as well. Because the time-varying rate of the magnetic flux density $\partial \vec{B}/\partial t$ is finite, the right side of Eq. (4.8) must vanish when $S \to 0$, given by

$$\int_S \frac{\partial \vec{B}}{\partial t} \cdot d\vec{s} = 0. \tag{4.10}$$

From Eqs. (4.8)–(4.10), we get

$$(E_{t1} - E_{t2}) \cdot L = 0 \quad \Rightarrow \quad E_{t1} = E_{t2}. \tag{4.11}$$

This is the 1st boundary condition of an E-field at the interface of dielectric media. It shows that E_{t1} must be equal to E_{t2}.

Now, we utilize Gauss's Law to find the boundary condition regarding the normal components E_{n1} and E_{n2}. We start from the electric flux density $\vec{D}$ at the interface and decompose it as

$$\vec{D} = D_t \hat{a}_t + D_n \hat{a}_n, \tag{4.12}$$

where D_t is the tangential component and D_n is the normal component to the interface. Since $\vec{D} = \epsilon \, \vec{E}$, we have

$$D_{n1} = \epsilon_1 \, E_{n1}, \tag{4.13}$$

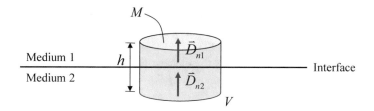

Fig. 4.6 Boundary condition of normal component of electric field

$$D_{n2} = \epsilon_2 \, E_{n2}. \tag{4.14}$$

Suppose V is an arbitrary volume and from Eq. (4.2) we have

$$\int_V \left(\nabla \cdot \vec{D} \right) dv = \int_V 0 \, dv = 0. \tag{4.15}$$

Then from divergence theorem, we obtain

$$\int_V \left(\nabla \cdot \vec{D} \right) dv = \oint_S \vec{D} \cdot d\vec{s} = 0, \tag{4.16}$$

where S is the surface of V. Equation (4.16) tells us that the surface integral of $\vec{D}$ on a volume V must be zero. We can readily apply Eq. (4.16) to the interface as shown in Fig. 4.6. Let V be a pillbox with the height h, and the area of the top face as well as the bottom face is M. Suppose M is very small so that D_{n1} at the top face is a constant, and so is D_{n2} at the bottom face. Furthermore, when the height h approaches zero, i.e., $h \to 0$, the top face and the bottom face approach the interface. In this case, the contribution to the surface integral from the sidewall of the box vanishes. Hence, we only need to consider the contribution from the top face and the bottom face, and the surface integral in Eq. (4.16) becomes

$$\oint_S \vec{D} \cdot d\vec{s} = D_{n1}M - D_{n2}M = 0. \tag{4.17}$$

From Eq. (4.17), we get

$$D_{n1} = D_{n2}. \tag{4.18}$$

From Eqs. (4.18), (4.13) and (4.14), we finally obtain

$$\epsilon_1 \, E_{n1} = \epsilon_2 \, E_{n2}. \tag{4.19}$$

This is the 2nd boundary condition of an electric field at the interface of dielectric media. From above, $\in_1 E_{n1}$ must be equal to $\in_2 E_{n2}$ at the interface. Note that because $\in_1 \neq \in_2$, E_{n1} is not equal to E_{n2}.

From Eqs. (4.11) and (4.19), for the electric field, we learn that the tangential component is continuous at the interface, i.e., $E_{t1} = E_{t2}$, but the normal component is discontinuous, i.e., $E_{n1} \neq E_{n2}$.

Example 4.1

Suppose medium 1 and medium 2 both are dielectric media. Their permittivities are $\in_1 = 2 \in_0$ and $\in_2 = 3 \in_0$. Let their interface lie on the $x-y$ plane and the electric field at the interface in medium 1 is $\vec{E}_1 = 2\hat{x} + 3\hat{y} + 5\hat{z}$. Please derive the electric field $\vec{E}_2$ at the interface in medium 2.

Solution

First, we can decompose $\vec{E}_1$ as

$$\vec{E}_1 = \vec{E}_{t1} + \vec{E}_{n1},$$

where $\vec{E}_{t1}$ is the tangential component and $\vec{E}_{n1}$ is the normal component to the interface. Because the interface lies on the $x-y$ plane, we have $\vec{E}_{t1}$ and $\vec{E}_{n1}$ given by

$$\vec{E}_{t1} = 2\hat{x} + 3\hat{y},$$

$$\vec{E}_{n1} = 5\hat{z}.$$

Suppose $\vec{E}_2$ is composed of a tangential component and a normal component, denoted by $\vec{E}_2 = \vec{E}_{t2} + \vec{E}_{n2}$. From the 1st boundary condition in Eq. (4.11) and 2nd boundary condition in Eq. (4.19), we have

$$E_{t1} = E_{t2} \quad \Rightarrow \quad \vec{E}_{t1} = \vec{E}_{t2},$$

$$\in_1 E_{n1} = \in_2 E_{n2} \quad \Rightarrow \quad \in_1 \vec{E}_{n1} = \in_2 \vec{E}_{n2}.$$

Immediately we get

$$\vec{E}_{t2} = \vec{E}_{t1} = 2\hat{x} + 3\hat{y},$$

$$\vec{E}_{n2} = \frac{\epsilon_1}{\epsilon_2} \vec{E}_{n1} = \frac{2}{3} \vec{E}_{n1} = \frac{10}{3} \hat{z}.$$

Finally, we obtain

$$\vec{E}_2 = \vec{E}_{t2} + \vec{E}_{n2} = 2\hat{x} + 3\hat{y} + \frac{10}{3}\hat{z}.$$

■

B Interface Between Dielectric Medium and Conductor

When we investigate EM wave behavior as it crosses an interface between a dielectric medium and a conductor, for example, between air and metal, we must consider the effect of abundant free electrons on the surface of the conductor. When an EM wave is incident on a conductor, the associated electric field makes these free charges redistribute and thus form **surface charges**. Suppose we have free moving charges attracted by the incident electric field on the surface as shown in Fig. 4.7. The charge density at the interface is not zero in this case.

In Fig. 4.7, we utilize Faraday's Law and Gauss's Law to get

$$\nabla \times \vec{E} = -\frac{\partial \vec{B}}{\partial t}, \tag{4.20}$$

$$\nabla \cdot \vec{D} = \rho, \tag{4.21}$$

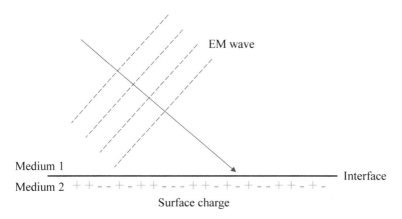

Fig. 4.7 Interface between dielectric medium and conductor

respectively, where ρ is the charge density. Comparing Eqs. (4.20) and (4.21) with Eqs. (4.1) and (4.2) for the interface between two dielectric media, we find that Eq. (4.20) is identical to Eq. (4.1) regarding Faraday's Law. The difference comes from the charge density in Gauss's Law as shown in Eqs. (4.21) and (4.2).

Therefore, similar to what we derived in Eqs. (4.3)–(4.11) based on Faraday's Law, we obtain the relation of the tangential components of electric field given by

$$E_{t1} = E_{t2}. \tag{4.22}$$

This is the 1st boundary condition of the interface between a dielectric medium and a conductor. It is identical to Eq. (4.11) for the case between two dielectric media.

Next, we derive the relation of the normal components E_{n1} and E_{n2}. Suppose V is an arbitrary volume and we do volume integral on both sides of Eq. (4.21) to get

$$\int_V (\nabla \cdot \vec{D}) dv = \int_V \rho dv. \tag{4.23}$$

Using divergence theorem, Eq. (4.23) can be rewritten as

$$\oint_S \vec{D} \cdot d\vec{s} = \int_V \rho dv, \tag{4.24}$$

where S is the surface of V.

Let V be a pillbox with the height h, and the area of the top face as well as the bottom face is M as shown in Fig. 4.8. Suppose M is very small so that D_{n1} at the top face is a constant, and so is D_{n2} at the bottom face. Furthermore, when the height h approaches zero, i.e., $h \rightarrow 0$, the top face and the bottom face will approach the interface. In this case, the contribution to the surface integral from the sidewall vanishes. Hence, we only need to consider the contribution from the top face and the bottom face, and the surface integral in the left side of Eq. (4.24) becomes

$$\oint_S \vec{D} \cdot d\vec{s} = D_{n1} \cdot M - D_{n2} \cdot M = (D_{n1} - D_{n2}) \cdot M. \tag{4.25}$$

In addition, when $h \rightarrow 0$, the surface charge at the interface remains in the pillbox. Hence the right side of Eq. (4.24) is given by

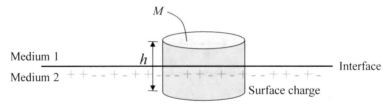

Fig. 4.8 Boundary condition of normal component of electric field at conductor interface

$$\int_V \rho \cdot dv = Q,$$ (4.26)

where Q is the total charge inside the pillbox. Note that when $h \to 0$, the volume V approaches zero. Since the surface charge remains, the charge density ρ will approach infinity so that Q is not zero.

Meanwhile, we can imagine that when $h \to 0$, the pillbox becomes an extremely thin disk with the area M. In order to properly describe the role of surface charge, we define a new parameter called **surface charge density** given by

$$\rho_s = \frac{Q}{M},$$ (4.27)

where ρ_s denotes the charge density per unit area and the unit is Coulomb/m^2.

From Eqs. (4.24)–(4.27), we obtain

$$(D_{n1} - D_{n2}) \cdot M = Q = \rho_s \cdot M$$ (4.28)

Hence we have

$$D_{n1} - D_{n2} = \rho_s \quad \Rightarrow \quad D_{n1} = D_{n2} + \rho_s$$ (4.29)

Finally, Eq. (4.29) can be rewritten as

$$\in_1 E_{n1} = \in_2 E_{n2} + \rho_s.$$ (4.30)

Equation (4.30) is the 2nd boundary condition of the interface between a dielectric medium and a conductor.

Comparing Eqs. (4.30) with (4.19), we can easily find the difference is the surface charge density ρ_s. In other words, the existence of surface charge makes the boundary condition of a conductor's interface different from that of the dielectric media.

Example 4.2

Suppose medium 1 is a dielectric medium and medium 2 is a conductor. The permittivity of medium 1 is $\in_1 = 2 \in_0$ and that of medium 2 is $\in_2 = \in_0 = 8.85 \times 10^{-12}$ (*Farad/m*). Suppose their interface lies on the x–y plane and the electric field in medium 1 is $\vec{E}_1 = 5\hat{x} - 3\hat{y} + 4\hat{z}$. Let the surface charge density be $\rho_s = 10^{-11}$ (*Coulomb/m^2*). Please derive the electric field $\vec{E}_2$ in medium 2.

Solution

First, we decompose the electric field $\vec{E}_1$ as

$$\vec{E}_1 = \vec{E}_{t1} + \vec{E}_{n1},$$

where $\vec{E}_{t1}$ is the tangential component and $\vec{E}_{n1}$ is the normal component. Because the interface lies on the $x-y$ plane, we have

$$\vec{E}_{t1} = 5\hat{x} - 3\hat{y},$$

$$\vec{E}_{n1} = 4\hat{z}.$$

Next, let

$$\vec{E}_2 = \vec{E}_{t2} + \vec{E}_{n2}.$$

From the boundary conditions in Eqs. (4.22) and (4.30), we have

$$E_{t1} = E_{t2},$$

$$\in_1 E_{n1} = \in_2 E_{n2} + \rho_s \quad \Rightarrow \quad E_{n2} = \frac{\in_1}{\in_2} E_{n1} - \frac{\rho_s}{\in_2}.$$

Hence

$$\vec{E}_{t2} = \vec{E}_{t1} = 5\hat{x} - 3\hat{y}$$

$$\vec{E}_{n2} = E_{n2} \cdot \hat{z} = (\frac{\in_1}{\in_2} E_{n1} - \frac{\rho_s}{\in_2}) \cdot \hat{z} = 6.87 \cdot \hat{z}.$$

Finally, we get

$$\vec{E}_2 = \vec{E}_{t2} + \vec{E}_{n2} = 5\hat{x} - 3\hat{y} + 6.87\hat{z}.$$

∎

C   Summary

It takes a lot of effort to derive the boundary conditions of electric fields. Fortunately, the results are quite simple and summarized as follows.

1. For the interface between two dielectric media:

$$E_{t1} = E_{t2} \tag{4.31}$$

$$\in_1 E_{n1} = \in_2 E_{n2} \tag{4.32}$$

2. For the interface between a dielectric medium and a conductor:

$$E_{t1} = E_{t2} \tag{4.33}$$

$$\in_1 E_{n1} = \in_2 E_{n2} + \rho_s \tag{4.34}$$

where ρ_s is the surface charge density.

In above, the tangential component of an electric field is always continuous when an EM wave crosses an interface. On the other hand, the associated normal component is discontinuous across the interface. Besides, the induced surface charge makes the boundary conditions between a dielectric medium and a conductor more complicated than those of two dielectric media as shown in Eqs. (4.34) and (4.32).

Finally, readers may wonder why we do not consider the interface between two conductors. The reason is simple: because EM waves decay rapidly after getting into a conductor, normally they have disappeared before having the chance to encounter another conductor. Therefore, it is unnecessary to consider the interface between two conductors since it rarely happens in practice.

4.2 Boundary Conditions of Magnetic Fields

When learning Electromagnetics, lots of beginners feel confused about the term "boundary conditions", and do not understand why it is necessary for us to learn these conditions. Actually, boundary conditions define behaviors of EM waves across an interface of different media. By learning theses conditions, we can change the behavior of EM waves such as propagation direction at the interface. This grants us the ability to control an EM wave properly for a specific application.

In the previous section, we learn the boundary conditions of E-field. In this section, we learn the boundary conditions of M-field.

A Interfaces Between Two Dielectric Media

Suppose medium 1 and medium 2 are dielectric media, and the associated permittivity and permeability are $(\in_1, \mu_1)$ for medium 1 and $(\in_2, \mu_2)$ for medium 2, where $\in_1 \neq \in_2$ and $\mu_1 = \mu_2 = \mu_0$.

First, Maxwell's equations hold across the interface. For the boundary conditions of magnetic fields, we apply **Ampere's Law** and **Gauss's Law of magnetic field**. Suppose the current density is zero in these two media, i.e., $\vec{J} = 0$. Then we have

$$\nabla \times \vec{H} = \frac{\partial \vec{D}}{\partial t}, \tag{4.35}$$

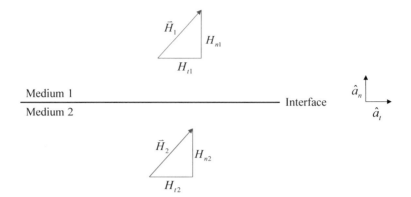

Fig. 4.9 Decomposition of magnetic field at interface

$$\nabla \cdot \vec{B} = 0, \tag{4.36}$$

where Eq. (4.35) is Ampere's Law and Eq. (4.36) is Gauss's Law of magnetic field.

For an arbitrary magnetic field $\vec{H}$ at the interface between medium 1 and medium 2 as shown in Fig. 4.9, it can be decomposed as **tangential component** $\vec{H}_t$ and **normal component** $\vec{H}_n$. Hence we have

$$\vec{H} = H_t \hat{a}_t + H_n \hat{a}_n, \tag{4.37}$$

where $\hat{a}_t$ is the unit vector parallel to the interface and $\hat{a}_n$ is the unit vector perpendicular to the interface, while H_t and H_n are their respective magnitudes.

The magnetic field $\vec{H}_1$ in medium 1 and the magnetic field $\vec{H}_2$ in medium 2 are shown in Fig. 4.9. They can be decomposed as

$$\vec{H}_1 = H_{t1} \hat{a}_t + H_{n1} \hat{a}_n, \tag{4.38}$$

$$\vec{H}_2 = H_{t2} \hat{a}_t + H_{n2} \hat{a}_n, \tag{4.39}$$

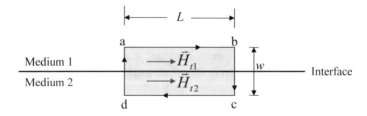

Fig. 4.10 Boundary condition of tangential component of magnetic field

where H_{t1} and H_{t2} are tangential components to the interface, and H_{n1} and H_{n2} are normal components to the interface. Similar to the case of electric fields in the previous section, H_{t1} and H_{t2} have a universal relation for different dielectric media, and so do H_{n1} and H_{n2}. In the following, we will find out the **boundary conditions of magnetic fields**.

First, we derive the relation between H_{t1} and H_{t2} based on Ampere's Law. Suppose S is an arbitrary surface and we do surface integral on both sides of Eq. (4.35). We obtain

$$\int_S \left(\nabla \times \vec{H} \right) \cdot d\vec{s} = \int_S \frac{\partial \vec{D}}{\partial t} \cdot d\vec{s}. \tag{4.40}$$

Applying Stokes' theorem on the left side of Eq. (4.40), we get

$$\int_S \left(\nabla \times \vec{H} \right) \cdot d\vec{s} = \oint_C \vec{H} \cdot d\vec{l}, \tag{4.41}$$

where C is the contour surrounding S. From Eqs. (4.40) and (4.41), we get

$$\oint_C \vec{H} \cdot d\vec{l} = \int_S \frac{\partial \vec{D}}{\partial t} \cdot d\vec{s} \tag{4.42}$$

Equation (4.42) means that the line integral of $\vec{H}$ along the surrounding contour of an arbitrary surface S is equal to the surface integral of $\frac{\partial \vec{D}}{\partial t}$ over S.

Next, we apply Eq. (4.42) at the boundary between medium 1 and medium 2. Let S be a rectangular surface with the length L and width w across the interface as shown in Fig. 4.10. Suppose L is very short so that H_{t1} and H_{t2} remain constants along L. Then we let w approach zero, i.e., $w \rightarrow 0$, so that the path ab and path cd of the contour of S approaches the interface. In Fig. 4.10, when $w \rightarrow 0$, the contribution of line integral along the path bc and the path da vanishes. Therefore, the left side of Eq. (4.42) becomes

$$\oint_C \vec{H} \cdot d\vec{l} = H_{t1}L - H_{t2}L = (H_{t1} - H_{t2}) \cdot L \tag{4.43}$$

On the other hand, when $w \rightarrow 0$, the area of S approaches zero as well. Because the time-varying rate of the electric flux density $\partial \vec{D}/\partial t$ is finite, the right side of Eq. (4.42) must vanish and is given by

$$\int_S \frac{\partial \vec{D}}{\partial t} \cdot d\vec{s} = 0. \tag{4.44}$$

From Eqs. (4.42)–(4.44), we get

$$(H_{t1} - H_{t2}) \cdot L = 0 \quad \Rightarrow \quad H_{t1} = H_{t2}. \tag{4.45}$$

This is the 1st boundary condition of a magnetic field at the interface of dielectric media.

Now, we utilize Gauss's Law of magnetic field to find the boundary condition regarding the normal components H_{n1} and H_{n2}. We start from the magnetic flux density $\vec{B}$ at the interface and decompose it as

$$\vec{B} = B_t \hat{a}_t + B_n \hat{a}_n \tag{4.46}$$

where B_t is the tangential component and B_n is the normal component to the interface. Since $\vec{B} = \mu \vec{H}$ and $\mu_1 = \mu_2 = \mu_0$, we have

$$B_{n1} = \mu_0 H_{n1}, \tag{4.47}$$

$$B_{n2} = \mu_0 H_{n2}. \tag{4.48}$$

Suppose V is an arbitrary volume and from Eq. (4.35) we have

$$\int_V \left(\nabla \cdot \vec{B} \right) dv = 0. \tag{4.49}$$

Then from divergence theorem, we obtain

$$\oint_V \left(\nabla \cdot \vec{B} \right) dv = \oint_S \vec{B} \cdot d\vec{s} = 0, \tag{4.50}$$

where S is the surface of V. Equation (4.50) tells us that the surface integral of $\vec{B}$ on a volume V must be zero. We can readily apply Eq. (4.50) to the interface as shown in Fig. 4.11. Let V be a pillbox with the height h, and the area of the top face as well as the bottom face is M. Suppose M is very small so that B_{n1} at the top face is a constant, and so is B_{n2} at the bottom face. Furthermore, when the height h approaches zero, i.e., $h \to 0$, the top face and the bottom face approach the interface. In this case, the contribution to the surface integral from the sidewall vanishes. Hence we only need

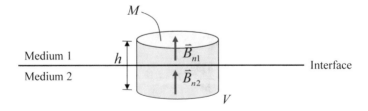

Fig. 4.11 Boundary condition of normal component of magnetic field

to consider the contribution from the top face and the bottom face, and the surface integral in Eq. (4.50) becomes

$$\oint_S \vec{B} \cdot d\vec{s} = (B_{n1} - B_{n2}) \cdot M = 0. \tag{4.51}$$

From Eq. (4.51), we get

$$B_{n1} = B_{n2}. \tag{4.52}$$

From Eqs. (4.52), (4.47) and (4.48), we finally get

$$H_{n1} = H_{n2}. \tag{4.53}$$

This is the 2nd boundary condition of the magnetic field at the interface of dielectric media.

In above, we learn that both the tangential component and the normal component of a magnetic field are continuous, i.e., $H_{t1} = H_{t2}$ and $H_{n1} = H_{n2}$. In other words, a magnetic field is continuous at the interface across two dielectric media, i.e., $\vec{H}_1 = \vec{H}_2$. Obviously in this case, the boundary conditions of a magnetic field is simpler than those of an electric field. Note that for a magnetic medium, because $\mu_1 \neq \mu_0$ or $\mu_2 \neq \mu_0$, we have $\vec{H}_1 \neq \vec{H}_2$.

Example 4.3

Suppose we have two dielectric media with $\mu_1 = \mu_2 = \mu_0$. The magnetic field in the medium 1 is given by $\vec{H}_1 = 4\hat{x} - 2\hat{y} + \hat{z}$. Please derive the magnetic field $\vec{H}_2$ in the medium 2.

Solution

According to the boundary conditions of two dielectric media (non-magnetic media), a magnetic field is continuous. Hence

$$\vec{H}_2 = \vec{H}_1 = 4\hat{x} - 2\hat{y} + \hat{z}$$

∎

B   Interface between Dielectric Medium and Conductor

As discussed in the previous section, when we investigate EM wave behavior across an interface between a dielectric medium and a conductor, we need to consider

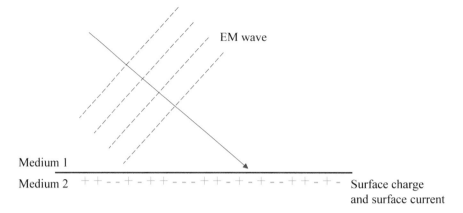

Fig. 4.12 Interface between dielectric medium and conductor

surface charges because abundant free electrons exist in a conductor. These surface charges will be attracted by the applied electric field and form a **surface current** as shown in Fig. 4.12.

At the interface, Ampere's Law and Gauss's Law of magnetic field are given by

$$\nabla \times \vec{H} = \vec{J} + \frac{\partial \vec{D}}{\partial t}, \tag{4.54}$$

$$\nabla \cdot \vec{B} = 0, \tag{4.55}$$

respectively, where $\vec{J}$ is the current density. Comparing Eqs. (4.54) and (4.55) with Eqs. (4.35) and (4.36), we can easily find the only difference comes from the current density $\vec{J}$ regarding Ampere's Law.

First, a magnetic field across the interface can be decomposed as a tangential component and a normal component as shown in Eqs. (4.38) and (4.39). We then utilize Ampere's Law to derive the relation between H_{t1} and H_{t2}. Suppose S is an arbitrary surface and we do surface integral on both sides of Eq. (4.54) to get

$$\int_S \left(\nabla \times \vec{H} \right) \cdot d\vec{s} = \int_S \vec{J} \cdot d\vec{s} + \int_S \frac{\partial \vec{D}}{\partial t} \cdot d\vec{s}. \tag{4.56}$$

Applying Stokes' theorem, Eq. (4.56) can be rewritten as

$$\oint_C \vec{H} \cdot d\vec{\ell} = \int_S \vec{J} \cdot d\vec{s} + \int_S \frac{\partial \vec{D}}{\partial t} \cdot d\vec{s}, \tag{4.57}$$

where C is the contour surrounding S.

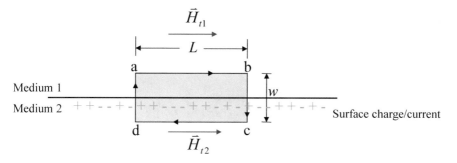

Fig. 4.13 Boundary condition of tangential component of magnetic field at conductor interface

Next, let S be a rectangular surface with the length L and width w across the interface as shown in Fig. 4.13. Suppose L is very short so that H_{t1} and H_{t2} remain constants along L. Then we let w approach zero, i.e., $w \to 0$, so that the path ab and the path cd of the contour of S approach the interface. In Fig. 4.13, when $w \to 0$, the contribution of line integral along the path bc and the path da vanishes. Therefore, the left side of Eq. (4.57) becomes

$$\oint_C \vec{H} \cdot d\vec{\ell} = H_{t1} \cdot L - H_{t2} \cdot L = (H_{t1} - H_{t2}) \cdot L. \tag{4.58}$$

On the other hand, when $w \to 0$, the area of S approaches zero as well. Because the time-varying rate of the electric flux density $\partial \vec{D}/\partial t$ is finite, the 2nd term of the right side of Eq. (4.57) vanishes. Hence from Eqs. (4.58) and (4.57), we have

$$(H_{t1} - H_{t2}) \cdot L = \int_S \vec{J} \cdot d\vec{s}. \tag{4.59}$$

Note that when $w \to 0$, the area of S approaches zero. Since the surface current remains in S, the current density $\vec{J}$ will approach infinity so that the surface current is not zero, i.e.,

$$I = \int_S \vec{J} \cdot d\vec{s}, \tag{4.60}$$

where I is the total surface current in S.

Meanwhile, we can imagine that when $w \to 0$, the rectangular surface approximates a line with the length L. In order to properly describe the role of surface current, we define a new parameter called **surface current density** given by

$$J_s = \frac{I}{L}, \tag{4.61}$$

where J_s denotes the current density per unit length and the unit is (Ampere/m). Note that J_s is actually a vector with the direction perpendicular to the magnetic field. Here to ease the problem and help readers get the insight of surface current, we take it as a scalar.

From Eqs. (4.59)–(4.61), we obtain

$$(H_{t1} - H_{t2}) \cdot L = I = J_s \cdot L. \tag{4.62}$$

Hence, the relation between H_{t1} and H_{t2} is given by

$$H_{t1} = H_{t2} + J_s. \tag{4.63}$$

This is the 1st boundary condition of the interface between a dielectric medium and a conductor.

Comparing Eqs. (4.63) with (4.45), we can easily find the difference is the surface current density J_s. In other words, the existence of surface current makes the boundary condition of a conductor's interface differ from that of the dielectric media regarding magnetic fields.

Finally, we derive the relation between the normal components H_{n1} and H_{n2} using Gauss's Law of magnetic field. As Eq. (4.55) is identical to Eq. (4.36), we get

$$H_{n1} = H_{n2}. \tag{4.64}$$

This is the 2nd boundary condition of the interface between a dielectric medium and a conductor. This condition is the same as what we have for the interface between two dielectric media.

Example 4.4

Suppose medium 1 is a dielectric medium and medium 2 is a conductor with $\mu_1 = \mu_2 = \mu_0$. Suppose their interface is the x–y plane and the magnetic field in medium 1 is given by $\vec{H}_1 = 3\hat{x} - 2\hat{z}$. Assume the surface current density $J_s = 0.5(A/m)$ with the direction perpendicular to the magnetic field. Please derive the magnetic field $\vec{H}_2$ in medium 2.

Solution

First, $\vec{H}_1$ can be decomposed as

$$\vec{H}_1 = \vec{H}_{t1} + \vec{H}_{n1},$$

where $\vec{H}_{t1}$ is parallel to the interface and $\vec{H}_{n1}$ is normal to the interface. Because the interface lies on the xy-plane, we have $\vec{H}_{t1}$ and $\vec{H}_{n1}$ given by

$$\vec{H}_{t1} = 3\hat{x},$$

$$\vec{H}_{n1} = -2\hat{z}.$$

We let

$$\vec{H}_2 = \vec{H}_{t2} + \vec{H}_{n2}.$$

From the boundary conditions in Eqs. (4.63) and (4.64), we obtain

$$H_{t1} = H_{t2} + J_s \quad \Rightarrow \quad H_{t2} = H_{t1} - J_s = 2.5,$$

$$H_{n1} = H_{n2} \quad \Rightarrow \quad H_{n2} = -2.$$

Finally, we get

$$\vec{H}_2 = \vec{H}_{t2} + \vec{H}_{n2} = 2.5\hat{x} - 2\hat{z}.$$

■

C Summary

The boundary conditions of magnetic fields are summarized as follows.

1. For the interface between two dielectric media:

$$H_{t1} = H_{t2}, \tag{4.65}$$

$$H_{n1} = H_{n2}. \tag{4.66}$$

Hence $\vec{H}_1 = \vec{H}_2$. In a word, a magnetic field is continuous across the interface.

2. For the interface between a dielectric medium and a conductor

$$H_{t1} = H_{t2} + J_s, \tag{4.67}$$

$$H_{n1} = H_{n2}, \tag{4.68}$$

where J_s is the surface current density. Obviously, $H_{t1} \neq H_{t2}$ and a magnetic field is not continuous in this case.

4.3   General Laws of Dielectric Interface

In the previous sections, we discussed the boundary conditions defining E-field and
M-field between two media. In this section, we discuss the relationships between the
incident angle, reflection angle, and refraction angle of dielectric interface. It is well
known that when an EM wave travels from a dielectric medium to another dielectric
medium, a portion is reflected and another portion is transmitted. For example, light
is the most commonly observed EM wave in daily life. When light propagates from
air to water, a portion of light will be reflected and the remaining will transmit into the
water with a change in direction (refraction). We discuss the relationships between
incidence, reflection and refraction of EM waves in the following.

A   Plane of Incidence

In Fig. 4.14, we have two dielectric media and their interface is the xy-plane. The
EM characteristic parameters of medium 1 and medium 2 are $(\in_1, \mu_0)$ and $(\in_2, \mu_0)$,
respectively, where $\in_1 \neq \in_2$. When an EM wave impinges on the interface between
medium 1 and medium 2, a portion is returned to medium 1 (reflection) and the
remaining transmits into medium 2 (refraction). In Fig. 4.14, the unit vector $\hat{a}_i$ denotes
the direction of the **incident wave**, the unit vector $\hat{a}_r$ denotes the direction of the
reflected wave, and the unit vector $\hat{a}_t$ denotes the direction of the **transmitted wave**
into the medium 2. For a plane wave, because the direction of propagation is also the

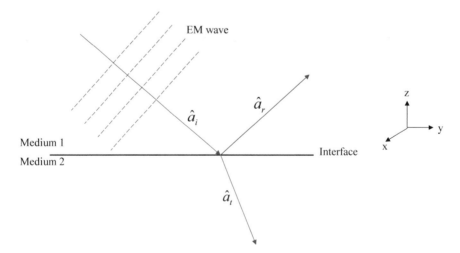

Fig. 4.14 Incidence of EM wave at dielectric interface (I)

direction of Poynting vector, these unit vectors $(\hat{a}_i, \hat{a}_r, \hat{a}_t)$ also show the direction of Poynting vectors in three cases, which imply the direction of energy transfer.

As shown in Fig. 4.14, these three important vectors $(\hat{a}_i, \hat{a}_r, \hat{a}_t)$ lie in the yz-plane and we call it the **plane of incidence**:

$$\text{Plane of incidence} \equiv \text{yz-plane}$$

Plane of incidence is very critical when we analyze the behaviors of an EM wave at an interface. It is perpendicular to the interface, i.e., xy-plane, in this case. Notice that unlike the physically real interface, the plane of incidence is an hypothetical plane defined for analyzing EM wave behavior at the interface.

For an arbitrary incident E-field, we can decompose it into two components given by

$$\vec{E} = \vec{E}_{//} + \vec{E}_{\perp}, \tag{4.69}$$

where $\vec{E}_{//}$ denotes the component parallel to the plane of incidence and $\vec{E}_{\perp}$ denotes the component perpendicular to the plane of incidence. In other words, $\vec{E}_{//}$ is the projection of $\vec{E}$ on the plane of incidence, and $\vec{E}_{\perp}$ is the projection of $\vec{E}$ on a plane perpendicular to the plane of incidence.

Example 4.5

Suppose the interface between two dielectric media is the xy-plane and the plane of incidence is the yz-plane as shown in Fig. 4.14. Let the incident E-field be $\vec{E} = 3\hat{x} - 2\hat{y} + 5\hat{z}$. Please derive the associated $\vec{E}_{//}$ and $\vec{E}_{\perp}$.

Solution

First, we can decompose the incident E-field as

$$\vec{E} = \vec{E}_{//} + \vec{E}_{\perp}.$$

Since the plane of incidence is the yz-plane, we have

$$\vec{E}_{//} = -2\hat{y} + 5\hat{z}, \quad \text{(parallel to yz-plane)}$$

$$\vec{E}_{\perp} = 3\hat{x}. \quad \text{(perpendicular to yz-plane)}$$

■

Example 4.6

Suppose the interface between two dielectric media is the xy-plane and the plane of incidence is the yz-plane. Please prove that for an arbitrary incident E-field $\vec{E} = a\hat{x} + b\hat{y} + c\hat{z}$, it can be decomposed as $\vec{E} = \vec{E}_{//} + \vec{E}_{\perp}$.

Solution

Because the plane of incidence is the yz-plane, the parallel component of $\vec{E}$ to the plane of incidence is obviously

$$\vec{E}_{//} = b\hat{y} + c\hat{z}.$$

On the other hand, the perpendicular component of $\vec{E}$ to the plane of incidence is

$$\vec{E}_{\perp} = a\hat{x}.$$

Therefore, the incident E-field $\vec{E}$ is a combination of the above two components and given by

$$\vec{E} = \vec{E}_{//} + \vec{E}_{\perp}.$$

This completes the proof.

∎

In general, both $\vec{E}_{//}$ and $\vec{E}_{\perp}$ are not zero in Eq. (4.69). In a specific case, when an incident E-field is parallel to the plane of incidence, we have $\vec{E}_{\perp} = 0$ and hence

$$\vec{E} = \vec{E}_{//}. \tag{4.70}$$

In this case, the incident EM wave is called **parallel polarized wave**. On the other hand, when an incident E-field is perpendicular to the plane of incidence, we have $\vec{E}_{//} = 0$ and hence

$$\vec{E} = \vec{E}_{\perp}. \tag{4.71}$$

In this case, the incident EM wave is called **perpendicular polarized wave**. Readers, especially beginners, may notice that parallel polarized wave and perpendicular polarized wave are defined based on the plane of incidence, instead of the interface. For example in Fig. 4.14, the E-field of a parallel polarized wave is parallel to the yz-plane, and the E-field of a perpendicular polarized wave is perpendicular to the yz-plane, i.e., its E-field is along $\hat{x}$ or $-\hat{x}$ direction.

From the experimental results, the above two polarized waves have different behaviors at the interface of two dielectric media. Hence we need to analyze them separately. However, they also share some common characteristics, which are called **general laws of dielectric interface**. This is the major topic of this section.

B Law of Reflection

In Fig. 4.15, we have an EM wave impinging on a dielectric interface and the plane of incidence is the yz-plane. As defined in the previous subsection, the plane of incidence contains three important unit vectors $(\hat{a}_i, \hat{a}_r, \hat{a}_t)$. Let the **incident angle** θ_i be the angle between z-axis (the normal line to the interface) and $\hat{a}_i$ (incident wave), the **reflection angle** θ_r be the angle between z-axis and $\hat{a}_r$ (reflected wave), and the **transmission angle** θ_t be the angle between z-axis and $\hat{a}_t$ (transmitted wave). The relationships between these three angles are important because they determine the propagation direction of reflected wave and transmitted wave.

First, we discuss the relationship between an incident angle θ_i and the associated reflection angle θ_r. In Fig. 4.16, suppose we have an incident plane wave. The point P and the point P' are in the same plane (wavefront). When the wavefront propagates from point P to O, the point P' also propagates to O' and their propagation distances are equal:

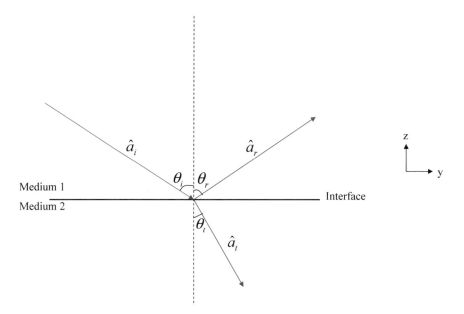

Fig. 4.15 Incidence of EM wave at dielectric interface (II)

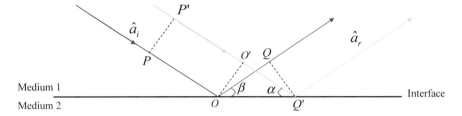

Fig. 4.16 Incident wave and reflected wave at dielectric interface

$$\overline{PO} = \overline{P'O'}. \tag{4.72}$$

Next, when the EM wave at point O is reflected to point Q, the wavefront at O' also propagates to point Q'. The distances of propagation, $\overline{OQ}$ and $\overline{O'Q'}$, are equal so that Q and Q' are in the same plane (wavefront).

Comparing Figs. 4.15 and 4.16, we get the relationship between the angle α and the incident angle θ_i given by

$$\alpha = 90° - \theta_i. \tag{4.73}$$

Besides, the relationship between the angle β and the reflection angle θ_r is given by

$$\beta = 90° - \theta_r. \tag{4.74}$$

From Fig. 4.16, we can easily get the following from trigonometric function:

$$\overline{OQ} = \overline{OQ'} \cdot \cos \beta = \overline{OQ'} \cdot \sin \theta_r, \tag{4.75}$$

$$\overline{O'Q'} = \overline{OQ'} \cdot \cos \alpha = \overline{OQ'} \cdot \sin \theta_i, \tag{4.76}$$

Because $\overline{OQ} = \overline{O'Q'}$, from Eqs. (4.75) and (4.76), we obtain

$$\sin \theta_i = \sin \theta_r. \tag{4.77}$$

Hence

$$\theta_i = \theta_r \tag{4.78}$$

Equation (4.78) is called **Law of Reflection**, which states that the reflection angle is equal to the incident angle. This law is important because it allows us to precisely predict the direction of a reflected wave.

Example 4.7

In Fig. 4.15, the unit vectors $\hat{a}_i$ and $\hat{a}_r$ denote the propagation direction of an incident wave and that of the reflected wave, respectively. Suppose $\hat{a}_i = \frac{1}{2}\hat{y} - \frac{\sqrt{3}}{2}\hat{z}$. Please derive $\hat{a}_r$.

Solution

From Law of Reflection, $\theta_i = \theta_r$. In Fig. 4.15, the y-component of $\hat{a}_r$ shall be equal to that of $\hat{a}_i$. Besides, the z-component of $\hat{a}_i$ and that of $\hat{a}_r$ have the same magnitude but opposite directions. Hence

$$\hat{a}_r = \frac{1}{2}\hat{y} + \frac{\sqrt{3}}{2}\hat{z}.$$

■

C Law of Refraction (Snell's Law)

Next, we discuss the relationship between an incident angle θ_i and the associated transmission angle θ_t. Figure 4.17 shows the plane of incidence, where we have an incident plane wave having point P and point P' in the same plane (wavefront). In medium 1, when the wavefront goes from point P to O, the wavefront at point P' also goes to O', and their distances of propagation are equal: $\overline{PO} = \overline{P'O'}$. After the plane wave transmits into medium 2, during Δt the wavefront at point O propagates

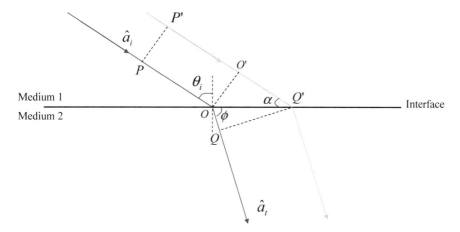

Fig. 4.17 Incident wave and transmitted wave at dielectric interface

to point Q. During the same time period, the wavefront at point O' propagates to point Q' in medium 1, where Q and Q' are in the same plane (wavefront.)

From Figs. 4.15 and 4.17, we get $\alpha = 90° - \theta_i$ and

$$\phi = 90° - \theta_t. \tag{4.79}$$

Using trigonometric identity, we obtain

$$\overline{OQ} = \overline{OQ'} \cdot \cos \phi = \overline{OQ'} \cdot \sin \theta_t, \tag{4.80}$$

$$\overline{O'Q'} = \overline{OQ'} \cdot \cos \alpha = \overline{OQ'} \cdot \sin \theta_i. \tag{4.81}$$

Note that an EM wave has different propagating speed in two dielectric media. Hence we have

$$v_{p1} = \frac{c}{n_1}, \tag{4.82}$$

$$v_{p2} = \frac{c}{n_2}, \tag{4.83}$$

where v_{p1} and v_{p2} are phase velocities in medium 1 and medium 2, respectively, and n_1 and n_2 are their respective refractive indexes. Because during Δt, the wave propagates for a distance $\overline{O'Q'}$ in medium 1 and for a distance $\overline{OQ}$ in medium 2, we have

$$\Delta t = \frac{\overline{O'Q'}}{v_{p1}} = \frac{\overline{OQ}}{v_{p2}}. \tag{4.84}$$

Inserting Eqs. (4.80)–(4.83) into Eq. (4.84), we finally get

$$n_1 \cdot \sin \theta_i = n_2 \cdot \sin \theta_t. \tag{4.85}$$

Equation (4.85) is called **Law of Refraction** or **Snell's Law**. It defines the relationship between an incident angle θ_i and the associated transmission angle θ_t. It enables us to precisely predict the direction of a transmitted EM wave at an interface between two dielectric media.

In above, the following laws always hold for an interface between two dielectric media:

$$\theta_i = \theta_r, \quad \text{(Law of Reflection)}$$

$$n_1 \cdot \sin \theta_i = n_2 \cdot \sin \theta_t. \quad \left(\text{Snell's Law}\right)$$

These two laws are very important in dealing with interface problems between two dielectric media since they are valid for any incident EM waves.

Example 4.8

In Fig. 4.15, the unit vectors $\hat{a}_i$ and $\hat{a}_t$ denote the propagation direction of an incident wave and that of the transmitted wave, respectively. Suppose $\epsilon_1 = 4 \epsilon_0$, $\epsilon_2 = 3 \epsilon_0$ and $\hat{a}_i = \frac{1}{2}\hat{y} - \frac{\sqrt{3}}{2}\hat{z}$. Please derive $\hat{a}_t$.

Solution

First, because the y-component of $\hat{a}_i$ is 1/2 and $|\hat{a}_i| = 1$, from Fig. 4.15, we get

$$|\hat{a}_i| \cdot \sin \theta_i = \sin \theta_i = \frac{1}{2}.$$

According to Snell's Law, we have

$$n_1 \sin \theta_i = n_2 \sin \theta_t.$$

Because $n_1 = \sqrt{4} = 2$ and $n_2 = \sqrt{3}$, we get

$$\sin \theta_t = \frac{n_1}{n_2} \sin \theta_i = \frac{1}{\sqrt{3}}$$

From trigonometric identity, we have

$$\cos \theta_t = \sqrt{1 - \sin \theta_t^2} = \sqrt{\frac{2}{3}}.$$

Because $|\hat{a}_t| = 1$, from Fig. 4.15, the y-component of $\hat{a}_t$ is $\sin \theta_t$ and z-component of $\hat{a}_t$ is $-\cos \theta_t$. Hence

$$\hat{a}_t = \sin \theta_t \cdot \hat{y} - \cos \theta_t \cdot \hat{z} = \frac{1}{\sqrt{3}} \cdot \hat{y} - \sqrt{\frac{2}{3}} \cdot \hat{z}.$$

■

D   Total Reflection

When an EM wave propagates from a medium to a less dense medium, i.e., $n_1 > n_2$, an interesting phenomenon called **total reflection** may occur. It can be verified by

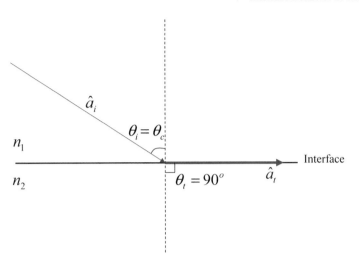

Fig. 4.18 Total reflection and critical angle

Snell's Law. From Eq. (4.85), we obtain

$$\sin \theta_t = \frac{n_1}{n_2} \cdot \sin \theta_i. \tag{4.86}$$

In Eq. (4.86), when the incident angle θ_i is $0°$, the transmission angle θ_t is also $0°$; when θ_i increases, θ_t also increases. Because $n_1 > n_2$, θ_t must be greater than θ_i. We can expect that when θ_i increases to a specific angle θ_c as shown in Fig. 4.18, θ_t will reach $90°$. In this case, no EM wave transmits into medium 2. From Snell's Law, we can derive this specific angle θ_c as follows. Let θ_t be $90°$ in Eq. (4.85), we have

$$n_1 \cdot \sin \theta_c = n_2 \cdot \sin 90° = n_2 \tag{4.87}$$

Hence θ_c is given by

$$\theta_c = \sin^{-1}\left(\frac{n_2}{n_1}\right). \tag{4.88}$$

Actually, when the incident angle is greater than or equal to θ_c, i.e., $\theta_i \geq \theta_c$, an incident EM wave is totally returned to the medium 1. This phenomenon is called **total reflection** and θ_c is called **critical angle** of total reflection.

In above, the total reflection occurs when the following two conditions hold:

1. An EM wave propagates from a medium having a greater refractive index to a less dense medium, i.e., $n_1 > n_2$.

2. The incident angle is greater than or equal to the critical angle, i.e., $\theta_i \geq \theta_c = \sin^{-1}\left(\frac{n_2}{n_1}\right)$.

One of the most important applications of total reflection is **optical fiber communications**. As shown in Fig. 4.19, an optical fiber consists of two parts. The inner core is a dielectric medium having the refractive index n_1 and it is surrounded by another dielectric medium called **cladding** having the refractive index n_2, where $n_1 > n_2$. When we transmit an optical signal into the inner core with a selected incident angle, we can make the optical signal totally reflected at the interface between these two dielectric media. In that case, total reflection occurs all the way and the optical signal is fully kept in the inner core. Hence, the energy loss is extremely limited and the signal can propagate for a very long distance. This technique prospers inter-continental communications and achieves global networking.

Example 4.9

We have an optical fiber as shown in Fig. 4.20. Let an optical signal transmit from air to the inner core having the refractive index n_1. Suppose $n_1 > n_2$ and the refractive index of air is $n = 1$. If we want to achieve total reflection (between inner core and cladding) in this fiber, please derive the maximal allowed incident angle of θ_i at the interface between air and the inner core.

Solution

In Fig. 4.20, when we have total reflection between inner core and cladding, the corresponding critical angle is given by

$$\theta_c = \sin^{-1} \frac{n_2}{n_1}$$

From Snell's Law, we get the following relationship at the interface between air ($n = 1$) and inner core:

$$1 \cdot \sin \theta_i = n_1 \cdot \sin \phi.$$

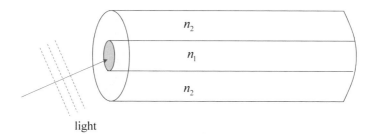

light

Fig. 4.19 Total reflection applied to optical fiber communications

Fig. 4.20 Plot of Example
4.9

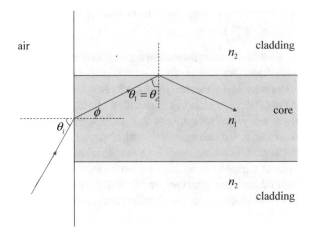

Because $\phi = 90° - \theta_c$, we have

$$\sin \theta_i = n_1 \cdot \sin \phi = n_1 \cdot \cos \theta_c = n_1 \cdot \sqrt{1 - \sin^2 \theta_c} = \sqrt{n_1^2 - n_2^2}$$

Hence the maximal allowed angle of θ_i is given by

$$\theta_{\max} = \sin^{-1}(\sqrt{n_1^2 - n_2^2}).$$

Note that from Fig. 4.20, when θ_i is greater than $\theta_{\max}$, we will not have total reflection at the interface between the inner core and the cladding.   ∎

4.4 Parallel Polarized Wave

In the previous section, we introduce two important laws at the interface between two dielectric media:

$$\theta_i = \theta_r \tag{4.89}$$

$$n_1 \cdot \sin \theta_i = n_2 \cdot \sin \theta_t \tag{4.90}$$

Equation (4.89) is Law of Reflection and Eq. (4.90) is Law of Refraction (Snell's Law). These two laws greatly simplify the analysis of an EM wave when impinging on an interface because they clearly identify the direction of the reflected wave and transmitted wave. Besides, an arbitrary incident wave can be decomposed into two components: a parallel polarized wave and a perpendicular polarized wave. In this section, we discuss the behaviors of a **parallel polarized wave** at an interface.

A E-field in Plane of Incidence

Suppose we have two dielectric media. The associated permittivity and permeability are (ϵ_1, μ_0) for medium 1 and (ϵ_2, μ_0) for medium 2, where $\epsilon_1 \neq \epsilon_2$. Let a plane wave go from medium 1 to medium 2 as shown in Fig. 4.21. The interface is the xy-plane and the **plane of incidence** is the yz-plane. These two planes are mutually perpendicular.

For an arbitrary incident EM wave, the electric field $\vec{E}$ can be decomposed as two components:

$$\vec{E} = \vec{E}_{//} + \vec{E}_{\perp}, \tag{4.91}$$

where $\vec{E}_{//}$ is parallel to the plane of incidence and $\vec{E}_{\perp}$ is perpendicular to the plane of incidence. When an incident EM wave has $\vec{E}_{\perp} = 0$, we have

$$\vec{E} = \vec{E}_{//}. \tag{4.92}$$

In this case, the EM wave is called parallel polarized wave whose electric field is parallel to the plane of incidence. For example in Fig. 4.21, the E-field of a parallel polarized wave is parallel to the $y-z$ plane and its x-component is zero.

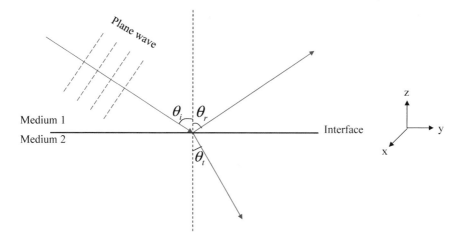

Fig. 4.21 Parallel-polarized wave at dielectric interface (I)

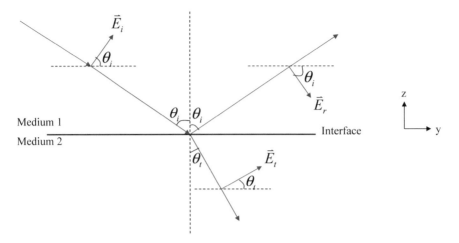

Fig. 4.22 Parallel polarized wave at dielectric interface (II)

A parallel polarized wave has an interesting feature: the incident electric field $\vec{E}_i$, the reflected electric field $\vec{E}_r$ and the transmitted electric field $\vec{E}_t$ all lie in the plane of incidence. As shown in Fig. 4.22, $\vec{E}_i$, $\vec{E}_r$ and $\vec{E}_t$ lie in the plane of incidence (y–z plane) and all of them are perpendicular to the propagation direction of the EM wave. In Fig. 4.22, the reflection angle θ_r is replaced by the incident angle θ_i because of the Law of Reflection. Besides, we draw three dashed lines which are parallel to the interface in order to facilitate our analysis.

For a parallel polarized wave, we focus on the E-field. Once the E-field is attained, we can readily derive the associated M-field. For example in Fig. 4.22, once $\vec{E}_r$ is attained, the magnitude of the reflected magnetic field $\vec{H}_r$ can be readily derived by $\left|\vec{H}_r\right| = \left|\vec{E}_r\right|/\eta_1$, where η_1 is the wave impedance of medium 1. In addition, the direction of the Poynting vector $\vec{P}_r$ of a reflected wave is its propagation direction. Because $\vec{P}_r = \vec{E}_r \times \vec{H}_r$, we can easily derive the direction of $\vec{H}_r$ from the directions of $\vec{E}_r$ and $\vec{P}_r$. Hence $\vec{H}_r$ can be readily derived once $\vec{E}_r$ is attained.

B Reflection Coefficient and Transmission Coefficient

From Sect. 4.1, an electric field at an interface of two dielectric media must satisfy the boundary conditions given by

$$E_{t1} = E_{t2}, \tag{4.93}$$

$$\epsilon_1\ E_{n1} = \epsilon_2\ E_{n2}, \tag{4.94}$$

where the subscript "t" denotes the tangential component parallel to the interface and the subscript "n" denotes the normal component perpendicular to the interface.

Suppose E_i, E_r and E_t denote the magnitude of an incident E-field, reflected E-field and transmitted E-field, respectively. From boundary conditions in Eqs. (4.93) and (4.94), we can derive their relationships. First, we consider the tangential components parallel to the interface. From the directions of $\vec{E}_i$ and $\vec{E}_r$ in Fig. 4.22, we get the tangential component in medium 1 given by

$$E_{t1} = E_i \cos\theta_i + E_r \cos\theta_i. \tag{4.95}$$

On the other hand, from the direction of $\vec{E}_t$, we get the tangential component in medium 2 given by

$$E_{t2} = E_t \cdot \cos\theta_t. \tag{4.96}$$

Because $E_{t1} = E_{t2}$, we have

$$(E_i + E_r) \cdot \cos\theta_i = E_t \cdot \cos\theta_t. \tag{4.97}$$

Next, we consider the normal components perpendicular to the interface. From Fig. 4.22, we get

$$E_{n1} = E_i \cdot \sin\theta_i - E_r \cdot \sin\theta_i, \tag{4.98}$$

$$E_{n2} = E_t \cdot \sin\theta_t. \tag{4.99}$$

Because $\epsilon_1\ E_{n1} = \epsilon_2\ E_{n2}$, we have

$$\epsilon_1\ (E_i - E_r) \cdot \sin\theta_i = \epsilon_2\ E_t \cdot \sin\theta_t. \tag{4.100}$$

In addition, from Snell's Law, we have

$$\sin\theta_t = \frac{n_1}{n_2}\sin\theta_i = \sqrt{\frac{\epsilon_1}{\epsilon_2}} \cdot \sin\theta_i. \tag{4.101}$$

Inserting Eq. (4.101) into Eq. (4.100), we obtain

$$E_i - E_r = \sqrt{\frac{\epsilon_2}{\epsilon_1}} \cdot E_t \quad \Rightarrow \quad E_i - E_r = \frac{n_2}{n_1} \cdot E_t. \tag{4.102}$$

Equations (4.97) and (4.102) are two independent equations, and we have two unknown variables E_r and E_t when E_i is given. Thus we attain the following solutions.

$$E_r = \frac{n_1 \cos\theta_t - n_2 \cos\theta_i}{n_1 \cos\theta_t + n_2 \cos\theta_i} \cdot E_i, \tag{4.103}$$

$$E_t = \frac{2n_1 \cos\theta_i}{n_1 \cos\theta_t + n_2 \cos\theta_i} \cdot E_i. \tag{4.104}$$

Obviously, Eq. (4.103) gives the relationship between reflected E-field and incident E-field; Eq. (4.104) gives the relationship between transmitted E-field and incident E-field.

Now, we are ready to define the **reflection coefficient** and the **transmission coefficient** for a parallel polarized wave. From Eqs. (4.13) and (4.104), we define the **reflection coefficient** given by

$$R_{//} = \frac{E_r}{E_i} = \frac{n_1 \cos\theta_t - n_2 \cos\theta_i}{n_1 \cos\theta_t + n_2 \cos\theta_i}, \tag{4.105}$$

and the **transmission coefficient** given by

$$T_{//} = \frac{E_t}{E_i} = \frac{2n_1 \cos\theta_i}{n_1 \cos\theta_t + n_2 \cos\theta_i}. \tag{4.106}$$

Both coefficients depend on (n_1, n_2) and $(\cos\theta_i, \cos\theta_t)$.

In order to explain the physical meanings of Eqs. (4.105) and (4.106) effectively, we choose a specific incident angle $\theta_i = 0°$ as an example. When $\theta_i = 0°$, from Snell's Law, we get $\theta_t = 0°$. Because $\cos\theta_i = \cos\theta_t = 1$, we have

$$R_{//} = \frac{n_1 - n_2}{n_1 + n_2} \tag{4.107}$$

$$T_{//} = \frac{2n_1}{n_1 + n_2} \tag{4.108}$$

In this case, we find

1. When $n_1 < n_2$, e.g., an EM wave propagates from air to water, we have $R_{//} < 0$ from Eq. (4.107). On the other hand, when $n_1 > n_2$, e.g., an EM wave propagates from water to air, we have $R_{//} > 0$.
2. From Eq. (4.108), we learn that $T_{//}$ must be positive. It means E_t and E_i always have the same sign.
3. From Eq. (4.107), we learn that when the difference between n_1 and n_2 is greater, $|R_{//}|$ is greater. On the other hand, when n_1 approaches n_2, $|R_{//}|$ vanishes, i.e., when $n_1 = n_2$, $R_{//} = 0$. This result is reasonable because reflection can be regarded as the response of an EM wave when encountering a change of medium. When the change is large, reflection is significant. When the change vanishes, i.e., $n_1 = n_2$, the EM wave propagates without any reflection.

Although the above principles are found at a specific case of $\theta_i = 0°$, they apply to other incident angles. In general, reflection and transmission occur between two dielectric media.

Example 4.10

For a parallel polarized wave as shown in Fig. 4.22, suppose $n_1 = 2$, $n_2 = 3$ and $\theta_i = 30°$. Please derive the reflection coefficient $R_{//}$ and the transmission coefficient $T_{//}$.

Solution

First, from Snell's Law, we have

$$n_1 \sin \theta_i = n_2 \sin \theta_t \quad \Rightarrow \quad \sin \theta_t = \frac{n_1}{n_2} \cdot \sin 30° = \frac{1}{3}$$

Because the incident angle $\theta_i = 30°$, we have

$$\cos \theta_i = \cos 30° = \frac{\sqrt{3}}{2}.$$

And

$$\cos \theta_t = \sqrt{1 - \sin^2 \theta_t} = \frac{\sqrt{8}}{3}.$$

Finally, from Eqs. (4.105) and (4.106) we get

$$R_{//} = \frac{n_1 \cos \theta_t - n_2 \cos \theta_i}{n_1 \cos \theta_t + n_2 \cos \theta_i} = \frac{2 \cdot \frac{\sqrt{8}}{3} - 3 \cdot \frac{\sqrt{3}}{2}}{2 \cdot \frac{\sqrt{8}}{3} + 3 \cdot \frac{\sqrt{3}}{2}} = -0.16,$$

$$T_{//} = \frac{2n_1 \cos \theta_i}{n_1 \cos \theta_t + n_2 \cos \theta_i} = \frac{2 \cdot 2 \cdot \frac{\sqrt{3}}{2}}{2 \cdot \frac{\sqrt{8}}{3} + 3 \cdot \frac{\sqrt{3}}{2}} = 0.77.$$

∎

Example 4.11

For a parallel polarized wave as shown in Fig. 4.22, suppose $n_1 = 4$, $n_2 = 3$ and $\theta_i = 30°$. Please derive the reflection coefficient $R_{//}$ and the transmission coefficient $T_{//}$.

Solution

From Snell's Law, we have

$$n_1 \sin \theta_i = n_2 \sin \theta_t \quad \Rightarrow \quad \sin \theta_t = \frac{n_1}{n_2} \cdot \sin 30° = \frac{2}{3}.$$

Because the incident angle $\theta_i = 30°$, we have

$$\cos \theta_i = \cos 30° = \frac{\sqrt{3}}{2}.$$

And

$$\cos \theta_t = \sqrt{1 - \sin^2 \theta_t} = \frac{\sqrt{5}}{3}.$$

Finally, from Eqs. (4.105) and (4.106), we get

$$R_{//} = \frac{n_1 \cos \theta_t - n_2 \cos \theta_i}{n_1 \cos \theta_t + n_2 \cos \theta_i} = \frac{4 \cdot \frac{\sqrt{5}}{3} - 3 \cdot \frac{\sqrt{3}}{2}}{4 \cdot \frac{\sqrt{5}}{3} + 3 \cdot \frac{\sqrt{3}}{2}} = 0.068,$$

$$T_{//} = \frac{2 n_1 \cos \theta_i}{n_1 \cos \theta_t + n_2 \cos \theta_i} = \frac{2 \cdot 4 \cdot \frac{\sqrt{3}}{2}}{4 \cdot \frac{\sqrt{5}}{3} + 3 \cdot \frac{\sqrt{3}}{2}} = 1.24.$$

∎

From the above two examples, the reflection coefficient $R_{//}$ might be positive or negative as discussed previously, and $|R_{//}| < 1$. Besides, the transmission coefficient $T_{//}$ must be positive. Note that $T_{//}$ may be greater than 1. This issue will be discussed later.

C   Brewster Angle

As we observe Eq. (4.105), the numerator of reflection coefficient $R_{//}$ vanishes when $n_1 \cos \theta_t = n_2 \cos \theta_i$. Actually when $R_{//} = 0$, no reflection occurs and a parallel polarized wave perfectly transmits through the interface into medium 2. From results of experiments, $R_{//} = 0$ only occurs at a specific angle called **Brewster angle**, which is named after the Scottish scientist, David Brewster.

From Eq. (4.105), when $R_{//} = 0$, we have

$$n_1 \cos \theta_t = n_2 \cos \theta_i. \tag{4.109}$$

From Snell's Law, we have

$$n_1 \sin \theta_i = n_2 \sin \theta_t. \tag{4.110}$$

From Eqs. (4.109) and (4.110), we get

$$\cos^2 \theta_t + \sin^2 \theta_t = \left(\frac{n_2}{n_1}\right)^2 \cos^2 \theta_i + \left(\frac{n_1}{n_2}\right)^2 \sin^2 \theta_i = 1. \tag{4.111}$$

Since $\cos^2 \theta_i = 1 - \sin^2 \theta_i$, from Eq. (4.111), we obtain

$$\sin \theta_i = \sqrt{\frac{n_2^2}{n_1^2 + n_2^2}}. \tag{4.112}$$

In addition, we have

$$\cos \theta_i = \sqrt{1 - \sin^2 \theta_i} = \sqrt{\frac{n_1^2}{n_1^2 + n_2^2}}. \tag{4.113}$$

Finally, from Eqs. (4.112) and (4.113), we get

$$\tan \theta_i = \frac{\sin \theta_i}{\cos \theta_i} = \frac{n_2}{n_1}. \tag{4.114}$$

According to Eq. (4.114), we learn that $R_{//} = 0$ is achieved at a specific incident angle which depends on the ratio of refractive indexes of two dielectric media. This angle is called Brewster angle and denoted by θ_B. From Eq. (4.114), we have

$$\theta_B = \tan^{-1}\left(\frac{n_2}{n_1}\right). \tag{4.115}$$

For arbitrary n_1 and n_2, we can always derive θ_B so that an incident wave perfectly transmits into medium 2 without any reflection.

As shown in Fig. 4.23, when $\theta_i = \theta_B$, an incident wave totally transmits into medium 2 and this phenomenon can be called **total transmission**. Comparing it with **total reflection** in the previous section, the conditions of total reflection seem more restrictive. When $n_1 > n_2$ and $\theta_i \geq \theta_c$, total reflection occurs. When $\theta_i = \theta_B$, total transmission occurs no matter $n_1 > n_2$ or $n_1 \leq n_2$. It can be seen in the following two examples.

Example 4.12

Suppose we have $n_1 = 3$ and $n_2 = 2$. Please derive the angle so that total transmission occurs, i.e., $\theta_i = \theta_B$. In addition, please derive the associated reflection coefficient $R_{//}$ and transmission coefficient $T_{//}$.

Fig. 4.23 Total transmission
and Brewster angle

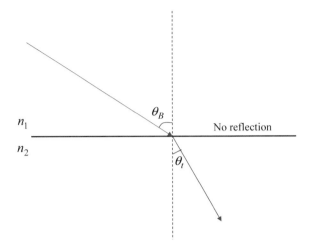

Solution

First, from Eq. (4.114), we can derive Brewster angle given by

$$\tan \theta_B = \frac{n_2}{n_1} = \frac{2}{3}.$$

Hence we have an angle θ_B in a triangle as shown in Fig. 4.24. When $\theta_i = \theta_B$, we obtain the following from trigonometric identities:

$$\sin \theta_i = \sin \theta_B = \frac{2}{\sqrt{13}},$$

$$\cos \theta_i = \cos \theta_B = \frac{3}{\sqrt{13}}.$$

Besides, from Snell's Law, we have

Fig. 4.24 Explaining
trigonometric identity in
Example 4.12

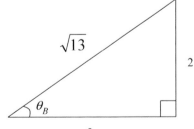

$$\sin \theta_t = \frac{n_1}{n_2} \sin \theta_i = \frac{3}{\sqrt{13}}$$

And thus

$$\cos \theta_t = \sqrt{1 - \sin^2 \theta_t} = \frac{2}{\sqrt{13}}.$$

Finally, from Eqs. (4.105) and (4.106), we get

$$R_{//} = \frac{n_1 \cos \theta_t - n_2 \cos \theta_i}{n_1 \cos \theta_t + n_2 \cos \theta_i} = \frac{3 \cdot \frac{2}{\sqrt{13}} - 2 \cdot \frac{3}{\sqrt{13}}}{3 \cdot \frac{2}{\sqrt{13}} + 2 \cdot \frac{3}{\sqrt{13}}} = 0,$$

$$T_{//} = \frac{2n_1 \cos \theta_i}{n_1 \cos \theta_t + n_2 \cos \theta_i} = \frac{2 \cdot 3 \cdot \frac{3}{\sqrt{13}}}{3 \cdot \frac{2}{\sqrt{13}} + 2 \cdot \frac{3}{\sqrt{13}}} = \frac{3}{2}.$$

∎

Example 4.13

Comparing with Example 4.12, conversely, suppose we have $n_1 = 2$ and $n_2 = 3$. Please derive the Brewster angle so that total transmission occurs. In addition, please derive the associated reflection coefficient $R_{//}$ and transmission coefficient $T_{//}$.

Solution

First, from Eq. (4.114), we can derive Brewster angle given by

$$\tan \theta_B = \frac{n_2}{n_1} = \frac{3}{2}.$$

Hence we have an angle θ_B in a triangle as shown in Fig. 4.25. When $\theta_i = \theta_B$, we obtain the following from trigonometric identities:

$$\sin \theta_i = \sin \theta_B = \frac{3}{\sqrt{13}},$$

$$\cos \theta_i = \cos \theta_B = \frac{2}{\sqrt{13}}.$$

Besides, from Snell's Law, we have

$$\sin \theta_t = \frac{n_1}{n_2} \sin \theta_i = \frac{2}{\sqrt{13}}$$

Fig. 4.25 Explaining
trigonometric identity in
Example 4.13

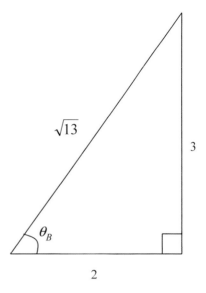

And thus

$$\cos \theta_t = \sqrt{1 - \sin^2 \theta_t} = \frac{3}{\sqrt{13}}.$$

Finally, from Eqs. (4.105) and (4.106), we get

$$R_{//} = \frac{n_1 \cos \theta_t - n_2 \cos \theta_i}{n_1 \cos \theta_t + n_2 \cos \theta_i} = \frac{2 \cdot \frac{3}{\sqrt{13}} - 3 \cdot \frac{2}{\sqrt{13}}}{2 \cdot \frac{3}{\sqrt{13}} + 3 \cdot \frac{2}{\sqrt{13}}} = 0,$$

$$T_{//} = \frac{2n_1 \cos \theta_i}{n_1 \cos \theta_t + n_2 \cos \theta_i} = \frac{2 \cdot 2 \cdot \frac{2}{\sqrt{13}}}{2 \cdot \frac{3}{\sqrt{13}} + 3 \cdot \frac{2}{\sqrt{13}}} = \frac{2}{3}.$$

∎

From Example 4.12 and Example 4.13, we find that total transmission occurs no matter $n_1 > n_2$ or $n_1 \le n_2$. For example in our daily life, no matter we have light transmitting from air to water or from water to air, each has a respective Brewster angle. Besides, in Example 4.13, because the reflection coefficient $R_{//}$ is zero, readers might expect that the transmission coefficient $T_{//}$ shall be one. But why do we get $T_{//} = \frac{2}{3}$? Actually when $R_{//} = 0$, $T_{//}$ is not necessary to be 1. It can be explained by **Law of Energy Conservation**, which is introduced in the following subsection.

D Law of Energy Conservation

First, we discuss the energy distribution at an interface between two dielectric media as shown in Fig. 4.26. Let $(\vec{P}_i, \vec{P}_r, \vec{P}_t)$ be Poynting vectors of incident wave, reflected wave and transmitted wave, respectively. From Sect. 3.3, we know that the direction of Poynting vector is exactly the propagation direction of the associated EM wave. Suppose the incident wave is a uniform plane wave and the associated Poynting vector is $\vec{P}_i$. Then we have

$$\left|\vec{P}_i\right| = \left|\vec{E}_i \times \vec{H}_i\right| = \frac{E_i^2}{\eta_1}, \tag{4.116}$$

where η_1 is the wave impedance of medium 1 and the unit of $|\vec{P}_i|$ is $watt/m^2$, which means the **power density** of the incident wave. In addition, we have the power density for reflected wave and transmitted wave given by

$$\left|\vec{P}_r\right| = \left|\vec{E}_r \times \vec{H}_r\right| = \frac{E_r^2}{\eta_1}, \tag{4.117}$$

$$\left|\vec{P}_t\right| = \left|\vec{E}_t \times \vec{H}_t\right| = \frac{E_t^2}{\eta_2}, \tag{4.118}$$

where η_2 is the wave impedance of medium 2.

In Fig. 4.27, let M be a tiny area on the interface. Because the incident angle is θ_i, the incident power of M is given by

Fig. 4.26 Power distribution at dielectric interface

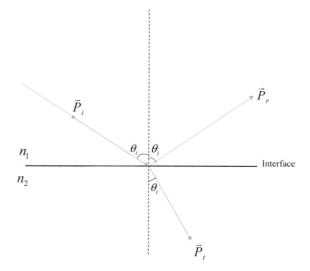

Fig. 4.27 Power distribution
at a tiny area of interface

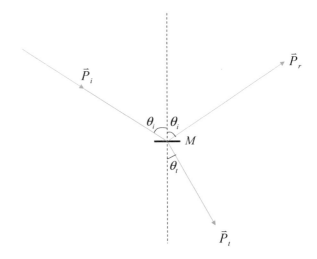

$$U_i = \left|\vec{P}_i\right| \cos\theta_i \cdot M. \tag{4.119}$$

In Eq. (4.119), $|\vec{P}_i| \cos\theta_i$ is the incident power density on M and U_i is the incident power. Similarly, the reflected power U_r and the transmitted power U_t on M are given by

$$U_r = \left|\vec{P}_r\right| \cos\theta_i \cdot M, \tag{4.120}$$

$$U_t = \left|\vec{P}_t\right| \cos\theta_t \cdot M, \tag{4.121}$$

respectively. From the Law of Energy Conservation, U_i must be equal to the sum of U_r and U_t, i.e.,

$$\left|\vec{P}_i\right| \cos\theta_i = \left|\vec{P}_r\right| \cos\theta_i + \left|\vec{P}_t\right| \cos\theta_t. \tag{4.122}$$

Inserting Eqs. (4.116)–(4.118) into Eq. (4.122), we have

$$\frac{E_i^2}{\eta_1} \cdot \cos\theta_i = \frac{E_r^2}{\eta_1} \cdot \cos\theta_i + \frac{E_t^2}{\eta_2} \cdot \cos\theta_t. \tag{4.123}$$

Equation (4.123) can be rewritten as

$$1 = \frac{E_r^2}{E_i^2} + \frac{E_t^2}{E_i^2} \cdot \frac{\eta_1}{\eta_2} \cdot \frac{\cos\theta_t}{\cos\theta_i} \quad \Rightarrow \quad 1 = R_{//}^2 + T_{//}^2 \cdot \frac{\eta_1}{\eta_2} \cdot \frac{\cos\theta_t}{\cos\theta_i} \tag{4.124}$$

Because

$$\eta_1 = \frac{\eta_0}{n_1}, \tag{4.125}$$

$$\eta_2 = \frac{\eta_0}{n_2}, \tag{4.126}$$

where η_0 is the wave impedance of free space. From Eq. (4.124), we have

$$R_{//}^2 + \frac{n_2 \cos \theta_t}{n_1 \cos \theta_i} \cdot T_{//}^2 = 1. \tag{4.127}$$

Equation (4.127) is attained based on the Law of Energy Conservation. If we check Eqs. (4.105) and (4.106) by inserting them into Eq. (4.127), we can find they satisfy the Law of Energy Conservation. It tells us that the behavior of an EM wave at an interface, are consistent with not only Maxwell's equations, but also the Law of Energy Conservation.

Finally, from Eq. (4.127), when $R_{//} = 0$, we have

$$T_{//} = \sqrt{\frac{n_1 \cos \theta_i}{n_2 \cos \theta_t}}. \tag{4.128}$$

Obviously in this case, $T_{//}$ might not be 1. It justifies the result we got in Example 4.12 and Example 4.13.

Example 4.14

In Example 4.11, we have $n_1 = 4$, $n_2 = 3$ and the incident angle $\theta_i = 30°$. Suppose the magnitude of incident E-field $|\vec{E}_i| = E_0$ and M is a tiny area at the interface. Please derive the incident power U_i, reflected power U_r and transmitted power U_t on M.

Solution

From the results in Example 4.11, we have

$$\cos \theta_i = \frac{\sqrt{3}}{2}, \quad \cos \theta_t = \frac{\sqrt{5}}{3}, \quad R_{//} = 0.068, \quad T_{//} = 1.24.$$

From Eqs. (4.116) and (4.119), we have the incident power given by

$$U_i = \frac{E_0^2}{\eta_1} \cdot \cos \theta_i \cdot M = \frac{E_0^2}{\eta_0/4} \cdot \frac{\sqrt{3}}{2} \cdot M = 3.46 \cdot \frac{E_0^2 M}{\eta_0}.$$

Similarly, from Eqs. (4.117) and (4.120), we have the reflected power given by

$$U_r = \frac{E_r^2}{\eta_1} \cdot \cos \theta_i \cdot M = \frac{(E_0 R_{//})^2}{\eta_0/4} \cdot \cos \theta_i \cdot M = 0.02 \cdot \frac{E_0^2 M}{\eta_0}.$$

And from Eqs. (4.118) and (4.121), we have the transmitted power given by

$$U_t = \frac{E_t^2}{\eta_2} \cdot \cos \theta_t \cdot M = \frac{(E_0 T_{//})^2}{\eta_0/3} \cdot \cos \theta_t \cdot M = 3.44 \cdot \frac{E_0^2 M}{\eta_0}.$$

From the above results, we find $U_i = U_r + U_t$, which is consistent with the Law of Energy Conservation.

∎

Example 4.15

Please prove that the result of Example 4.13 is consistent with the Law of Energy Conservation.

Solution

From the results of Example 4.13, we have

$$\cos \theta_i = \frac{2}{\sqrt{13}}, \; \cos \theta_t = \frac{3}{\sqrt{13}}, \; R_{//} = 0, T_{//} = \frac{2}{3}.$$

Assume M is a tiny area on the interface and the magnitude of incident E-field is $|\vec{E}_i| = E_0$. From Eqs. (4.116)–(4.121), the incident power on M is given by

$$U_i = \frac{E_0^2}{\eta_1} \cdot \cos \theta_i \cdot M = \frac{E_0^2}{\eta_0/2} \cdot \frac{2}{\sqrt{13}} \cdot M = \frac{4}{\sqrt{13}} \cdot \frac{E_0^2 M}{\eta_0}.$$

The reflected power and transmitted power on M are given by

$$U_r = \frac{E_r^2}{\eta_1} \cdot \cos \theta_i \cdot M = 0,$$

$$U_t = \frac{E_t^2}{\eta_2} \cdot \cos \theta_t \cdot M = \frac{(E_0 T_{//})^2}{\eta_0/3} \cdot \frac{3}{\sqrt{13}} \cdot M = \frac{4}{\sqrt{13}} \cdot \frac{E_0^2 M}{\eta_0}.$$

It can be verified that $U_i = U_r + U_t$, which is consistent with the Law of Energy Conservation.

∎

4.5 Perpendicular Polarized Wave

In the previous section, we have learned the behaviors of parallel polarized waves at dielectric interface. In this section, we will introduce the behaviors of **perpendicular polarized waves** whose E-field is perpendicular to the plane of incidence. Because the interface is also perpendicular to the plane of incidence, the E-field of a perpendicular polarized wave is parallel to the interface.

As the E-field of a perpendicular polarized wave is parallel to the interface without any component normal to the interface, we have only one boundary condition for the E-field, i.e., we have $E_{t1} = E_{t2}$, but do not have $\epsilon_1 E_{n1} = \epsilon_2 E_{n2}$. Unfortunately, one boundary condition is not sufficient to analyze the associated behaviors at the interface. Therefore, when discussing behaviors of perpendicular polarized waves, we focus on the associated M-field in order to derive the reflection coefficient and transmission coefficient.

A M-field in Plane of Incidence

Suppose we have two dielectric media. The associated permittivity and permeability are (ϵ_1, μ_0) for medium 1 and (ϵ_2, μ_0) for medium 2, where $\epsilon_1 \neq \epsilon_2$. For an incident EM wave at the interface, the electric field $\vec{E}$ can be decomposed as two components:

$$\vec{E} = \vec{E}_{//} + \vec{E}_{\perp}, \tag{4.129}$$

where $\vec{E}_{//}$ is parallel to the plane of incidence and $\vec{E}_{\perp}$ is perpendicular to the plane of incidence. When an incident EM wave has $\vec{E}_{//} = 0$, we have

$$\vec{E} = \vec{E}_{\perp}. \tag{4.130}$$

In this case, the EM wave is called perpendicular polarized wave whose E-field is perpendicular to the plane of incidence.

Because the M-field of a perpendicular polarized wave is perpendicular to the E-field, it is parallel to the plane of incidence. Hence the incident M-field $\vec{H}_i$, reflected M-field $\vec{H}_r$ and transmitted M-field $\vec{H}_t$ all lie in the plane of incidence. As illustrated in Fig. 4.28, $\vec{H}_i$, $\vec{H}_r$ and $\vec{H}_t$ all lie in the plane of incidence ($y-z$ plane) and the interface is the $x-y$ plane. Note that the direction of these M-fields are so arranged that the corresponding incident E-field $\vec{E}_i$, reflected E-field $\vec{E}_r$ and transmitted E-field $\vec{E}_t$ have the reference direction inserting the paper. In addition, the reflection angle θ_r is replaced by the incident angle θ_i because of the Law of Reflection. Besides, we draw three dashed lines which are parallel to the interface in order to facilitate

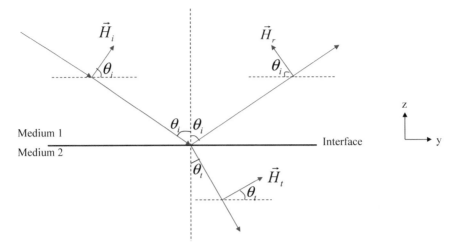

Fig. 4.28 Perpendicular-polarized wave at dielectric interface

our analysis. In the following, we derive the reflection coefficient and transmission coefficient of a perpendicular polarized wave.

B Reflection Coefficient and Transmission Coefficient

From Sect. 4.2, a magnetic field at an interface of two dielectric media must satisfy the boundary conditions given by

$$H_{t1} = H_{t2}, \tag{4.131}$$

$$H_{n1} = H_{n2} \tag{4.132}$$

where the subscript "t" denotes the tangential component parallel to the interface and the subscript "n" denotes the normal component perpendicular to the interface.

Suppose H_i, H_r and H_t denote the magnitude of an incident M-field, reflected M-field and transmitted M-field, respectively. From the directions of $\vec{H}_i$ and $\vec{H}_r$ in Fig. 4.28, we get the tangential component in medium 1 given as

$$H_{t1} = H_i \cdot \cos\theta_i - H_r \cdot \cos\theta_i. \tag{4.133}$$

On the other hand, from the direction of $\vec{H}_t$ in medium 2, we get

$$H_{t2} = H_t \cdot \cos\theta_t. \tag{4.134}$$

From Eqs. (4.131), (4.133) and (4.134), we attain

$$(H_i - H_r) \cdot \cos \theta_i = H_t \cdot \cos \theta_t. \tag{4.135}$$

Next, when considering normal components to the interface in Fig. 4.28, we have

$$H_{n1} = H_i \cdot \sin \theta_i + H_r \cdot \sin \theta_i, \tag{4.136}$$

$$H_{n2} = H_t \cdot \sin \theta_t. \tag{4.137}$$

From Eqs. (4.132), (4.136) and (4.137), we attain

$$(H_i + H_r) \cdot \sin \theta_i = H_t \cdot \sin \theta_t. \tag{4.138}$$

Moreover, from Snell's Law, we have

$$\sin \theta_t = \frac{n_1}{n_2} \sin \theta_i. \tag{4.139}$$

Inserting Eq. (4.139) into Eq. (4.138), we get

$$n_2 \cdot (H_i + H_r) = n_1 \cdot H_t. \tag{4.140}$$

Equations (4.135) and (4.140) are two independent equations. Suppose H_i is given. From Eqs. (4.135) and (4.140), we can derive H_r and H_t given by

$$H_r = \frac{n_1 \cos \theta_i - n_2 \cos \theta_t}{n_1 \cos \theta_i + n_2 \cos \theta_t} \cdot H_i, \tag{4.141}$$

$$H_t = \frac{2n_2 \cos \theta_i}{n_1 \cos \theta_i + n_2 \cos \theta_t} \cdot H_i. \tag{4.142}$$

Equations (4.141) and (4.142) show the relationship between H_i, H_r and H_t. Obviously, their relationship depends on the angles (θ_i, θ_t) and the property of dielectric media, i.e., (n_1, n_2).

In order to be consistent with those of parallel polarized waves, we still define **reflection coefficient and transmission coefficient** by using E-field for perpendicular polarized waves. First, let E_i, E_r and E_t be the magnitude of incident, reflected, and transmitted E-field, respectively. Suppose η is the wave impedance and thus $E = \eta H$. The reflection coefficient and transmission coefficient are given by

$$R_\perp = \frac{E_r}{E_i} = \frac{\eta_1 H_r}{\eta_1 H_i} = \frac{H_r}{H_i}, \tag{4.143}$$

$$T_\perp = \frac{E_t}{E_i} = \frac{\eta_2 H_t}{\eta_1 H_i} = \frac{n_1}{n_2} \cdot \frac{H_t}{H_i}, \tag{4.144}$$

respectively. Inserting Eqs. (4.141) and (4.142) into Eqs. (4.143) and (4.144), we finally get

$$R_\perp = \frac{n_1 \cos \theta_i - n_2 \cos \theta_t}{n_1 \cos \theta_i + n_2 \cos \theta_t}, \tag{4.145}$$

$$T_\perp = \frac{2n_1 \cos \theta_i}{n_1 \cos \theta_i + n_2 \cos \theta_t}. \tag{4.146}$$

Comparing Eqs. (4.145) and (4.146) with the results for parallel polarized waves, i.e., Eqs. (4.105) and (4.106) in Sect. 4.4, we can find that they are different. This is why we need to discuss these two cases separately.

Finally, from Eqs. (4.145) and (4.146), we have

$$T_\perp = 1 + R_\perp. \tag{4.147}$$

It shows for a perpendicular polarized wave, we have a simple relationship between $R_\perp$ and $T_\perp$ as shown in Eq. (4.147). However, for a parallel polarized wave, we do not have similar relationship. In addition, from Eq. (4.147), we can immediately derive $R_\perp$ when $T_\perp$ is given, and vice versa.

Example 4.16

In Fig. 4.28, suppose $n_1 = 2$, $n_2 = 3$ and $\theta_i = 30°$. Please derive the reflection coefficient $R_\perp$ and the transmission coefficient $T_\perp$.

Solution

First, from Snell's Law, we have

$$n_1 \sin \theta_i = n_2 \sin \theta_t \quad \Rightarrow \quad \sin \theta_t = \frac{2}{3} \cdot \sin 30° = \frac{1}{3}$$

Next, we have

$$\cos \theta_i = \cos 30° = \frac{\sqrt{3}}{2},$$

$$\cos \theta_t = \sqrt{1 - \sin^2 \theta_t} = \frac{\sqrt{8}}{3}.$$

From Eq. (4.145), we get the reflection coefficient given by

$$R_\perp = \frac{n_1 \cos \theta_i - n_2 \cos \theta_t}{n_1 \cos \theta_i + n_2 \cos \theta_t} = \frac{2 \cdot \frac{\sqrt{3}}{2} - 3 \cdot \frac{\sqrt{8}}{3}}{2 \cdot \frac{\sqrt{3}}{2} + 3 \cdot \frac{\sqrt{8}}{3}} = -0.24.$$

Since $R_\perp$ is known, we can utilize Eq. (4.147) and immediately get the transmission coefficient given by

$$T_\perp = 1 + R_\perp = 0.76.$$

∎

Example 4.17

In Fig. 4.28, suppose $n_1 = 4$, $n_2 = 3$ and $\theta_i = 30°$. Please derive the reflection coefficient $R_\perp$ and the transmission coefficient $T_\perp$.

Solution

First, from Snell's Law, we have

$$n_1 \sin \theta_i = n_2 \sin \theta_t \quad \Rightarrow \quad \sin \theta_t = \frac{4}{3} \cdot \sin 30° = \frac{2}{3}.$$

Next, we have

$$\cos \theta_i = \cos 30° = \frac{\sqrt{3}}{2},$$

$$\cos \theta_t = \sqrt{1 - \sin^2 \theta_t} = \frac{\sqrt{5}}{3}.$$

From Eq. (4.145), we get the reflection coefficient given by

$$R_\perp = \frac{n_1 \cos \theta_i - n_2 \cos \theta_t}{n_1 \cos \theta_i + n_2 \cos \theta_t} = \frac{4 \cdot \frac{\sqrt{3}}{2} - 3 \cdot \frac{\sqrt{5}}{3}}{4 \cdot \frac{\sqrt{3}}{2} + 3 \cdot \frac{\sqrt{5}}{3}} = 0.215$$

Since $R_\perp$ is known, we can utilize Eq. (4.147) and immediately get the transmission coefficient given by

$$T_\perp = 1 + R_\perp = 1.215$$

∎

From above two examples, we find that reflection coefficient $R_\perp$ might be positive or negative. But the transmission coefficient $T_\perp$ is always positive and might be greater than 1. This is similar to what we have for a parallel polarizad wave.

C Polarizer

From Sect. 4.4, we know that total transmission exists for a parallel polarized wave when the incident angle equals to Brewster angle, i.e., $\theta_i = \theta_B$. In this case, no reflection occurs ($R_{//} = 0$) and the wave perfectly transmits through interface into the next medium. It is useful for a number of applications. In the following, we want to check if total transmission also occurs for a perpendicular polarizad wave.

First, suppose total transmission occurs for a perpendicular polarizad wave. In this case, the corresponding reflection coefficient $R_\perp = 0$, and from Eq. (4.145), we have

$$n_1 \cos \theta_i = n_2 \cos \theta_t. \tag{4.148}$$

In addition, from Snell's Law, we have

$$n_1 \sin \theta_i = n_2 \sin \theta_t. \tag{4.149}$$

From Eqs. (4.148) and (4.149), we attain

$$\sin^2 \theta_i + \cos^2 \theta_i = \left(\frac{n_2}{n_1}\right)^2 (\sin^2 \theta_t + \cos^2 \theta_t) \Rightarrow \quad 1 = \left(\frac{n_2}{n_1}\right)^2 \tag{4.150}$$

Because $n_1 \neq n_2$, Eq. (4.150) can not hold. Hence total transmission ($R_\perp = 0$) does not occur for a perpendicular polarized wave.

Fortunately, we can take advantage of this excellent property—**total transmission exists only for a parallel polarized wave, but not for a perpendicular polarized wave**—to extract a specific polarized wave. As shown in Fig. 4.29, suppose an incident EM wave ($\vec{E}_i$) is decomposed as a parallel polarized wave ($\vec{E}_{//}$) and a perpendicular polarized wave ($\vec{E}_\perp$), where $\vec{E}_i = \vec{E}_{//} + \vec{E}_\perp$. When we purposely select the incident angle so that $\theta_i = \theta_B$ and thus $R_{//} = 0$, then the parallel polarized wave perfectly transmits into medium 2. In this case, the reflected E-field $\vec{E}_r$ is given by

$$\vec{E}_r = R_{//}\vec{E}_{//} + R_\perp \vec{E}_\perp = R_\perp \vec{E}_\perp. \tag{4.151}$$

Hence the reflected wave consists of solely the perpendicular polarized wave. From a user's perspective, what we achieve in Fig. 4.29 is a **polarizer** as shown in Fig. 4.30, where the input is the incident wave ($\vec{E}_i$) and the output is the reflected wave ($\vec{E}_r$). The polarizer completely removes the parallel polarized wave and the output consists of only the perpendicular polarized wave. Furthermore, because the E-field of a perpendicular polarized wave is parallel to the interface, the reflected E-field $\vec{E}_r$ is parallel to the interface. Hence no matter what the original $\vec{E}_i$ may be,

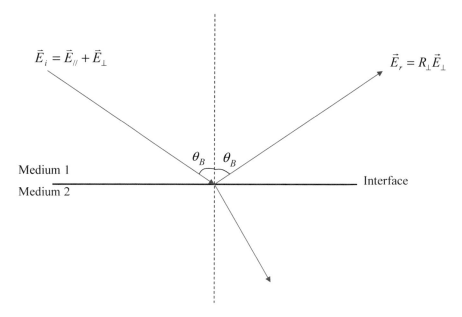

Fig. 4.29 Explaining principle of polarizer

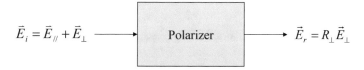

Fig. 4.30 Explaining function of polarizer

after a polarizer, the output is a well polarized wave where E-field is parallel to the interface. In practice, polarizers are widely used in optical applications.

Example 4.18

In Fig. 4.29, suppose $n_1 = 2$, $n_2 = 3$, the plane of incidence is the yz-plane and the incident E-field is given by $\vec{E}_i = 2\hat{x} + 4\hat{y} + 3\hat{z}$.

(a). Let $\vec{E}_i = \vec{E}_{//} + \vec{E}_\perp$. Please derive $\vec{E}_{//}$ and $\vec{E}_\perp$.
(b). Derive Brewster angle θ_B.
(c). When $\theta_i = \theta_B$, derive the reflected E-field $\vec{E}_r$.

Solution

(a) Because $\vec{E}_{//}$ is parallel to the plane of incidence (yz-plane), we have

$$\vec{E}_{//} = 4\hat{y} + 3\hat{z}.$$

On the other hand, $\vec{E}_\perp$ is perpendicular to the plane of incidence and given by

$$\vec{E}_\perp = 2\hat{x}.$$

(b)   The Brewster angle is given by

$$\tan \theta_B = \frac{n_2}{n_1} = \frac{3}{2} \quad \Rightarrow \quad \theta_B = \tan^{-1}(\frac{3}{2}) = 56.4°.$$

(c)   According to the previous discussions, when $\theta_i = \theta_B$, the reflected wave consists of only the perpendicular polarized wave $\vec{E}_\perp$. Let's simply focus on $\vec{E}_\perp$. First, when $\theta_i = \theta_B$, we have

$$\tan \theta_i = \tan \theta_B = \frac{3}{2}.$$

As we plot the associated triangle as shown in Fig. 4.31, we have

$$\sin \theta_i = \frac{3}{\sqrt{13}},$$

Fig. 4.31 Explaining trigonometric identity in Example 4.18

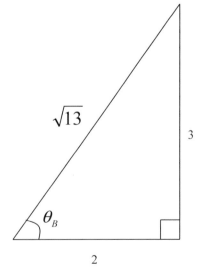

$$\cos\theta_i = \frac{2}{\sqrt{13}}.$$

Next, from Snell's Law, we have

$$\sin\theta_t = \frac{n_1}{n_2}\sin\theta_i = \frac{2}{\sqrt{13}}$$

and thus

$$\cos\theta_t = \sqrt{1 - \sin^2\theta_t} = \frac{3}{\sqrt{13}}.$$

Then from Eq. (4.145), the reflection coefficient can be obtained by

$$R_\perp = \frac{n_1\cos\theta_i - n_2\cos\theta_t}{n_1\cos\theta_i + n_2\cos\theta_t} = \frac{2\cdot\frac{2}{\sqrt{13}} - 3\cdot\frac{3}{\sqrt{13}}}{2\cdot\frac{2}{\sqrt{13}} + 3\cdot\frac{3}{\sqrt{13}}} = -\frac{5}{13}.$$

Finally, the reflected E-field is given by

$$\vec{E}_r = R_\perp \cdot \vec{E}_\perp = -\frac{10}{13}\hat{x}.$$

■

4.6 Wave Behavior at Conductor Interface

Different from dielectric media, a conductor contains a lot of free electrons. These free electrons grant a conductor much better electric conductivity than dielectric media. When an EM wave impinges on a conductor, free electrons are attracted or expelled and form **surface charges**. These surface charges will move due to the external electric field and dynamically form **surface current**. Owing to the existence of surface charges and surface current, the analysis of an EM wave at conductor interface is more complicated than that at dielectric interface.

A Interface of a Conductor

In Fig. 4.32, suppose medium 1 is a dielectric medium and medium 2 is a conductor. From the above discussion, we know that surface charge and surface current exist on medium 2. Let the associated permittivity and permeability of these two media be $(\in_1, \mu_1)$ and $(\in_2, \mu_2)$, respectively, where $\in_1 \neq \in_2$. From Sect. 4.1, when a uniform

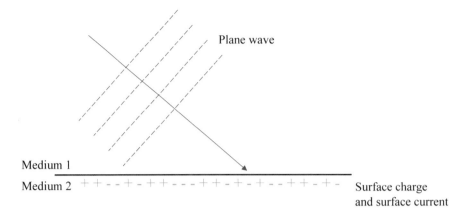

Medium 1

Medium 2   Surface charge and surface current

Fig. 4.32 Incidence of EM wave at conductor interface

plane wave impinges on the interface of a conductor, the associated E-field must satisfy the following boundary conditions:

$$E_{t1} = E_{t2}, \tag{4.152}$$

$$\in_1 E_{n1} = \in_2 E_{n2} + \rho_s, \tag{4.153}$$

where ρ_s denotes **surface charge density**. In addition, from Sect. 4.2, the associated M-field must satisfy the following boundary conditions:

$$H_{t1} = H_{t2} + J_s, \tag{4.154}$$

$$H_{n1} = H_{n2}, \tag{4.155}$$

where J_s is **surface current density**. In Eqs. (4.152)–(4.155), the left side shows the E-field and M-field of medium 1; the right side shows those of medium 2. Besides, the subscript "t" denotes the tangential component parallel to the interface and the subscript "n" denotes the normal component perpendicular to the interface.

Owing to the existence of surface charge and surface current, as shown in Eqs. (4.153) and (4.154), the analysis of an EM wave at conductor interface is very complicated. In order to help readers catch the useful insights effectively, we skip the complicated analysis and introduce the behaviors in a more qualitative way.

As illustrated in Fig. 4.32, what will happen when an EM wave impinges on such an interface?

1. **Law of Reflection still holds.**

When discussing dielectric interface in Sect. 4.3, Law of Reflection holds, i.e., $\theta_i = \theta_r$, because the incident wave and the reflected wave must propagate the same distance at a given time duration. Similarly in Fig. 4.32, because both the incident wave and the reflected wave propagate in medium 1, they must follow the same rule. Hence, Law of Reflection also holds for an interface of a conductor.

2. **Snell's Law does not hold.**

When discussing dielectric interface in Sect. 4.3, both media are dielectric. The phase velocity of an EM wave in medium 1 is $v_{p1} = \frac{c}{n_1}$ and that in medium 2 is $v_{p2} = \frac{c}{n_2}$, where n_1 and n_2 are the respective refractive indexes. Because of different phase velocities, the propagation distance of incident wave and that of transmitted wave are different during a specific time period. Based on this feature, we can derive Snell's Law, given by

$$n_1 \cdot \sin \theta_i = n_2 \cdot \sin \theta_t \tag{4.156}$$

However, for an interface of a conductor as shown in Fig. 4.32, medium 2 is a conductor, instead of a dielectric medium. From Sect. 3.4, the phase velocity of an EM wave in a conductor is much slower than $\frac{c}{n_2}$. Hence Snell's Law does **NOT** hold at an interface of a conductor. This is a major difference between a dielectric interface and a conductor interface.

3. **Reflection coefficient $R \approx 1$ and transmission coefficient $T \approx 0$**

From both analytical and experimental results, an interface of a conductor will reflect most incident EM wave and only very few will get into the conductor. For instance, a copper is a good conductor with conductivity $5.8 \times 10^7 (S/m)$. For most microwave

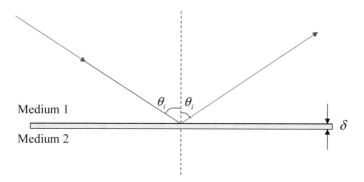

Fig. 4.33 Law of Reflection and skin effect at conductor interface

spectrum, the reflection coefficient may go up to 0.999. It means almost all the EM waves are reflected back to the medium 1.

For an interface of a conductor, suppose E_i, E_r and E_t denote the magnitude of incident, reflected and transmitted E-fields, respectively. Because most EM waves are reflected back, we have $E_r \approx E_i$ and $E_t << E_i$. Hence reflection coefficient $R \approx 1$ and transmission coefficient $T \approx 0$.

Because the transmission coefficient is small, very limited EM wave is getting into a conductor. Since an EM wave decays very quickly in a conductor, a transmitted EM wave propagates for a very short distance and then vanishes.

4. **Surface resistance**

When an EM wave impinges on a conducting surface, surface charges are induced in the conductor and mainly bound within the distance of skin depth (δ) as shown in Fig. 4.33. Because skin depth δ is usually very small and free electrons repel each other in this small area, the bound surface current encounter resistance. This phenomenon is similar to what happens when free electrons go through a resistor. We can imagine that a resistor exists in the surface of a conductor and call it **surface resistance**. Like common resistors, a surface resistance also consumes energy.

From above, we can regard a conductor as a good "EM wave reflector." Most incident EM waves will be reflected and very limited energy will be consumed at the surface of the conductor.

B   *Surface Resistance*

Surface resistance is an important feature of a conductor. As shown in the left side of Fig. 4.34, a conductor has a length L, width W and thickness H. When an EM wave impinges on the conductor, the transmitted wave will concentrate on a short distance of skin depth δ and form a surface current. From Sect. 3.4, the skin depth is given by

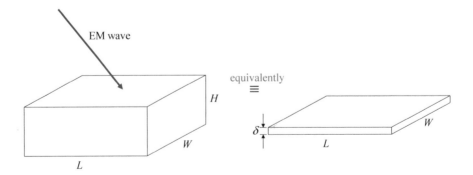

Fig. 4.34 Skin effect and surface resistance

$$\delta = \sqrt{\frac{1}{\pi f \mu \sigma}}, \tag{4.157}$$

where f is frequency, σ is conductivity and μ is permeability. Because all the prominent behaviors of the transmitted wave occur within the depth of δ, we can regard this conductor as a thin plate having the thickness δ as shown in the right side of Fig. 4.34. It greatly simplifies our analysis when calculating the associated surface resistance.

From Fundamental Electronics, when a medium has length L, cross-sectional area A, and conductivity σ, the associated resistance is proportional to L, and inversely proportional to A and σ, given by

$$R = \frac{L}{\sigma A}. \tag{4.158}$$

In Fig. 4.34, suppose a surface current flows along L. Because the cross-sectional area is $A = W \cdot \delta$, the surface resistance is given by

$$R = \frac{1}{\sigma} \cdot \frac{L}{W \delta}. \tag{4.159}$$

Equation (4.159) is typically rewritten as

$$R = \frac{1}{\sigma \delta} \cdot \frac{L}{W} = R_S \cdot \frac{L}{W}, \tag{4.160}$$

where R_S is the surface resistance per unit length and unit width of the conductor, given by

$$R_S = \frac{1}{\sigma \delta} = \sqrt{\frac{\pi f \mu}{\sigma}}. \tag{4.161}$$

Obviously, R_S increases when σ decreases.

The above analytical technique can be used to solve EM problems involving a conductor. For example in Fig. 4.35, we have a copper line with a length L and radius a. When a high-frequency EM wave (signal) is transmitted by this copper line, the induced electric current does not uniformly distribute on the cross-section (πa^2). Instead, all the prominent EM effects concentrate within the depth of δ from

Fig. 4.35 A copper line with a given radius

Fig. 4.36 Skin effect of the
copper line

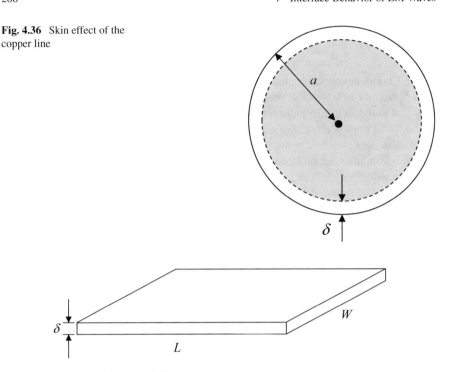

Fig. 4.37 Equivalent thin plate of skin effect

the surface as shown in Fig. 4.36. Hence, in order to simplify the problem, we can
regard the copper line as a thin plate as shown in Fig. 4.37, where the length is L, the
thickness is δ, and the width W is equal to the circumstance of the original copper
line, i.e., $W = 2\pi a$. Hence, the surface resistance is given by

$$R = R_S \cdot \frac{L}{W} = R_S \cdot \frac{L}{2\pi a}. \tag{4.162}$$

Once the surface resistance is obtained, we can derive the power consumption of
the copper line.

Example 4.19

Suppose a copper line has a length $L = 0.1\,m$, radius $a = 2$ mm, and conductivity
$\sigma = 5.8 \times 10^7 (S/m)$. Please derive the surface resistance when an EM wave has the
frequency $f = 100\,MHz$.

Solution

First, from Eq. (4.161), we have

$$R_S = \sqrt{\frac{\pi f \mu}{\sigma}} = \sqrt{\frac{\pi \cdot (10^8) \cdot (4\pi \times 10^{-7})}{5.8 \times 10^7}} = 2.6 \times 10^{-3}(\Omega).$$

From Eq. (4.162), we have the surface resistance given by

$$R = R_S \cdot \frac{L}{2\pi a} = (2.6 \times 10^{-3}) \cdot \frac{0.1}{2\pi \cdot (2 \times 10^{-3})} = 0.021(\Omega).$$

∎

Example 4.20

Suppose we transmit a signal using the copper line in Example 4.19 and the surface current is given by $I(t) = I_A \cdot \cos 2\pi f t$, where $I_A = 2(mA)$. Please derive the average power consumption.

Solution

From the result in Example 4.19, we have the surface resistance given by

$$R = 0.021(\Omega).$$

Next, we regard this copper line as a resistor having the resistance $R = 0.021(\Omega)$. Because the current is $I(t) = I_A \cdot \cos 2\pi f t$, at a specific time t, the power consumption is given by

$$P(t) = I^2(t) \cdot R = I_A^2 R \cdot \cos^2 2\pi f t.$$

For a sinusoidal period, the average of $\cos^2 2\pi f t$ is equal to $\frac{1}{2}$. Hence the average power consumption is given by

$$P_{avg} = \frac{1}{2} I_A^2 R = \frac{1}{2} \cdot (2 \times 10^{-3})^2 \cdot (0.021) = 4.2 \times 10^{-8}(W).$$

∎

C Perfect Conductor

Suppose we have an ideal conductor whose conductivity approaches infinity, i.e., $\sigma \to \infty$. This conductor is called a **perfect conductor**. From Eq. (4.157), we know that when $\sigma \to \infty$, the skin depth $\delta = 0$. The null skin depth is an important feature of a perfect conductor.

Because of the null skin depth ($\delta = 0$), any incident EM wave can not penetrate into a perfect conductor. Hence, both E-field and M-field are absent in a perfect conductor. From the boundary conditions in Eqs. (4.152)–(4.155), since $E_{t2} = E_{n2} = 0$ and $H_{t2} = H_{n2} = 0$, we have

$$E_{t1} = 0, \tag{4.163}$$

$$\in_1 E_{n1} = \rho_s, \tag{4.164}$$

$$H_{t1} = J_s, \tag{4.165}$$

$$H_{n1} = 0. \tag{4.166}$$

Obviously, the boundary conditions of a perfect conductor are simpler than those of an ordinary conductor. These conditions will be used in the next chapter when we introduce transmission lines.

Besides, from Eq. (4.161), the surface resistance of a perfect conductor is zero ($R_S = 0$) because $\sigma \to \infty$. It implies that a perfect conductor does not consume power, which can be used to produce an ideal transmission line.

Summary

We learn boundary conditions and behaviors of EM waves at different interfaces in this chapter, which includes six sections:

4.1:   Boundary conditions of E-field

- For the interface between two dielectric media:

$$E_{t1} = E_{t2},$$

$$\in_1 E_{n1} = \in_2 E_{n2}.$$

- For the interface between a dielectric medium and a conductor:

$$E_{t1} = E_{t2},$$

$$\in_1 E_{n1} = \in_2 E_{n2} + \rho_s,$$

where ρ_s is the surface charge density.

4.2:   Boundary conditions of M-field

- For the interface between two dielectric media:

$$H_{t1} = H_{t2},$$

$$H_{n1} = H_{n2}.$$

Hence $\vec{H}_1 = \vec{H}_2$.

- For the interface between a dielectric medium and a conductor:

$$H_{t1} = H_{t2} + J_s,$$

$$H_{n1} = H_{n2},$$

where J_s is the surface current density.

4.3: General laws at dielectric interface

- Law of Reflection: $\theta_i = \theta_r$,
- Snell's Law: $n_1 \cdot \sin \theta_i = n_2 \cdot \sin \theta_t$,
- Total reflection occurs when $n_1 > n_2$ and $\theta_i \geq \theta_c = \sin^{-1}(\frac{n_2}{n_1})$,

where θ_i is the incident angle, θ_r is the reflection angle, θ_t is the transmission angle and θ_c is the critical angle.

4.4: We learn behaviors of parallel polarized waves and Brewster angle.
4.5: We learn behaviors of perpendicular polarized waves and polarizer.
4.6: We learn behaviors of EM waves at a conductor interface.

The above knowledge makes us understand behaviors of an EM wave when it encounters an interface. Furthermore, we can utilize these knowledge to **control an EM wave because it must satisfy the boundary conditions**. Hence we can arrange the boundaries where an EM wave encounters and enforce an EM wave to adjust its behaviors to what we desire. This concept is used in a lot of applications. In the remaining of the book, we will learn applications of EM waves.

Exercises

1. Please derive boundary conditions of electric fields between

    (a). two dielectric media,
    (b). a dielectric medium and a conductor,

    in your own way. (Hint: Refer to Sect. 4.1).

2. Assume that a dielectric medium A has $\epsilon_1 = 2 \epsilon_0$, another dielectric medium B has $\epsilon_2 = 4 \epsilon_0$, and their interface is the xy-plane. Let $\vec{E}_1$ be the E-field of medium

A at the interface and $\vec{E}_2$ be that of medium B. Please calculate $\vec{E}_2$ when $\vec{E}_1$ is given in the following cases.

(a). $\vec{E}_1 = 5\hat{x} + 4\hat{y} - 2\hat{z}$,
(b). $\vec{E}_1 = 4\cos(\omega t + \frac{\pi}{3}) \cdot \hat{x} - 3\sin(\omega t - \frac{\pi}{4}) \cdot \hat{y} + 2\cos\omega t \cdot \hat{z}$,
(c). $\vec{E}_1 = 6e^{j\frac{\pi}{5}} \cdot \hat{x} + 8e^{j\frac{\pi}{4}} \cdot \hat{y} - 5e^{j\frac{2\pi}{3}} \cdot \hat{z}$ (phasor representation).

(Hint: Example 4.1).

3. Repeat Exercise 2(a) when the interface is perpendicular to the unit vector $\hat{n} = \frac{1}{\sqrt{2}}\hat{x} + \frac{1}{\sqrt{2}}\hat{y}$.

4. Assume A is a dielectric medium and B is a conductor, and their interface is the yz-plane. Let $\vec{E}_1$ be the E-field of medium A at the interface and $\vec{E}_2$ be that of medium B. When A has $\epsilon_1 = 3\ \epsilon_0$, B has $\epsilon_2 = \epsilon_0$ and ρ_s is the surface charge density, please calculate $\vec{E}_2$ when $\vec{E}_1$ is given in each following case.

(a). $\vec{E}_1 = 3\hat{x} - 2\hat{y} + 7\hat{z}$,
(b). $\vec{E}_1 = 6e^{j\frac{\pi}{5}} \cdot \hat{x} + 8e^{j\frac{\pi}{4}} \cdot \hat{y} - 5e^{j\frac{2\pi}{3}} \cdot \hat{z}$ (phasor representation).

(Hint: Example 4.2).

5. Repeat Exercise 4(a) when the interface is perpendicular to the unit vector $\hat{n} = \frac{1}{\sqrt{2}}\hat{x} + \frac{1}{\sqrt{2}}\hat{y}$.

6. Please derive boundary conditions of magnetic fields between

   (a). two dielectric media,
   (b). a dielectric medium and a conductor,

   in your own way. (Hint: Refer to Sect. 4.2).

7. Assume A and B are dielectric media, where A has $\epsilon_1 = 2\ \epsilon_0$, B has $\epsilon_2 = 4\ \epsilon_0$, and their interface is the xy-plane. Suppose $\vec{H}_1$ and $\vec{H}_2$ are M-fields of medium A and B at the interface, respectively. Please calculate $\vec{H}_2$ when $\vec{H}_1$ is given in each following case.

   (a). $\vec{H}_1 = 3\hat{x} + \hat{y} - 5\hat{z}$,
   (b). $\vec{H}_1 = \cos(\omega t + \frac{\pi}{3}) \cdot \hat{x} - 2\sin(\omega t - \frac{\pi}{4}) \cdot \hat{y} - 4\cos\omega t \cdot \hat{z}$.

   (Hint: Example 4.3).

8. Assume that A is a dielectric medium and B is a conductor, and their interface is the xy-plane. Let A have $\epsilon_1 = 3\ \epsilon_0$ and B have $\epsilon_2 = \epsilon_0$. Suppose $\vec{H}_1$ is the M-field of medium A at the interface, $\vec{H}_2$ is that of medium B, and J_s is the surface current density. If $\vec{H}_1 = 3\hat{x} + 4\hat{y} + 6\hat{z}$, please calculate $\vec{H}_2$.

   (Hint: Example 4.4).

9. Assume that the interface between two dielectric media A and B is the xy-plane, where A has $\epsilon_1 = 2\ \epsilon_0$, and B has $\epsilon_2 = 4\ \epsilon_0$. A plane wave propagating in medium A impinges on the interface between media A and B. Because

of reflection, the incident wave and the reflected wave coexist in medium A, while the transmitted wave exists in medium B. At the interface, assume B has $\vec{E}_2 = 3\cos(\omega t + \theta) \cdot \hat{x}$ and $\vec{H}_2 = H_2 \cdot \hat{y}$. The transmitted wave propagates along $+z$ direction.

(a). Please calculate H_2.

(b). Please derive E-field and M-field of medium A at the interface, i.e. $\vec{E}_1 = E_1 \cdot \hat{x}$ and $\vec{H}_1 = H_1 \cdot \hat{y}$.

(c). Is $E_1/H_1 = \eta_1$ (wave impedance) valid in medium A? Please explain why or why not.

10. Assume the interface between two dielectric media is the xy-plane, and yz-plane is the plane of incidence. Let $\vec{E} = \vec{E}_{//} + \vec{E}_{\perp}$, where $\vec{E}_{//}$ is parallel to the plane of incidence and $\vec{E}_{\perp}$ is perpendicular to that. Please derive $\vec{E}_{//}$ and $\vec{E}_{\perp}$ in the following cases.

(a). $\vec{E} = 3\hat{x} - 2\hat{y} + 5\hat{z}$,

(b). $\vec{E} = 2\cos(\omega t + \frac{\pi}{4}) \cdot \hat{x} + 3\sin(\omega t - \frac{\pi}{4}) \cdot \hat{y} + 5\sin\omega t \cdot \hat{z}$,

(c). $\vec{E} = 2e^{j\frac{\pi}{5}} \cdot \hat{x} - 3e^{j\frac{\pi}{4}} \cdot \hat{y} - 3e^{j\frac{2\pi}{3}} \cdot \hat{z}$.

(Hint: Refer to Example 4.5).

11. In Exercise 10, let $\vec{E} = \vec{E}_t + \vec{E}_n$, where $\vec{E}_t$ is parallel to the interface and $\vec{E}_n$ is perpendicular to the interface. Please derive $\vec{E}_t$ and $\vec{E}_n$.

12. A wave is incident on the interface of dielectric medium 1 and dielectric medium 2, where $n_1 = 2$ and $n_2 = 3$. Assume the incident angle is $\theta_i = 30°$.

(a). Please derive the reflection angle θ_r and transmission angle θ_t.

(b). If we increase θ_i, then θ_t increases too. What is the maximum of θ_t?

(Hint: Refer to Sect. 4.3).

13. A wave is incident on the interface of two dielectric media, where $n_1 = 4$ and $n_2 = 2$. Derive the critical angle θ_c and describe what happens when $\theta_i > \theta_c$?

14. A parallel-polarized wave is incident on the interface between two dielectric media, where $n_1 = 3$ and $n_2 = 5$.

(a). Please derive the reflection coefficient $(R_{//})$.

(b). Please calculate $R_{//}$ when $\theta_i = 0°$, $30°$, $60°$ or $90°$.

(c). Please derive the transmission coefficient $(T_{//})$.

(d). Please calculate $T_{//}$ for $\theta_i = 0°$, $30°$, $60°$ or $90°$.

(Hint: Example 4.10).

15. Repeat Exercise 14 when $n_1 = 5$ and $n_2 = 3$.

16. Prove the equality, $R_{//}^2 + \frac{n_2 \cos\theta_t}{n_1 \cos\theta_i} \cdot T_{//}^2 = 1$, by using the formulas of $R_{//}$ and $T_{//}$.

17. A parallel-polarized wave is incident on the interface between two dielectric media, where $n_1 = 2$ and $n_2 = 2\sqrt{3}$.

(a). Derive the Brewster angle (θ_B).

(b).   When $\theta_i = \theta_B$, calculate $R_{//}$ and $T_{//}$.

(c).   When $\theta_i = \theta_B$, prove that $\tan \theta_t = \frac{n_1}{n_2}$.

(Hint: Example 4.12).

18.   A wave is incident on the interface of two dielectric media, where $n_1 = 2$ and $n_2 = 3$. Plot $R_{//}$ and $T_{//}$ when $0 \le \theta_i \le 90°$ by using computers.

19.   Repeat Exercise 18 when $n_1 = 3$ and $n_2 = 2$.

20.   A perpendicular polarized wave is incident on the interface between two dielectric media, where $n_1 = 3$ and $n_2 = 5$.

(a).   Derive the reflection coefficient ($R_\perp$).

(b).   Calculate $R_\perp$ when $\theta_i = 0°, 30°, 60°$ or $90°$.

(c).   Derive the transmission coefficient ($T_\perp$).

(d).   Calculate $T_\perp$ when $\theta_i = 0°, 30°, 60°$ or $90°$.

(Hint: Example 4.16).

21.   Repeat Exercise 20 when $n_1 = 5$ and $n_2 = 3$.

22.   A wave is incident on the interface of two dielectric media, where $n_1 = 2$ and $n_2 = 3$. Plot $R_\perp$ and $T_\perp$ when $0 \le \theta_i \le 90°$ by using computers.

23.   Repeat Exercise 22 when $n_1 = 3$ and $n_2 = 2$.

24.   Assume the plane of incidence of two dielectric media is the yz-plane, where $n_1 = 3$ and $n_2 = 4$. Suppose the incident E-field is $\vec{E}_i = 2\hat{x} + \hat{y} + 5\hat{z}$ and $\vec{E}_r$ is the reflected E-field.

(a).   Let $\vec{E}_i = \vec{E}_{//} + \vec{E}_\perp$. Please find $\vec{E}_{//}$ and $\vec{E}_\perp$.

(b).   Derive the Brewster angle θ_B.

(c).   When $\theta_i = \theta_B$, derive $\vec{E}_r$.

(Hint: Example 4.18).

25.   Prove the equality, $R_\perp^2 + \frac{n_2 \cos \theta_t}{n_1 \cos \theta_i} \cdot T_\perp^2 = 1$, by using the formulas of $R_\perp$ and $T_\perp$.

26.   Using the results in Example 4.16, suppose $n_1 = 4$, $n_2 = 3$ and $\theta_i = 30°$. Let $M = 1$ (m^2) be a unit area on the interface and the incident E-field $|\vec{E}_i| = E_0$. Please derive the incident power (U_i), reflected power (U_r) and transmitted power (U_t).

(Hint: Example 4.14).

27.   Please prove that the results in Exercise 26 is consistent with the Law of Energy Conservation.

28.   A refractive index of lenses for eyeglasses is typically between 1.5 and 1.74. Suppose a light is incident on the interface between the air ($n_1 = 1$) and a lens ($n_2 = 1.586$). Consider a parallel polarized wave.

(a).   When $\theta_i = 0°$, please derive the reflection coefficient $R_{//}$ and the transmission coefficient $T_{//}$.

(b).   When $\theta_i = 90°$, please derive the reflection coefficient $R_{//}$ and the transmission coefficient $T_{//}$.

(c). Plot $R_{//}$ and $T_{//}$ when $0 \le \theta_i \le 90°$ by using computers.

(d). Verify the Law of Energy Conservation by checking the equality $R_{//}^2 +$ $\frac{n_2 \cos \theta_t}{n_1 \cos \theta_i} \cdot T_{//}^2 = 1$ when $\theta_i = 30°$.

29. Repeat Exercise 28 when considering perpendicular polarized component as follows.

(a). When $\theta_i = 0°$, please derive the reflection coefficient $R_\perp$ and the transmission coefficient $T_\perp$.

(b). When $\theta_i = 90°$, please derive the reflection coefficient $R_\perp$ and the transmission coefficient $T_\perp$.

(c). Plot $R_\perp$ and $T_\perp$ when $0 \le \theta_i \le 90°$ by using computers.

30. An EM wave is incident on a rectangular conductor, whose width $W = 2\ m$, length $L = 10\ m$, and its thickness is much larger than the skin depth. Assume that $\sigma = 5 \times 10^7 (S/m)$, $f = 200 MHz$, and the surface current flows along the direction of L.

(a). Please calculate the skin depth.

(b). Derive the surface resistance.

(c). If the voltage drop between two ends of the conductor is $V(t) = (0.2V) \cdot \sin 2\pi f t$, please calculate the average power loss of this conductor.

(Hint: Examples 4.19 and 4.20).

31. A wave is incident on a copper wire, whose radius $r = 1\ cm$ and length $L = 100\ m$. Assume $\sigma = 5 \times 10^8 (S/m)$ and $f = 1 MHz$.

(a). Please calculate the skin depth.

(b). Derive the surface resistance along the wire.

(c). If the voltage drop between two ends of the wire is $V(t) = (0.1V) \cdot \sin 2\pi f t$, calculate its average power loss.

32. The interface between free space and a perfect conductor is the xy-plane. Assume the incident E-field at the interface is $\vec{E}_i = 5\hat{x} + 2\hat{y} + 3\hat{z}$ and $\vec{E}_r$ is the reflected E-field. Let $\vec{E}_i = \vec{E}_{i,t} + \vec{E}_{i,n}$, where $\vec{E}_{i,t}$ is parallel to the interface and $\vec{E}_{i,n}$ is perpendicular to the interface.

(a). Please calculate $\vec{E}_{i,t}$ and $\vec{E}_{i,n}$.

(b). Assume $\vec{E}_r = \vec{E}_{r,t} + \vec{E}_{r,n}$, where $\vec{E}_{r,t}$ is parallel to the interface and $\vec{E}_{r,n}$ is perpendicular to the interface. If the surface charge density is ρ_s, please derive $\vec{E}_{r,t}$ and $\vec{E}_{r,n}$.

(c). Calculate E-field of the free space at the interface.

(d). Explain why E-field is always perpendicular to the surface of perfect conductor.

Chapter 5
Transmission Lines

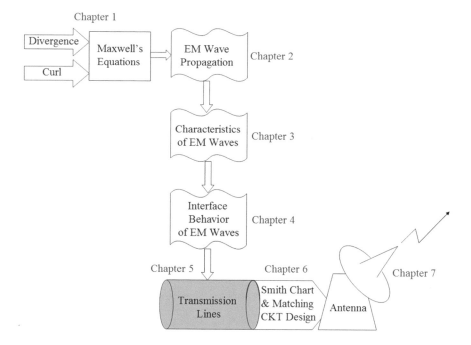

© Springer Nature Switzerland AG 2020

M.-S. Kao and C.-F. Chang, *Understanding Electromagnetic Waves*,

https://doi.org/10.1007/978-3-030-45708-2_5

Abstract Transmission line can deliver EM waves to a far destination. It is interesting and useful to learn how EM waves propagate in a transmission line and why it can propagate for a long distance. We answer these questions step by step with physical intuition plus simple mathematics as follows:

Step 1. Deriving EM fields in a transmission line,
Step 2. Transferring EM fields into voltages and currents which are much easier and more familiar for readers to deal with.
Step 3. Introducing critical parameters of transmission lines, such as characteristic impedance, propagation constant and reflection coefficient.
Step 4. Introducing how to produce a high-frequency circuit component using transmission lines and measure useful parameters of a standing wave.

In each step, we try to guide readers in an intuitive way to catch the idea behind each critical parameter and principle. For example, we start from circuit perspective so that the transmission line equations can be easily derived. We further show that the reflection—the most important phenomenon in transmission line—is actually a smart solution for EM waves to resolve the problem of impedance mismatch. In addition, many illustrations and examples are provided to help readers get the ideas. Hence this chapter forms an important basis for applications and advanced study such as high-frequency circuit design.

Keywords Transmission line · Parallel plate transmission line · Two-wire transmission line · Coaxial transmission line · TEM wave · Transmission line equations · Propagation constant · Attenuation constant · Phase constant · Characteristic impedance · Reflection coefficient · Impedance matching circuit · Transmission line circuit component · Characteristic resistance · Input impedance · Impedance inverter · Standing wave · Voltage standing wave ratio (VSWR)

5.1 Principles of TX Lines

From the previous chapters, we learn many features of EM waves and they can propagate in various mediums including free space. Of course we would like to use them to deliver messages. However, without proper guidance, they might radiate in undesired directions in open space. For example, in communications, we usually want to deliver messages to a specific destination. Therefore, we need to figure out a way to make EM waves propagate to a specific direction, so that the destination can receive sufficient EM energy and extract the desired messages.

A way to satisfy this need is the invention of **transmission lines (TX lines)**. A TX line is a pair of conductors that can convey and guide EM waves from one place to another. In this chapter, we will introduce the principles and applications of

TX lines. This chapter consists of six sections, and each section will introduce an important concept or a critical parameter of TX lines. Learning these will build a solid background for a microwave engineer to design high-frequency circuits.

A   Categories of TX Lines

In electrical engineering, TX lines are used in many applications. For example, the power distribution lines, telecommunication lines, coaxial cable of measurement instruments, and the connection lines in circuits. All of them are TX lines. Obviously, TX lines are critical components in electrical engineering because most electrical signals are transmitted through TX lines.

There are three major types of TX lines as follows.

1. Parallel plate TX line

In Fig. 5.1, a parallel plate TX line consists of two parallel conducting plates. The operation principle is easy to understand: suppose we have an EM wave propagating between these two conducting plates. Because a conducting plate (metal) can reflect the EM wave effectively, the EM wave will bounce back and forward between these two plates. As shown in Fig. 5.1, the EM wave will be restricted in a specific space and directed to the desired destination.

The principle of parallel plate TX lines is widely used in a **printed circuit board (PCB)**. For a PCB, it has at least two layers. A layer consists of circuit components having conducting lines in between, and a layer of ground plane as illustrated in Fig. 5.2. Similar to what occurs in two parallel plates, EM waves propagate between the conducting line and the ground layer.

Here we would like to remind readers an important concept: because EM waves must satisfy the boundary conditions we learned in Chap. 4, we can utilize them to guide EM waves so that our goal can be achieved. For example, in this case, we place two parallel metal plates in open space so that the EM waves originally supposed to

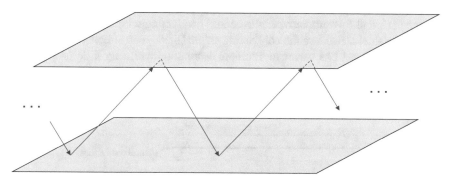

Fig. 5.1 Propagation of EM wave in parallel plate TX line

Fig. 5.2 Illustrating TX line applied in printed circuit board

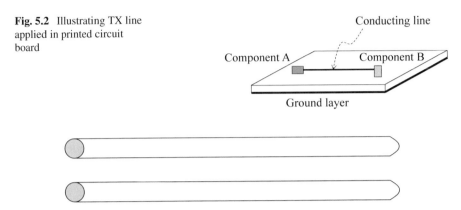

Fig. 5.3 Two-wire TX line

radiate omnidirectionally are thus restricted and directed to the destination we desire. This concept also applies to the following two types of TX lines.

2. Two-wire TX line

Two parallel conducting wires can be used as a transmission lines for EM waves as shown in Fig. 5.3. The working principle is similar to that of two parallel conducting plates: EM waves bounce back and forward between two conducting (metal) wires and hence are directed to the remote destination. Intuitively, the propagation loss of two parallel wires is greater than that of two parallel plates because the latter has better shielding. Generally, two-wire transmission lines only apply to short-distance transmission. For example, they are widely used as power lines of household electric appliances.

3. Coaxial TX line

A coaxial transmission line consists of a conducting wire and a surrounding tubular conducting shield with insulating medium in between as shown in Fig. 5.4. The working principle is also similar to that of parallel plate TX lines, that EM waves bounce back and forward between the central conducting wire and the surrounding conducting shield. Because the conducting shield (metal) can effectively reflect EM waves, it can direct EM waves to a remote destination with very little propagation loss. Intuitively, a coaxial TX line provides the better shielding than the previous two TX lines for EM wave propagation.

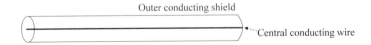

Fig. 5.4 Coaxial TX line

B   TEM Wave

In order to realize the principles of TX lines, we consider parallel plate TX lines and explain how EM waves propagate between two metal plates. First, in Fig. 5.5, suppose two plates locate at $x = 0$ and $x = d$, respectively. Assume both plates extend infinitely along the y-axis and EM wave propagates along the z-axis. An insulating medium is filled between two plates in order to separate them.

Although parallel plate TX lines can support many kinds of EM waves, we consider a simple and critical case, called **Transverse Electromagnetic wave(TEM wave)**. The term "transverse" means "perpendicular", and it emphasizes that the E-field and the M-field are both perpendicular to the propagation direction. A schematic plot of a TEM wave is shown in Fig. 5.6, where the plane wave propagates along the z-axis, while the E-field and M-field are perpendicular to the propagation direction (z-axis). On the other hand, we show another example of a non-TEM wave in Fig. 5.7. In Fig. 5.7, the propagation direction is along the z-axis, but the associated E-field and M-field may not be perpendicular to the propagation direction.

Next, we investigate how a TEM wave propagates. Suppose the parallel plates are both **perfect conductors**. From Sect. 4.6, when an EM wave impinges on a perfect conductor from an insulating medium, the E-field and the M-field must satisfy the

Fig. 5.5 Parallel plate TX line with insulating medium between two plates

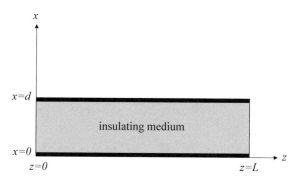

Fig. 5.6 Schematic plot of TEM wave propagating in parallel plate TX line

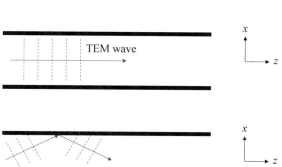

Fig. 5.7 Schematic plot of non-TEM wave propagating in parallel plate TX line

following boundary conditions:

$$E_{t1} = 0, \tag{5.1}$$

$$\epsilon_1 \, E_{n1} = \rho_s, \tag{5.2}$$

$$H_{t1} = J_s, \tag{5.3}$$

$$H_{n1} = 0. \tag{5.4}$$

In the left side of the above formulas, we have E-field and M-field in the insulating medium at the interface between the insulating medium and the metal plate. The subscript "t" denotes the tangential component parallel to the interface and the subscript "n" denotes the normal component perpendicular to the interface. ρ_s denotes the surface charge density and J_s denotes the surface current density. From Eqs. (5.1) and (5.4), we learn that the component of E-field parallel to the interface (E_{t1}) is null, and so is the component of M-field perpendicular to the interface (H_{n1}).

In Fig. 5.5, suppose a TEM wave propagates along the z-axis. It is a uniform plane wave and hence all the points on an x-y plane have identical E-field phasor ($\vec{E}$) and M-field phasor ($\vec{H}$). Thus $\vec{E}$ and $\vec{H}$ are functions of z and can be denoted by $\vec{E} = \vec{E}(z)$ and $\vec{H} = \vec{H}(z)$, respectively. Because E-field and M-field of a TEM wave are perpendicular to the propagation direction (along z-axis), $\vec{E}$ and $\vec{H}$ do not have z-component and given by

$$\vec{E} = E_x \cdot \hat{x} + E_y \cdot \hat{y} \tag{5.5}$$

$$\vec{H} = H_x \cdot \hat{x} + H_y \cdot \hat{y}. \tag{5.6}$$

Because both plates are perfect conductors, the boundary conditions of Eqs. (5.1) and (5.4) must be satisfied at $x = 0$ and $x = d$. In Fig. 5.5, the x-axis is perpendicular to the interface and y-axis is parallel to the interface. Hence at $x = 0$ and $x = d$, we have

$$E_y = 0 \tag{5.7}$$

$$H_x = 0. \tag{5.8}$$

Because it is a uniform plane wave, all the points of an x-y plane have identical E_y and H_x as Eqs. (5.7) and (5.8). Hence Eqs. (5.5) and (5.6) can be simplified as

Fig. 5.8 EM fields in parallel plate TX line

$$\vec{E} = E_x \cdot \hat{x}, \tag{5.9}$$

$$\vec{H} = H_y \cdot \hat{y}. \tag{5.10}$$

The above formulas imply that the E-field of this TEM wave has only x-component and the M-field of this TEM wave has only y-component. Both fields are functions of z, i.e., $E_x = E_x(z)$ and $H_y = H_y(z)$. The schematic plot is shown in Fig. 5.8. Note that the propagation of TEM wave is bounded between two parallel plates.

In Fig. 5.8, TEM wave propagates in the insulating medium between two parallel plates and the associated electric field must satisfy the wave equation. Suppose the frequency is ω and the medium has permeability μ and permittivity ϵ. Then the wave equation is given by

$$\nabla^2 E_x + \omega^2 \mu \epsilon \cdot E_x = 0. \tag{5.11}$$

Because E_x is a function of z as discussed previously, Eq. (5.11) can be simplified as

$$\frac{d^2 E_x}{dz^2} + \beta^2 E_x = 0, \tag{5.12}$$

where

$$\beta = \omega \sqrt{\mu \epsilon} = \frac{n\omega}{c}, \tag{5.13}$$

and n is the refractive index of the dielectric medium between two plates. Note that β has the same form as the wave number k of a plane wave propagating in a dielectric medium.

Next, we can derive two solutions for Eq. (5.12) as follows:

① Forward wave
The first solution is the forward wave, given by

$$E_x(z) = E_a e^{-j\beta z}, \tag{5.14}$$

where E_a is the phasor of electric field at $z = 0$. Suppose $E_a = A e^{j\theta}$. Then the time-varying electric field $E_x(z, t)$ is given by

$$E_x(z, t) = \text{Re}\{E_x(z) \cdot e^{j\omega t}\} = A \cos(\omega t - \beta z + \theta), \qquad (5.15)$$

where A is the amplitude and β is the **phase constant** which determines the varying rate of phase along propagation direction (+z).

In Eq. (5.15), the formula stands for a **forward wave** toward +z with the phase velocity given by

$$v_p = \frac{\omega}{\beta} = \frac{1}{\sqrt{\mu\,\epsilon}} = \frac{c}{n}. \qquad (5.16)$$

And the wavelength is given by

$$\lambda = \frac{2\pi}{\beta} = \frac{\lambda_0}{n}, \qquad (5.17)$$

where λ_0 is the wavelength in vacuum.

② Backward wave

The second solution is the backward wave, give by

$$E_x(z) = E_b e^{j\beta z}, \qquad (5.18)$$

where E_b is the phasor of electric field at $z = 0$. Suppose $E_b = Be^{j\phi}$. Then the time-varying electric field is given by

$$E_x(z, t) = \text{Re}\{E_x(z) \cdot e^{j\omega t}\} = B \cos(\omega t + \beta z + \phi), \qquad (5.19)$$

where B is the amplitude and β is the phase constant. In Eq. (5.19), the formula stands for a **backward wave** along –z. This backward wave has identical phase velocity and wavelength of the forward wave.

Example 5.1

Suppose a TEM wave propagates in a parallel plate TX line toward z-direction. The E-field is given by $\vec{E} = E_x(z) \cdot \hat{x}$, the refractive index of the insulating medium is $n = 2$, and the frequency $f = 1\,\text{GHz}$. (a). Suppose the E-field of the forward wave at $z = 0$ is $E_a = 10$ (phasor). Please derive the E-field at $z = 5m$. (b). Suppose the E-field of the backward wave at $z = 0$ is $E_b = 15e^{j\frac{\pi}{6}}$. Please derive the E-field at $z = 5m$.

Solution

(a) First, the phase constant is given by

$$\beta = \frac{n\omega}{c} = \frac{2 \cdot (2\pi \times 10^9)}{3 \times 10^8} = \frac{40\pi}{3}.$$

Because $E_a = 10$, the E-field at $z = 5m$ is given by

$$E_x = 10 \cdot e^{-j\left(\frac{40\pi}{3} \times 5\right)} = 10 \cdot e^{-j\frac{200}{3}\pi} = 10 \cdot e^{-j\frac{2\pi}{3}}.$$

(b)   Because $E_b = 15e^{j\frac{\pi}{6}}$, the E-field at $z = 5m$ is given by

$$E_x = 15 \cdot e^{j\frac{\pi}{6}} \cdot e^{j\left(\frac{40\pi}{3} \times 5\right)} = 15 \cdot e^{j\frac{\pi}{6}} \cdot e^{j\frac{2}{3}\pi} = 15 \cdot e^{j\frac{5}{6}\pi}.$$

■

Example 5.2

Continued from Example 5.1, please derive the corresponding time-varying E-field at $z = 5m$ in both cases.

Solution

(a)   From Example 5.1, the phasor of E-field of the forward wave at $z = 5m$ is given by

$$E_x = 10e^{-j\frac{2\pi}{3}}.$$

Thus the corresponding time-varying E-field is given by

$$E_x(t) = \text{Re}\left\{10e^{-j\frac{2\pi}{3}} \cdot e^{j\omega t}\right\} = 10 \cdot \cos\left(\omega t - \frac{2}{3}\pi\right) = 10 \cdot \cos\left((2\pi \times 10^9)t - \frac{2}{3}\pi\right).$$

(b)   From Example 5.1, the phasor of E-field of the backward wave at $z = 5m$ is given by

$$E_x = 15e^{j\frac{5\pi}{6}}.$$

Thus the corresponding time-varying E-field is given by

$$E_x(t) = \text{Re}\left\{15e^{j\frac{5\pi}{6}} \cdot e^{j\omega t}\right\} = 15 \cdot \cos\left((2\pi \times 10^9)t + \frac{5}{6}\pi\right).$$

■

C Voltage of TX Lines

From previous results, we can derive the E-field of parallel plate TX lines by using wave equation. Once the E-field is attained, we can derive the voltage between two plates. Furthermore, we can find an interesting phenomenon in a TX line:

When we transmit a low-frequency signal (EM wave) in a TX line, the voltages at different positions of that line are almost the same. However, when we transmit a high-frequency signal in a TX line, the voltages at different positions are varying.

This phenomenon implies the transmission of high-frequency signal is more complicated than that of low-frequency signal. In the following, we explain the cause of this phenomenon by revealing the relevant mechanism in TX lines.

In Fig. 5.9, suppose a parallel plate TX line has the length L and the associated E-field at some point z is given by $\vec{E} = E_x(z) \cdot \hat{x}$, where $0 \leq z \leq L$. From fundamental electricity, the voltage between two points can be derived by the line integral of the E-field between these two points. Hence the voltage at some position z between two parallel plates is given by

$$V(z) = - \int_0^d E_x(z) \cdot dx. \tag{5.20}$$

When we consider a forward wave, the E-field is $E_x(z) = E_a e^{-j\beta z}$. Because $E_x(z)$ is a function of z only, $E_x(z)$ is a constant in Eq. (5.20). Hence we have

$$V(z) = -E_x(z) \cdot d = -E_a e^{-j\beta z} \cdot d. \tag{5.21}$$

Furthermore, suppose $E_a = -A$, where A is a real number. The time-dependent voltage $V(z, t)$ is expressed by

$$V(z, t) = \mathrm{Re}\{V(z)e^{j\omega t}\} = \mathrm{Re}\{Ad \cdot e^{-j\beta z} \cdot e^{j\omega t}\} = Ad \cdot \cos(\omega t - \beta z). \tag{5.22}$$

From Eq. (5.22), the voltage between two parallel plates changes with z, where $0 \leq z \leq L$.

In Eq. (5.22), the phase constant β can be represented by

$$\beta = \frac{2\pi}{\lambda}, \tag{5.23}$$

Fig. 5.9 Schematic plot of voltage along TX line

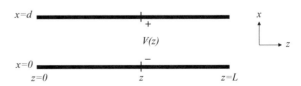

where λ is the wavelength in the dielectric medium and

$$\lambda = \frac{\lambda_0}{n}, \tag{5.24}$$

where λ_0 is the wavelength in vacuum.

It is well known that for an EM wave, the associated wavelength is shorter when the frequency is higher. For example, when the frequency $f = 1$ KHz, the corresponding wavelength in vacuum is given by $\lambda_0 = c/f = 300$ km. Suppose the refractive index of the propagation medium is $n = 3$. Then the wavelength of the EM wave in this medium is $\lambda = \lambda_0/n = 100$ km. Next, when the frequency increases to $f = 1$ MHz, the corresponding wavelength reduces to $\lambda = 100$ m. As the frequency keeps increasing to $f = 1$ GHz, the wavelength reduces to $\lambda = 0.1$ m $= 10$ cm. From the examples, the frequency of an EM wave determines the corresponding wavelength.

Now, we would like to explain the impact of the frequency to the voltage in a TX line. Suppose we have a parallel plate TX line with the length $L = 0.5$ m and the refractive index of the dielectric medium between two plates is $n = 3$. Then at the frequency $f = 1$ KHz, the associated wavelength $\lambda = 100$ km. Obviously in this case, $L \ll \lambda$. Hence we have

$$\beta z = \frac{2\pi}{\lambda} \cdot z \approx 0 \tag{5.25}$$

for $0 \leq z \leq L$. Hence from Eq. (5.22), the corresponding voltage at $t = 0$ is

$$V(z, t) = Ad \cdot \cos(-\beta z) \approx Ad, \tag{5.26}$$

where $0 \leq z \leq L$. Equation (5.26) shows that the voltage along the line is almost the same. It means that the voltage does not vary with the position.

However, when we transmit a high-frequency signal in the TX line, the voltage may significantly change with position. For example, suppose we have an EM wave with frequency $f = 100$ MHz in the above TX line. The corresponding wavelength λ is 1 m. Because $L = 0.5m = \lambda/2$, we have

$$\beta z = 0 \text{ when } z = 0, \tag{5.27}$$

$$\beta z = \frac{\pi}{2} \text{ when } z = \frac{L}{2}, \tag{5.28}$$

$$\beta z = \pi \text{ when } z = L. \tag{5.29}$$

From Eq. (5.22), the corresponding voltages at $t = 0$ are given by

$$V(0) = Ad, \tag{5.30}$$

$$V(\frac{L}{2}) = Ad \cdot \cos\frac{\pi}{2} = 0, \tag{5.31}$$

$$V(L) = Ad \cdot \cos\pi = -Ad. \tag{5.32}$$

Obviously, the voltage significantly varies with position! If we assume that the voltage remains almost the same along the TX line as what we did for low-frequency signal, we may make a big error. Hence **at different positions, we have different voltages when transmitting a high-frequency signal**.

What we discussed above can be applied to wide applications of electronic circuits. For example, suppose we transmit a signal by using a TX line on a circuit board. The length of the TX line is L. When the frequency of the signal is low, i.e., $\lambda \gg L$, the voltage can be regarded as the same along the TX line. In this case, we can greatly simplify the circuit analysis. On the other hand, when the signal frequency is high, the wavelength λ may be on the same order as L. In this case, we must consider different voltages at different positions of the TX. This is why the analysis of a high-frequency circuit is much more complicated than that of a low-frequency circuit.

Example 5.3

Suppose a parallel plate TX line has length $L = 2m$ and the refractive index of the dielectric medium is $n = 2$. The voltage is given by $V(z, t) = 3 \cdot \cos(\omega t - \beta z)$. In the following three cases:

(a) $f = 1$ kHz.
(b) $f = 1$ MHz.
(c) $f = 1$ GHz,

   please derive the voltage at $z = 0$ and $z = L$ when $t=0$.

Solution

(a) At $f = 1$ kHz, we have the corresponding wavelength given by

$$\lambda = \frac{\lambda_0}{n} = \frac{c}{n \cdot f} = \frac{3 \times 10^8}{2 \times 10^3} = 1.5 \times 10^5 (m).$$

And the phase constant is given by
$$\beta = \frac{2\pi}{\lambda} = \frac{4\pi}{3} \times 10^{-5}.$$

When $t = 0$, the voltages at $z = 0$ and $z = L = 2m$ are given by

$$V(0) = 3 \cdot \cos(0) = 3(V),$$

$$V(L) = 3 \cdot \cos\left(-\frac{4\pi}{3} \times 10^{-5} \times 2\right) \cong 3(V),$$

respectively.

(b) At $f = 1$ MHz, we have the corresponding wavelength given by

$$\lambda = \frac{c}{n \cdot f} = \frac{3 \times 10^8}{2 \times 10^6} = 150(m).$$

And the phase constant is given by

$$\beta = \frac{2\pi}{\lambda} = \frac{\pi}{75}.$$

When $t = 0$, the voltages at $z = 0$ and $z = L = 2m$ are given by

$$V(0) = 3 \cdot \cos(0) = 3(V),$$

$$V(L) = 3 \cdot \cos\left(-\frac{\pi}{75} \times 2\right) = 2.99(V),$$

respectively.

(c) At $f = 1$ GHz, we have the corresponding wavelength given by

$$\lambda = \frac{c}{n \cdot f} = \frac{3 \times 10^8}{2 \times 10^9} = \frac{3}{20}(m).$$

And the phase constant is given by

$$\beta = \frac{2\pi}{\lambda} = \frac{40\pi}{3}.$$

When $t = 0$, the voltages at $z = 0$ and $z = L = 2m$ are given by

$$V(0) = 3 \cdot \cos(0) = 3(V),$$

$$V(L) = 3 \cdot \cos\left(-\frac{40\pi}{3} \times 2\right) = -1.5(V).$$

■

From this example, we see the voltage along a TX line may vary significantly when the frequency of the transmitted EM wave is high.

5.2   Transmission Line Equations

In the previous section, we learned how a TEM wave propagates in a parallel plate TX line and derived the associated electric field and voltage. Although different types of TX lines may have different structures, TEM waves are the major ones we use for transmission. In addition, because the associated E-field and M-field are perpendicular to the propagation direction, it is convenient for us to analyze the propagation behavior.

Furthermore, when scientists tried to analyze the propagation of TEM waves in different types of TX lines, they found that it is easier to deal with voltage and current, instead of E-field and M-field. Hence they develop a formula, called **transmission line equation**, describing the variation of the voltage and current in TX lines. In this section, we introduce the TX line equation and the associated solutions.

A   *Electric Current in TX Lines*

In Sect. 5.1, we derived the E-field of TEM waves, and then obtained the voltage from the E-field between two parallel plates. Similarly, we can derive the M-field of TEM waves, and then obtain the current from the M-field. Because the derivation is a little bit tedious, we skip it and only introduce the critical results. First, as shown in Fig. 5.10, we have a TX line with the length L. The current in a TX line has two salient features:

1. When the frequency is high, the current may be different at different positions, where $0 \leq z \leq L$. In other words, the current is a function of position z. Suppose the current in the upper line is $\vec{I}_1$ and the current in the lower line is $\vec{I}_2$. Then both of them depend on z and can be represented by $\vec{I}_1 = \vec{I}_1(z)$ and $\vec{I}_2 = \vec{I}_2(z)$.
2. At a given position z, **the magnitude of** $\vec{I}_1(z)$ **is equal to that of** $\vec{I}_2(z)$**, but they have opposite directions**. In other words, $\vec{I}_1(z) = -\vec{I}_2(z)$, where $0 \leq z \leq L$.

Because of the symmetrical structure of a TX line, we can expect that $\vec{I}_1(z)$ and $\vec{I}_2(z)$ have identical magnitude. However, why do they have opposite directions? We explain this phenomenon from the perspective of M-field. First, from Ampere's law,

Fig. 5.10 Schematic plot of current along TX line

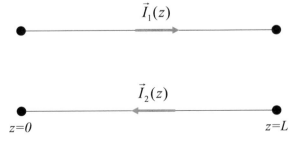

the current will generate M-field as shown in Fig. 5.11. The generated M-field is stronger when it is closer to $\vec{I}$.

In Fig. 5.12, suppose $\vec{I}_1 = \vec{I}_2$ at position z and both have the same direction. Let $\vec{H}_1$ and $\vec{H}_2$ be the generated M-fields of $\vec{I}_1$ and $\vec{I}_2$, respectively. The overall M-field $\vec{H}$ is their sum and given by

$$\vec{H} = \vec{H}_1 + \vec{H}_2. \tag{5.33}$$

From Fig. 5.12, when $\vec{I}_1 = \vec{I}_2$, it is easy to see that $\vec{H}_1$ and $\vec{H}_2$ have different directions between $x = 0$ and $x = d$, which will mutually cancel. For example, at

Fig. 5.11 Current and the induced M-field

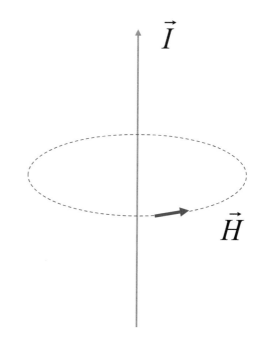

Fig. 5.12 Currents having the same direction and the induced M-fields along TX line

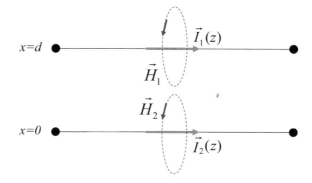

Fig. 5.13 Currents having opposite directions and the induced M-fields along TX line

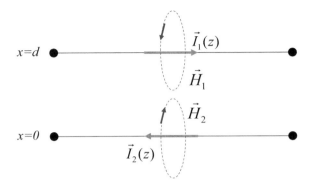

$x = 0$, because $\vec{H}_2$ is greater than $\vec{H}_1$, we have $\vec{H} \neq 0$. At $x = \frac{d}{2}$, obviously $\vec{H}_1 = -\vec{H}_2$ and thus $\vec{H} = 0$. Hence $\vec{H}$ is not a uniform M-field. This contradicts with the feature of TEM waves with a uniform M-field.

On the other hand, in Fig. 5.13, suppose $\vec{I}_1 = -\vec{I}_2$, which means they have opposite directions. From Fig. 5.13, it is easy to see that $\vec{H}_1$ and $\vec{H}_2$ have the same direction between $x = 0$ and $x = d$, and they are summed constructively to form a uniform M-field. For example, at $x = 0$, $\vec{H}_1$ is smaller than $\vec{H}_2$. At $x = \frac{d}{2}$, we have $\vec{H}_1 = \vec{H}_2$. At $x = d$, $\vec{H}_1$ is greater than $\vec{H}_2$. Hence for a specific position z, the M-field is uniform (the summed $\vec{H}$ remains constant for arbitrary x, where $0 \leq x \leq d$). This result is consistent with the feature of TEM waves.

In above, the current in one conducting line of a TX line has the same magnitude as that of another conducting line, but they flow in opposite directions. Hence when analyzing the TX line, we only need to consider the current of one conducting line and the other one can be readily obtained.

B Transmission Line Equations

The motivation of transmission line equations is to convert E-field and M-field of EM waves into electric voltage and current. This conversion greatly simplifies the analysis of a TX line because EM fields are 3-D quantities, which are much more complicated than a 1-D voltage and current. Moreover, lots of concepts we are familiar with in circuitry can be applied.

In Fig. 5.14, we have a TX line with the length L. First, we focus on a fraction of this TX line between z and $z + \Delta z$. The voltages at z and $z + \Delta z$ are $V(z)$ and $V(z + \Delta z)$, respectively. The corresponding currents are $I(z)$ and $I(z + \Delta z)$, respectively. Suppose the reference direction of the current is $+z$. For example, $I(z) = 2mA$ means that $2mA$ current flows toward $+z$ and $I(z) = -3mA$ means that $3mA$ current flows toward $-z$.

Next, we can regard this fraction of TX line between z and $z + \Delta z$ as a circuit. Then by principles and techniques of circuitry, we can derive voltage and current in

Fig. 5.14 Schematic plot of voltage and current along a segment of TX line

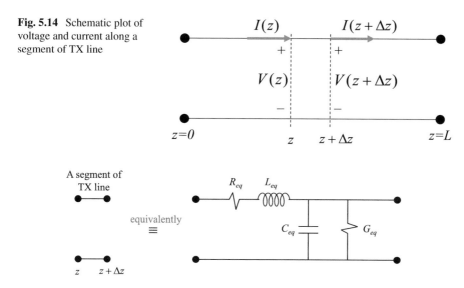

Fig. 5.15 Equivalent circuit for a segment of TX line

this circuit. The **equivalent circuit** of this fraction of TX line is shown in Fig. 5.15. This equivalent circuit is consistent with the physical properties of the TX line as explained below.

R_{eq}: Because both parallel lines are conductors, when an EM wave propagates between them, the surface resistance of conductors consumes a portion of energy. This effect is represented by an **equivalent resistance** R_{eq}.

L_{eq}: When an EM wave propagates in a TX line, the M-field between two conducting lines changes with time. From Faraday's law, the variation of M-field will induce an E-field. The E-field then causes the change of voltage. The overall effect is similar to an inductor and is thus represented by an **equivalent inductance** L_{eq}.

C_{eq}: In Fig. 5.15, an insulating medium is filled between two conducting lines. From electronics, this is exactly the structure of a capacitor. This property is represented by an **equivalent capacitance** C_{eq}.

G_{eq}: Between two conducing lines, the dielectric medium is not a perfect insulator. Suppose the equivalent resistance of the dielectric medium is R_a. Then the corresponding **equivalent conductance** is given by $G_{eq} = 1/R_a$ and the unit is *Simon* (1 Simon $= 1\ \Omega^{-1}$).

In Fig. 5.15, we convert a fraction of TX line with the length Δz to an equivalent circuit consisting of R_{eq}, L_{eq}, C_{eq}, and G_{eq}. As the length Δz increases, all of these four components increase proportionally. Hence they can be expressed as

$$R_{eq} = R \cdot \Delta z, \qquad (5.34)$$

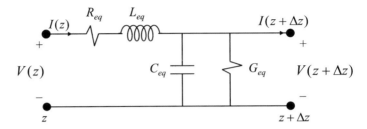

Fig. 5.16 Equivalent circuit diagram of a segment of TX line

$$L_{eq} = L \cdot \Delta z, \tag{5.35}$$

$$C_{eq} = C \cdot \Delta z, \tag{5.36}$$

$$G_{eq} = G \cdot \Delta z, \tag{5.37}$$

where

R: The equivalent resistance per unit length of a TX line. The unit is Ω/m.
L: The equivalent inductance per unit length of a TX line. The unit is $Henry/m$,
C: The equivalent capacitance per unit length of a TX line. The unit is $Farad/m$.
G: The equivalent conductance per unit length of a TX line. The unit is $Simon/m$.

After developing the equivalent circuit, we analyze the voltage and the current as shown in Fig. 5.16. Suppose ω is the frequency of the transmitted signal. The input voltage and current are $V(z)$ and $I(z)$, respectively. The output voltage and current are $V(z + \Delta z)$ and $I(z + \Delta z)$, respectively. In Fig. 5.16, the impedance of the equivalent inductor is $j\omega L_{eq}$. From Kirchhoff's voltage law, we have

$$\begin{aligned} V(z + \Delta z) &= V(z) - I(z) \cdot (R_{eq} + j\omega L_{eq}) \\ &= V(z) - I(z) \cdot (R + j\omega L) \cdot \Delta z. \end{aligned} \tag{5.38}$$

Next, because the admittance of the equivalent capacitor is $j\omega C_{eq}$, from Kirchhoff's current law, we have

$$\begin{aligned} I(z + \Delta z) &= I(z) - V(z + \Delta z) \cdot (G_{eq} + j\omega C_{eq}) \\ &= I(z) - V(z + \Delta z) \cdot (G + j\omega C) \cdot \Delta z. \end{aligned} \tag{5.39}$$

We can rewrite Eqs. (5.38) and (5.39) as

$$\frac{V(z + \Delta z) - V(z)}{\Delta z} = -I(z) \cdot (R + j\omega L), \tag{5.40}$$

$$\frac{I(z + \Delta z) - I(z)}{\Delta z} = -V(z + \Delta z) \cdot (G + j\omega C). \tag{5.41}$$

When $\Delta z \to 0$, we have $V(z + \Delta z) \to V(z)$ and $I(z + \Delta z) \to I(z)$. Hence Eqs. (5.40) and (5.41) can be approximated by the following differential equations:

$$\frac{dV(z)}{dz} = -I(z) \cdot (R + j\omega L), \tag{5.42}$$

$$\frac{dI(z)}{dz} = -V(z) \cdot (G + j\omega C). \tag{5.43}$$

Equations (5.42) and (5.43) are called **TX line equations**. They formulate the behavior of voltage and current in a TX line. The four featuring components R, L, C, and G may vary for different TX lines. However, they share the same form of TX line equations as shown in Eqs. (5.42) and (5.43).

C   Solutions of TX Line Equations

Equations (5.42) and (5.43) are both first-order differential equations and it is simple to derive the associated solutions. First, we take differentiation on both sides of Eq. (5.42) and we get

$$\frac{d^2 V(z)}{dz^2} = -\frac{dI(z)}{dz} \cdot (R + j\omega L). \tag{5.44}$$

Inserting Eqs. (5.43) into (5.44), we have

$$\frac{d^2 V(z)}{dz^2} = (R + j\omega L)(G + j\omega C) \cdot V(z). \tag{5.45}$$

Now, we define a parameter γ, given by

$$\gamma = \sqrt{(R + j\omega L)(G + j\omega C)}. \tag{5.46}$$

Then Eq. (5.45) can be rewritten as

$$\frac{d^2 V(z)}{dz^2} = \gamma^2 V(z). \tag{5.47}$$

This second-order differential equation formulates the variation of voltage in TX lines.

Similarly, we can get the second-order differential equation formulating the behavior of current in TX lines given by

$$\frac{d^2 I(z)}{dz^2} = \gamma^2 I(z). \tag{5.48}$$

From Eqs. (5.47) and (5.48), we learn that both voltage $V(z)$ and current $I(z)$ have similar form of a second-order differential equation characterized by the parameter γ. This parameter is called **propagation constant** of a TX line.

Furthermore, both Eqs. (5.47) and (5.48) have two solutions and can be simply derived by using exponential functions. The two solutions are given as follows:

① Forward wave

The first solution is given by

$$V(z) = V_a e^{-\gamma z}, \tag{5.49}$$

$$I(z) = I_a e^{-\gamma z}, \tag{5.50}$$

where $V(z)$ and $I(z)$ propagates in the form of exponential function $e^{-\gamma z}$ along z-axis. The terms V_a and I_a are the voltage phasor and the current phasor at $z = 0$, respectively. This solution stands for the **forward wave** in a TX line.

② Backward wave

The second solution is given by

$$V(z) = V_b e^{\gamma z}, \tag{5.51}$$

$$I(z) = I_b e^{\gamma z}, \tag{5.52}$$

where $V(z)$ and $I(z)$ propagates in the form of exponential function $e^{\gamma z}$ along z-axis. The terms V_b and I_b are the voltage phasor and the current phasor at $z = 0$, respectively. This solution stands for the **backward wave** in a TX line.

Generally, the propagation constant γ is a complex number and given by

$$\gamma = \sqrt{(R + j\omega L)(G + j\omega C)} = \alpha + j\beta, \tag{5.53}$$

where α and β are real numbers. Suppose

$$R + j\omega L = a \cdot e^{j\varphi_1}, \tag{5.54}$$

$$G + j\omega C = b \cdot e^{j\varphi_2}. \tag{5.55}$$

Because $R, L, G,$ and C are all positive numbers, the magnitudes a and b are both positive with $0 \le \varphi_1, \varphi_2 \le \frac{\pi}{2}$. From Eqs. (5.53)–(5.55), we have

$$\alpha + j\beta = \sqrt{ab \cdot e^{j(\varphi_1 + \varphi_2)}} = \sqrt{ab} \cdot e^{j\frac{\varphi_1 + \varphi_2}{2}}. \tag{5.56}$$

Hence

$$\alpha = \sqrt{ab} \cdot \cos \frac{\varphi_1 + \varphi_2}{2}, \tag{5.57}$$

$$\beta = \sqrt{ab} \cdot \sin \frac{\varphi_1 + \varphi_2}{2}. \tag{5.58}$$

Because $0 \le \varphi_1, \varphi_2 \le \frac{\pi}{2}$, the parameters α and β are both positive. From above, given R, L, G, C, we can derive α and β.

From Eq. (5.53), we can rewrite Eq. (5.49) as

$$V(z) = V_a \cdot e^{-\alpha z} \cdot e^{-j\beta z}. \tag{5.59}$$

Hence

$$|V(z)| = |V_a| \cdot e^{-\alpha z}. \tag{5.60}$$

We plot $|V(z)|$ in Fig. 5.17. From Fig. 5.17, we find $|V(z)|$ decreases along z. This is because the energy decays when the forward wave propagates toward $+z$. The decay is proportional to $e^{-\alpha z}$.

From Eq. (5.59), the parameter α characterizes the decay of $|V(z)|$ and is thus called **attenuation constant**. The greater the parameter α, the greater the decay rate of $V(z)$ along z. The other parameter β is called **phase constant** since it characterizes the behavior of phase variation. Both α and β are critical parameters for a TX line.

In Eq. (5.59), $V(z)$ is the phasor of voltage and we let $V_a = A \cdot e^{j\theta}$. Then the time-varying voltage can be derived by

$$V(z, t) = \text{Re}\{V(z)e^{j\omega t}\} = Ae^{-\alpha z} \cdot \cos(\omega t - \beta z + \theta). \tag{5.61}$$

Fig. 5.17 Voltage distribution of forward wave along a TX line

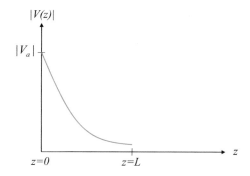

Fig. 5.18 Voltage
distribution of backward
wave along a TX line

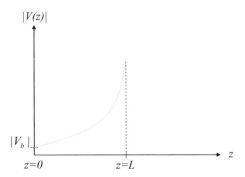

Hence the voltage in a TX line depends on z and t. Also, α determines the decay rate and β determines the varying rate of phase.

On the other hand, for the backward wave in Eq. (5.51), we can rewrite it as

$$V(z) = V_b \cdot e^{\alpha z} \cdot e^{j\beta z}. \tag{5.62}$$

Hence

$$|V(z)| = |V_b| \cdot e^{\alpha z}. \tag{5.63}$$

Suppose a backward wave propagates from $z = L$ to $z = 0$. We plot $|V(z)|$ in Fig. 5.18 where $|V(z)|$ attains the maximum when $z = L$ and the minimum when $z = 0$. We interpret $V(z)$ as the backward wave propagating along $-z$. The energy decays during propagation from $z = L$ to $z = 0$ and thus attains the minimum when $z = 0$.

In Eq. (5.62), $V(z)$ is the phasor of voltage and we let $V_b = B \cdot e^{j\phi}$. Then the time-varying voltage is given by

$$V(z, t) = \mathrm{Re}\{V(z)e^{j\omega t}\} = Be^{\alpha z} \cdot \cos(\omega t + \beta z + \phi). \tag{5.64}$$

Hence the voltage depends on z and t. Similar to the forward wave in Eq. (5.61), the attenuation constant α determines the decay rate and the phase constant β determines the varying rate of phase.

Example 5.4

Suppose we have a forward wave and a backward wave propagating in a TX line. The voltage of the forward wave is $V^+(z, t) = Ae^{-\alpha z} \cdot \cos(\omega t - \beta z)$ and the voltage of the backward wave is $V^-(z, t) = Be^{\alpha z} \cdot \cos(\omega t + \beta z - \pi)$. Suppose $A = B = 10(V)$, $\alpha = 0.01(m^{-1})$, and $\beta = \frac{\pi}{3}(m^{-1})$. When $t = 0$ and at $z = 5m$, please derive

(a) the voltage of the forward wave,
(b) the voltage of the backward wave,
(c) the measured voltage.

Solution

(a) When $t = 0$, we have

$$V^+(z) = Ae^{-\alpha z} \cdot \cos(-\beta z).$$

Hence at $z = 5m$, we have the voltage of the forward wave given by

$$V^+(z) = 10 \cdot e^{-(0.01) \times 5} \cdot \cos(-\frac{\pi}{3} \times 5) = 4.76(V).$$

(b) When $t = 0$, we have

$$V^-(z) = Be^{\alpha z} \cdot \cos(\beta z - \pi).$$

Hence at $z = 5m$, we have the voltage of the backward wave given by

$$V^-(z) = 10 \cdot e^{(0.01) \times (5)} \cdot \cos(\frac{\pi}{3} \times 5 - \pi) = -5.26(V).$$

(c) Finally, the voltage measured at $z = 5m$ is the sum of the voltages of both waves. It is given by

$$V(z) = V^+(z) + V^-(z) = 4.76 + (-5.26) = -0.5 \text{ (V)}.$$

■

D Attenuation Constant and Phase Constant

$R, L, G,$ and C are important quantities composing the equivalent circuit for a TX line. They actually determine the attenuation constant α and the phase constant β. Suppose we have a lossless TX line, which means that no loss occurs in the TX line. Hence the associated $R = 0$ and $G = 0$. In this case, from Eq. (5.46), we have

$$\gamma = \sqrt{(j\omega L)(j\omega C)} = j\omega\sqrt{LC}. \tag{5.65}$$

Since $\gamma = \alpha + j\beta$, we obtain

$$\alpha = 0, \tag{5.66}$$

$$\beta = \omega\sqrt{LC}. \tag{5.67}$$

Hence for a lossless TX line, the attenuation constant $\alpha = 0$ and the phase constant β increases with the frequency.

However, for a practical TX line, the associated R and G are not zeroes. We can rewrite Eq. (5.46) as

$$\gamma = \sqrt{(j\omega L)(j\omega C)(1 + \frac{R}{j\omega L})(1 + \frac{G}{j\omega C})}$$

$$= j\omega\sqrt{LC} \cdot \sqrt{(1 + \frac{R}{j\omega L})(1 + \frac{G}{j\omega C})}. \tag{5.68}$$

For a high-frequency circuit, we typically have $R << \omega L$ and $G << \omega C$. Since $\sqrt{1 + x} \approx 1 + \frac{x}{2}$ when $x << 1$, Eq. (5.68) can be approximated by

$$\gamma \approx j\omega\sqrt{LC} \cdot (1 + \frac{R}{2j\omega L})(1 + \frac{G}{2j\omega C})$$

$$\approx j\omega\sqrt{LC} \cdot \left[1 + \frac{1}{2j\omega}(\frac{R}{L} + \frac{G}{C})\right]$$

$$= \frac{\sqrt{LC}}{2}(\frac{R}{L} + \frac{G}{C}) + j\omega\sqrt{LC}. \tag{5.69}$$

Since $\gamma = \alpha + j\beta$, we have

$$\alpha = \frac{\sqrt{LC}}{2} \cdot (\frac{R}{L} + \frac{G}{C}), \tag{5.70}$$

$$\beta = \omega\sqrt{LC}. \tag{5.71}$$

Hence the attenuation constant α of a TX line can be derived using Eq. (5.70). Note that the associated phase parameter β in Eq. (5.71) has the same form as that of a lossless TX line in Eq. (5.67).

Example 5.5

Suppose we have a lossless TX line with the equivalent circuit parameters $R = 0$, $L = 4$ (μH/m), $C = 1$ (pF/m), and $G = 0$. Please derive the attenuation constant α and the phase constant β when the frequency $f = 100$ MHz.

Solution

Since it is a lossless TX line, from Eqs. (5.66) and (5.67), we have

$$\alpha = 0,$$

$$\beta = \omega\sqrt{LC} = (2\pi \times 10^8) \cdot \sqrt{(4 \times 10^{-6})(1 \times 10^{-12})} = \frac{2\pi}{5} \ (m^{-1}).$$

■

Example 5.6

Suppose we have a TX line with the equivalent circuit parameters $R = 0.1 (\Omega/m)$, $L = 10 \ (\mu H/m)$, $C = 10 \ (pF/m)$, and $G = 10^{-6} \ (S/m)$. Please derive the attenuation constant α and the phase constant β when the frequency $f = 100$ MHz.

Solution

Since $R << \omega L$ and $G << \omega C$, we can apply Eqs. (5.70) and (5.71) to attain

$$\alpha = \frac{\sqrt{LC}}{2} \cdot \left(\frac{R}{L} + \frac{G}{C}\right) = \frac{\sqrt{10^{-5} \cdot 10^{-11}}}{2} \cdot \left(\frac{0.1}{10^{-5}} + \frac{10^{-6}}{10^{-11}}\right) = 5.5 \times 10^{-4} \ (m^{-1}),$$

$$\beta = \omega\sqrt{LC} = (2\pi \times 10^8) \cdot \sqrt{(10^{-5})(10^{-11})} = 2\pi \ (m^{-1}).$$

■

5.3 Characteristic Impedance

We live in a fascinating world and many phenomena can be formulated by physical laws. For example, Newton's law of universal gravitation formulates the attracting force between two objects and Coulomb's law formulates the electrical force between two charged particles. With these physical laws, we can use mathematical formulas to explore the mechanism behind each phenomenon and even predict the associated behavior.

Similarly, although different TX lines may have different structures, the resulted voltage and current can be formulated by **transmission line equations (TX line equations)**. Transmission line equations are characterized by a number of parameters and each parameter represents a property of TX line. In this section, we are going to introduce an important property of TX lines: characteristic impedance.

A *Forward Wave*

First, we investigate the relationship between voltage and current of a forward wave. In Fig. 5.19, suppose we have a forward wave propagating toward $+z$. The associated voltage and current are given by

$$V(z) = V_a e^{-\gamma z}, \tag{5.72}$$

$$I(z) = I_a e^{-\gamma z}, \tag{5.73}$$

where γ is the **propagation constant**, V_a and I_a are the phasors at $z = 0$. It is easy to find that in Eqs. (5.72) and (5.73), $V(z)$ is proportional to $I(z)$. Their relationship is given by

$$\frac{V(z)}{I(z)} = \frac{V_a}{I_a} = \text{constant}. \tag{5.74}$$

Equation (5.74) implies that although $V(z)$ and $I(z)$ change with z, the ratio between them is constant. This ratio is called **characteristic impedance** and given by

$$Z_0 = \frac{V(z)}{I(z)}. \tag{5.75}$$

The name comes from circuitry, where the ratio of the voltage to the current is defined as the **impedance**. Since Z_0 characterizes the behavior of a TX line deeply, it is called the characteristic impedance. As will become clear later, Z_0 stands for an important property of a TX line.

From the previous section, we know that the voltage and the current for a TX line satisfy the following differential equation:

$$\frac{dV(z)}{dz} = -I(z) \cdot (R + j\omega L). \tag{5.76}$$

On the other hand, from Eq. (5.72), we get

Fig. 5.19 Forward wave and the associated voltage and current along a TX line

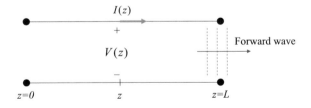

$$\frac{dV(z)}{dz} = -\gamma \cdot V_a e^{-\gamma z} = -\gamma \cdot V(z). \tag{5.77}$$

Inserting Eqs. (5.77) into (5.76), we get

$$-\gamma \cdot V(z) = -I(z) \cdot (R + j\omega L). \tag{5.78}$$

Hence

$$\frac{V(z)}{I(z)} = \frac{R + j\omega L}{\gamma} = Z_0. \tag{5.79}$$

Furthermore, because

$$\gamma = \sqrt{(R + j\omega L)(G + j\omega C)}, \tag{5.80}$$

we have

$$Z_0 = \frac{R + j\omega L}{\gamma} = \sqrt{\frac{R + j\omega L}{G + j\omega C}}. \tag{5.81}$$

Hence Z_0 is determined by R, L, C, G and frequency ω. Because different TX lines usually have different $R, L, C,$ and G, they may have a different characteristic impedance.

For a high-frequency circuit, we usually have $R << \omega L$ and $G << \omega C$. In this case, Eq. (5.81) becomes

$$Z_0 \approx \sqrt{\frac{j\omega L}{j\omega C}} = \sqrt{\frac{L}{C}}. \tag{5.82}$$

The characteristic impedance is approximated by the square root of the ratio between the equivalent inductance and the equivalent capacitance. In practice, we usually design a TX line so that Z_0 is a positive number. For example, we usually have a TX line with $Z_0 = 50\Omega$ in a high-frequency circuit or $Z_0 = 75\Omega$ for transmitting TV signal.

Example 5.7

Suppose we have a TX line with the equivalent circuit parameters $R = 0.1$ (Ω/m), $L = 2$ $(\mu H/m)$, $G = 1$ $(\mu S/m)$, and $C = 100$ (pF/m). If the frequency $f = 100\,MHz$, please derive the associated characteristic impedance Z_0.

Solution

From Eq. (5.81), we have

$$Z_0 = \sqrt{\frac{R + j\omega L}{G + j\omega C}} = \sqrt{\frac{0.1 + j(2\pi \times 10^8) \cdot (2 \times 10^{-6})}{10^{-6} + j(2\pi \times 10^8) \cdot (100 \times 10^{-12})}}$$

$$= \sqrt{\frac{0.1 + j(4\pi \times 10^2)}{10^{-6} + j(2\pi \times 10^{-2})}} \approx \sqrt{2 \times 10^4}$$

$$= 100\sqrt{2}$$

$$= 141.4 (\Omega).$$

∎

Example 5.8

Suppose we have a TX line with the characteristic impedance $Z_0 = 100\Omega$ and propagation constant $\gamma = 0.01 + j\frac{\pi}{4}$. Please derive the equivalent circuit parameters (R, L, C, G) when $f = 100\,MHz$.

Solution

From Eqs. (5.80) and (5.81), we have

$$\gamma = \sqrt{(R + j\omega L)(G + j\omega C)},$$

$$Z_0 = \sqrt{\frac{R + j\omega L}{G + j\omega C}}.$$

Hence

$$\gamma \cdot Z_0 = R + j\omega L$$

and then

$$(0.01 + j\frac{\pi}{4}) \cdot (100) = R + j\omega L.$$

Therefore, the equivalent resistance and the equivalent inductance are given by

$$R = (0.01) \cdot (100) = 1(\Omega/m),$$

$$L = \frac{(\frac{\pi}{4}) \cdot 100}{\omega} = 1.25 \times 10^{-7}(H/m),$$

respectively. In addition, from above we have

$$\frac{\gamma}{Z_0} = G + j\omega C.$$

Hence

$$\frac{0.01 + j\frac{\pi}{4}}{100} = G + j\omega C$$

and then

$$G = \frac{0.01}{100} = 1 \times 10^{-4}(S/m),$$

$$C = \frac{\frac{\pi}{4}}{\omega \cdot 100} = 1.25 \times 10^{-11}(F/m).$$

■

B Backward Wave

Next, we investigate the relationship between the voltage and the current of a backward wave. In Fig. 5.20, suppose we have a backward wave propagating along $-z$. The associated voltage and current are given by

Fig. 5.20 Backward wave and the associated voltage and current along a TX line

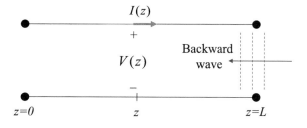

$$V(z) = V_b e^{\gamma z} \tag{5.83}$$

$$I(z) = I_b e^{\gamma z}, \tag{5.84}$$

where γ is the propagation constant, V_b and I_b are the phasors at $z = 0$. From Eqs. (5.83) and (5.84), we have

$$\frac{V(z)}{I(z)} = \frac{V_b}{I_b} = \text{constant}. \tag{5.85}$$

Hence the ratio of $V(z)$ to $I(z)$ is also a constant for a backward wave in a TX line. Now, a question arises: *Is this constant equal to the ratio Z_0 for a forward wave in* Eq. (5.75)?

From the previous section, we learn that the voltage $V(z)$ and the current $I(z)$ for a backward wave also satisfy the differential equation in Eq. (5.76). Inserting the differentiation of Eqs. (5.83) into (5.76), we have

$$\gamma \cdot V(z) = -I(z) \cdot (R + j\omega L). \tag{5.86}$$

Hence

$$\frac{V(z)}{I(z)} = -\frac{(R + j\omega L)}{\gamma} = -Z_0. \tag{5.87}$$

Obviously, the ratio of $V(z)$ to $I(z)$ of a backward wave is $-Z_0$, which is the negative of that of a forward wave. Readers shall pay attention to this opposite sign, because it makes a big difference in analysis.

In above, we summarize as follows:

① For a forward wave, we have

$$V(z) = V_a e^{-\gamma z},$$

$$I(z) = I_a e^{-\gamma z},$$

$$\frac{V(z)}{I(z)} = Z_0.$$

② For a backward wave, we have

$$V(z) = V_b e^{\gamma z},$$

$$I(z) = I_b e^{\gamma z},$$

$$\frac{V(z)}{I(z)} = -Z_0.$$

Hence for a forward wave or a backward wave in a TX line, we can easily derive the voltage from the known current, and vice versa.

Example 5.9

Suppose we have a TX line with the characteristic impedance $Z_0 = 50\Omega$. At some point of the line, a forward wave has the voltage $V^+ = 10$ V and the backward wave has the voltage $V^- = 4$ V. Please derive (a). the currents of the forward wave and the backward wave, (b). the measured voltage and current at that point.

Solution

(a) From Eq. (5.75), we can derive the forward current I^+ given by

$$I^+ = \frac{V^+}{Z_0} = \frac{10}{50} = 0.2(A).$$

On the other hand, from Eq. (5.85), we can derive the backward current I^- given by

$$I^- = -\frac{V^-}{Z_0} = -\frac{4}{50} = -0.08(A).$$

(b) Because the measured voltage V_m is the sum of V^+ and V^-, we have
$$V_m = V^+ + V^- = 14(V).$$

In addition, because the measured current I_m is the sum of I^+ and I^-, we have

$$I_m = I^+ + I^- = 0.2 + (-0.08) = 0.12(A).$$

■

C Forward Wave and Backward Wave Coexist

When we have only forward wave or backward wave in a TX line, the relationship between voltage and current is simple and can be characterized by characteristic impedance. However, when forward wave and backward wave coexist in a TX line, the relationship becomes much more complicated.

Suppose $V(z)$ and $I(z)$ denote the voltage and the current at the position z in a TX line, respectively. When a forward wave and a backward wave coexist, we have $V(z)$ and $I(z)$ given by

$$V(z) = V^+(z) + V^-(z), \tag{5.88}$$

$$I(z) = I^+(z) + I^-(z), \tag{5.89}$$

where $V^+(z)$ and $I^+(z)$ are the voltage and the current of the forward wave, and $V^-(z)$ and $I^-(z)$ are the voltage and the current of the backward wave. We call $V^+(z)$ **forward voltage** and $I^+(z)$ **forward current**. Similarly, we call $V^-(z)$ **backward voltage** and $I^-(z)$ **backward current**.

Equations (5.88) and (5.89) tell us that the resultant voltage $V(z)$ is the sum of the forward voltage and the backward voltage. The resultant current $I(z)$ is the sum of the forward current and the backward current. Because neither a voltage meter nor a current meter can distinguish the direction, all we can measure is the resultant $V(z)$ and $I(z)$, respectively.

Suppose at $z = 0$, the forward voltage is V_a and the backward voltage is V_b. Then for an arbitrary point z in a TX line, we have the forward voltage and the backward voltage given by

$$V^+(z) = V_a e^{-\gamma z}, \tag{5.90}$$

$$V^-(z) = V_b e^{\gamma z}. \tag{5.91}$$

Hence the resultant voltage is given by

$$V(z) = V^+(z) + V^-(z) = V_a e^{-\gamma z} + V_b e^{\gamma z}. \tag{5.92}$$

On the other hand, the forward current and the backward current are given by

$$I^+(z) = \frac{V^+(z)}{Z_0} = \frac{V_a e^{-\gamma z}}{Z_0}, \tag{5.93}$$

$$I^-(z) = -\frac{V^-(z)}{Z_0} = -\frac{V_b e^{\gamma z}}{Z_0}, \tag{5.94}$$

respectively. Hence the resultant current is given by

$$I(z) = I^+(z) + I^-(z) = \frac{V_a e^{-\gamma z} - V_b e^{\gamma z}}{Z_0}. \tag{5.95}$$

When a forward wave and a backward wave coexist in a TX line, from Eqs. (5.92) and (5.95), we have

$$\frac{V(z)}{I(z)} = Z_0 \cdot \frac{V_a e^{-\gamma z} + V_b e^{\gamma z}}{V_a e^{-\gamma z} - V_b e^{\gamma z}} \neq constant. \tag{5.96}$$

Hence the ratio of $V(z)$ to $I(z)$ is not a constant and it depends on the position z.
In above, the ratio of the voltage to the current is not a constant when a forward wave and a backward wave coexist in a TX line. However, we can still analyze them by separating the analysis into a case of forward waves and a case of backward waves. This is shown in the following example.

Example 5.10

Suppose we have a TX line with the characteristic impedance $Z_0 = 50\Omega$ and propagation constant $\gamma = j\frac{\pi}{3}$. If the forward voltage at $z = 0$ is $V_a = 10e^{j\frac{\pi}{6}}$ and the backward voltage is $V_b = 2e^{j\frac{\pi}{4}}$. Please derive

(a) the voltage and the current at $z = 5m$,
(b) the ratio of the voltage to the current at $z = 5m$.

Solution

(a) Suppose V^+ and V^- denote the forward voltage and the backward voltage at $z = 5m$, respectively. Then from Eqs. (5.90) and (5.91), we have

$$V^+ = V_a e^{-\gamma z} = \left(10e^{j\frac{\pi}{6}}\right) \cdot \left(e^{-j\frac{\pi}{3} \times 5}\right) = 10e^{-j\frac{3}{2}\pi} = j10,$$

$$V^- = V_b e^{\gamma z} = \left(2e^{j\frac{\pi}{4}}\right) \cdot \left(e^{j\frac{\pi}{3} \times 5}\right) = 2e^{j\frac{23}{12}\pi}.$$

Next, from Eqs. (5.93) and (5.94), the forward current and the backward current at $z = 5m$ are given by

$$I^+ = \frac{V^+}{Z_0} = \frac{j10}{50} = \frac{j}{5},$$

$$I^- = -\frac{V^-}{Z_0} = -\frac{2e^{j\frac{23}{12}\pi}}{50} = -\frac{e^{j\frac{23}{12}\pi}}{25},$$

respectively. Hence the voltage and the current at $z = 5m$ are given by

$$V = V^+ + V^- = j10 + 2e^{j\frac{23}{12}\pi},$$

$$I = I^+ + I^- = \frac{j}{5} - \frac{e^{j\frac{23}{12}\pi}}{25},$$

respectively.

(b)   The ratio of V to I is given by

$$\frac{V}{I} = \frac{j10 + 2e^{j\frac{23}{12}\pi}}{\frac{j}{5} - \frac{e^{j\frac{23}{12}\pi}}{25}} = \frac{j250 + 50e^{j\frac{23}{12}\pi}}{j5 - e^{j\frac{23}{12}\pi}}.$$

Obviously, the ratio is not equal to Z_0.

∎

5.4   Reflection

When you throw a ball forward, you can easily predict its movement and position. However, when you throw a ball to a wall and it bounces back, it is usually more difficult to predict its movement and position. Similarly in TX lines, when we only have forward waves, the relationship between voltage and current is simple and easy to handle. However, when we also have backward waves, it becomes more difficult as we have discussed in Sect. 5.3. In this section, we will investigate the cause of backward waves in TX lines and their impact. This is a critical issue in TX lines because it brings a lot of interesting and challenging results in applications of TX lines, particularly in high-frequency circuit design.

A   Occurrence of Reflection

Suppose we have a TX line of the length L and the load impedance is Z_L as shown in Fig. 5.21. Assume we have an EM wave propagating from $z = 0$ to $z = L$. Initially, we only have a forward wave in the TX line and the associated voltage and current

Fig. 5.21   EM wave along a TX line with load impedance

are simply given by

$$V(z) = V_a e^{-\gamma z}, \tag{5.97}$$

$$I(z) = I_a e^{-\gamma z}, \tag{5.98}$$

where V_a and I_a are the phasors at $z = 0$, and γ is the propagation constant. Because in the beginning, we only have forward wave so that the ratio of $V(z)$ to $I(z)$ is equal to the characteristic impedance Z_0, hence

$$\frac{V(z)}{I(z)} = Z_0. \tag{5.99}$$

When the forward wave travels to $z = L$, the forward voltage V_+ and the forward current I_+ are given by

$$V_+ = V_a e^{-\gamma L}, \tag{5.100}$$

$$I_+ = I_a e^{-\gamma L}. \tag{5.101}$$

Based on the TX line characteristic impedance Z_0 and the load impedance Z_L, we need to consider the following two cases.

Case 1: $Z_L \neq Z_0$

When the forward wave arrives $z = L$, the forward voltage V_+ and the forward current I_+ satisfy the following equality:

$$\frac{V_+}{I_+} = Z_0. \tag{5.102a}$$

On the other hand, according to circuit theory, at $z=L$ the voltage across Z_L and the current passing through Z_L must satisfy the following relationship:

$$\frac{V(z)}{I(z)} = Z_L. \tag{5.102b}$$

Now, we encounter a problem that not only the condition of the load impedance Z_L in Eq. (5.102b) needs to be satisfied, but also the condition of the characteristic impedance Z_0 in Eq. (5.102a) shall be met, but $Z_L \neq Z_0$. How can these two equations be satisfied simultaneously?

From the measured results, we find that the EM wave finds a clever way to resolve the problem, i.e., it resolves the problem by generating a reflected wave (backward wave) at $z = L$ and propagates toward $z = 0$. Hence when $Z_L \neq Z_0$, the reflection

Fig. 5.22 Schematic plot of forward wave and backward wave when reflection occurs

occurs and a forward wave and a backward wave exist simultaneously. In this case, the voltage and current across Z_L are not simply V_+ and I_+.

Now, we go further into the details as shown in Fig. 5.22. Suppose V_L is the voltage across Z_L and I_L is the current passing through Z_L. When the reflection occurs, we have the following relationships at $z = L$:

$$V_L = V_+ + V_-, \tag{5.103}$$

$$I_L = I_+ + I_-, \tag{5.104}$$

where V_+ is the forward voltage, V_- is the backward voltage, I_+ is the forward current, and I_- is the backward current. Meanwhile, V_+ and I_+ satisfy the property of the forward wave:

$$\frac{V_+}{I_+} = Z_0. \tag{5.105}$$

Also, V_- and I_- satisfy the property of the backward wave:

$$\frac{V_-}{I_-} = -Z_0. \tag{5.106}$$

On the other hand, V_L and I_L satisfy the condition of Z_L, given by

$$\frac{V_L}{I_L} = \frac{V_+ + V_-}{I_+ + I_-} = Z_L. \tag{5.107}$$

Equations (5.105)–(5.107) show that with the reflection, we can satisfy both the condition of the load impedance Z_L and the condition of the characteristic impedance Z_0. Furthermore, when the reflection occurs, a forward wave and a backward wave coexist in a TX line. For an arbitrary point in the TX line, the voltage and the current are given by

$$V(z) = V^+(z) + V^-(z), \tag{5.108}$$

$$I(z) = I^+(z) + I^-(z), \tag{5.109}$$

where $V^+(z)$ is the forward voltage, $I^+(z)$ is the forward current, $V^-(z)$ is the backward voltage, and $I^-(z)$ is the backward current. They satisfy the following equalities:

$$\frac{V^+(z)}{I^+(z)} = Z_0, \tag{5.110}$$

$$\frac{V^-(z)}{I^-(z)} = -Z_0. \tag{5.111}$$

From Eqs. (5.110) and (5.111), when the reflection occurs, the condition of the characteristic impedance for a TX line still holds. That is, the ratio of the forward voltage to the forward current remains a constant, and the ratio of the backward voltage to the backward current remains a constant, too. These two conditions provide a powerful tool when we try to analyze the reflection in a TX line.

Case 2: $Z_L = Z_0$

When the load impedance Z_L is equal to the characteristic impedance Z_0, we call it **impedance matched**. In this case, when a forward wave travels from $z = 0$ to $z = L$, $V(z)$ and $I(z)$ satisfy the condition of the characteristic impedance as formulated in Eq. (5.99). Meanwhile, at $z = L$, because $Z_L = Z_0$, the forward voltage V_+ and the forward current I_+ also satisfy the condition of the load impedance as formulated in Eq. (5.102b). Hence no reflection occurs. In other words, when impedance is matched, only forward waves exist in a TX line.

In above, when $Z_L \neq Z_0$, the reflection occurs at the load so that forward waves and the backward waves exist simultaneously in a TX line. On the other hand, when $Z_L = Z_0$, no reflection occurs at the load and only forward waves exist in a TX line.

B Reflection Coefficient

When $Z_L \neq Z_0$, the reflection occurs in a TX. We define the **reflection coefficient** as

$$\Gamma = \frac{V_-}{V_+}, \tag{5.112}$$

where V_+ and V_- are the forward voltage and the backward voltage at $z = L$, respectively. Readers shall pay attention that V_+ and V_- are both phasors and thus complex numbers. Hence the reflection coefficient Γ is also a complex number and can be represented by

$$\Gamma = |\Gamma| \cdot e^{j\theta}, \tag{5.113}$$

where $|\Gamma|$ is the magnitude of reflection and θ is the associated phase. Furthermore, by intuition, the reflected voltage shall be smaller than or equal to the input voltage. Therefore from Eq. (5.112), we have

$$|\Gamma| \leq 1. \tag{5.114}$$

This is an important property of a reflection coefficient Γ. Next, from Eqs. (5.105)–(5.107), we have

$$Z_L = \frac{V_+ + V_-}{I_+ + I_-} = Z_0 \cdot \frac{V_+ + V_-}{V_+ - V_-}. \tag{5.115}$$

Because $\Gamma = V^-/V^+$, Eq. (5.115) can be rewritten as

$$Z_L = Z_0 \cdot \frac{1 + \Gamma}{1 - \Gamma}. \tag{5.116}$$

From (5.116), the reflection coefficient can be derived by

$$\Gamma = \frac{Z_L - Z_0}{Z_L + Z_0}. \tag{5.117}$$

Hence given Z_L and Z_0, we can derive the reflection coefficient immediately. Equation (5.117) is an important and useful formula for a TX line, and will be applied in the following sections.

Example 5.11

Suppose a TX line has the following property: the length $L = 3$ m, the characteristic impedance $Z_0 = 50\Omega$, the load impedance $Z_L = 100\Omega$, and the propagation constant $\gamma = j\frac{\pi}{3}$. Assume the forward voltage at $z = 0$ is $V_a = 10e^{j\frac{\pi}{4}}$. At $z = L$, please derive

(a) forward voltage and forward current,
(b) backward voltage and backward current,
(c) measured voltage and current across the load.

Solution

(a) First, from Eq. (5.100), the forward voltage at $z = L$ is given by

$$V_+ = V_a \cdot e^{-\gamma L} = \left(10e^{j\frac{\pi}{4}}\right) \cdot e^{-\left(j\frac{\pi}{3} \times 3\right)} = -10e^{j\frac{\pi}{4}}.$$

Then from Eq. (5.105), the forward current at $z = L$ is given by

$$I_+ = \frac{V_+}{Z_0} = -\frac{1}{5}e^{j\frac{\pi}{4}}.$$

(b) Next, we derive the reflection coefficient given by

$$\Gamma = \frac{Z_L - Z_0}{Z_L + Z_0} = \frac{100 - 50}{100 + 50} = \frac{1}{3}.$$

Then from Eq. (5.112) and the result in (a), we can derive the backward voltage at $z = L$ as

$$V_- = \Gamma V_+ = -\frac{10}{3}e^{j\frac{\pi}{4}}.$$

From Eq. (5.106), the backward current at $z = L$ is given by

$$I_- = -\frac{V_-}{Z_0} = \frac{1}{15} \cdot e^{j\frac{\pi}{4}}.$$

(c) The measured voltage and current across the load are given by

$$V_L = V_+ + V_- = -\frac{40}{3}e^{j\frac{\pi}{4}},$$

$$I_L = I_+ + I_- = -\frac{2}{15}e^{j\frac{\pi}{4}}.$$

Note that the current can be derived from Eq. (5.107) as well, given by

$$I_L = \frac{V_L}{Z_L} = -\frac{2}{15}e^{j\frac{\pi}{4}}.$$

■

Example 5.12

Please repeat Example 5.11 but take the propagation constant $\gamma = \frac{1}{30} + j\frac{\pi}{5}$.

Solution

(a) First, from Eq. (5.100), the forward voltage at $z = L = 3m$ is given by

$$V_+ = V_a \cdot e^{-\gamma L} = \left(10e^{j\frac{\pi}{4}}\right) \cdot e^{-\left(\frac{1}{30}+j\frac{\pi}{5}\right)\times 3} = 10e^{-0.1} \cdot e^{j\left(\frac{\pi}{4}-\frac{3\pi}{5}\right)} \approx 9e^{-j\frac{7\pi}{20}}.$$

Then from Eq. (5.105), the forward current at $z = L$ is given by

$$I_+ = \frac{V_+}{Z_0} = \frac{9}{50}e^{-j\frac{7\pi}{20}}.$$

(b) Next, we derive the reflection coefficient given by

$$\Gamma = \frac{Z_L - Z_0}{Z_L + Z_0} = \frac{100 - 50}{100 + 50} = \frac{1}{3}.$$

Then from Eq. (5.112) and the result in (a), we can derive the backward voltage at $z = L$ as

$$V_- = \Gamma V_+ = 3e^{-j\frac{7\pi}{20}}.$$

From Eq. (5.106), the backward current at $z = L$ is given by

$$I_- = -\frac{V_-}{Z_0} = -\frac{3}{50} \cdot e^{-j\frac{7\pi}{20}}.$$

(c) The measured voltage and current across the load are given by

$$V_L = V_+ + V_- = 12e^{-j\frac{7\pi}{20}},$$

$$I_L = I_+ + I_- = \frac{3}{25}e^{-j\frac{7\pi}{20}}.$$

Note that the load current can be derived from Eq. (5.107) as well, given by

$$I_L = \frac{V_L}{Z_L} = \frac{3}{25}e^{-j\frac{7\pi}{20}}.$$

■

C Special Cases of Load Impedance

Reflection coefficient Γ is an important parameter which is closely related to the characteristic impedance Z_0 and load impedance Z_L. In the following, we investigate the reflection coefficient corresponding to three special cases of Z_L.

Case 1: $Z_L = 0$

In Fig. 5.23, we use a conducting line as the load so that $Z_L = 0$. Then from Eq. (5.117), we have

$$\Gamma = \frac{-Z_0}{Z_0} = -1. \qquad (5.118)$$

Hence when the load is a **short circuit**, the corresponding reflection coefficient $\Gamma = -1$. This is a reasonable result because when $Z_L = 0$, no matter what I_L is, the load voltage V_L is given by

$$V_L = I_L \cdot Z_L = 0. \qquad (5.119)$$

Since $V_L = V_+ + V_- = 0$, we have

$$V_+ = -V_- \quad \Rightarrow \quad \Gamma = \frac{V_-}{V_+} = -1. \qquad (5.120)$$

In circuitry, we learned that when $Z_L = 0$, the load voltage $V_L = I_L \cdot Z_L = 0$ according to Ohm's law. From the principles of TX lines, we can even go further and realize that $V_L = 0$ is the result of reflection, i.e., the forward voltage V_+ and the backward voltage V_- cancel each other out. It not only interprets the phenomenon ($V_L = 0$) at $z = L$, but also provides the voltage at an arbitrary point between $0 \leq z < L$. Hence, the reflection of TX lines gives a more general interpretation in this case.

Case 2: $Z_L \to \infty$

In Fig. 5.24, we leave it open at $z = L$ so that $Z_L \to \infty$. Then from Eq. (5.117), we have

Fig. 5.23 Case of short-circuit termination

$Z_L = 0$

$z = 0$ $z = L$

Fig. 5.24 Case of
open-circuit termination

$$\Gamma = \frac{Z_L - Z_0}{Z_L + Z_0} = 1. \qquad (5.121)$$

Hence for the case of **open circuit**, the corresponding reflection coefficient $\Gamma = 1$. This is a reasonable result because for an open circuit, we must have $I_L = 0$. Therefore

$$I_L = I_+ + I_- = \frac{V_+}{Z_0} - \frac{V_-}{Z_0} = 0. \qquad (5.122)$$

From Eq. (5.122), we get

$$V_+ = V_- \quad \Rightarrow \quad \Gamma = \frac{V_+}{V_-} = 1. \qquad (5.123)$$

In circuitry, we learned that when $Z_L \to \infty$, $I_L = V_L/Z_L = 0$ according to Ohm's law. From the principles of TX lines, we realize that $I_L = 0$ is the result of reflection, i.e., the forward current I_+ and the backward current I_- cancel each other out. It also gives a more general interpretation in this case.

Case 3: $Z_L = Z_0$

When we have impedance matched, i.e., $Z_L = Z_0$ as shown in Fig. 5.25. From Eq. (5.117), we have

$$\Gamma = \frac{Z_L - Z_0}{Z_L + Z_0} = 0. \qquad (5.124)$$

Thus when impedance is matched, no reflection occurs. In this case, only forward waves exist in the TX line.

Fig. 5.25 Case of
matched-impedance
termination

The above three special cases of load impedance correspond to different values of reflection coefficient. When $Z_L = 0$ or $Z_L \to \infty$, the magnitude of reflection coefficient achieves maximum, i.e., $|\Gamma| = 1$. On the other hand, when $Z_L = Z_0$, the corresponding reflection coefficient $\Gamma = 0$. In general cases, a reflection coefficient satisfies $0 \le |\Gamma| \le 1$. When Z_L approaches Z_0, the reflection coefficient approaches zero, and vice versa. It is an important property of a TX line.

Example 5.13

Suppose a TX line has the characteristic impedance $Z_0 = 50\Omega$ and the forward voltage at the load is $V_+ = 5e^{j\frac{\pi}{4}}$. Please derive the load current I_L when the load impedance (a). $Z_L = 0$ (b). $Z_L \to \infty$.

Solution

(a) When $Z_L = 0$, from Eq. (5.117), the reflection coefficient is given by

$$\Gamma = \frac{Z_L + Z_0}{Z_L - Z_0} = -1.$$

From Eq. (5.112), we have the backward voltage given by

$$V_- = \Gamma V_+ = -5e^{j\frac{\pi}{4}}.$$

Hence the forward current and the backward current are given by

$$I_+ = \frac{V_+}{Z_0} = \frac{1}{10}e^{j\frac{\pi}{4}},$$

$$I_- = -\frac{V_-}{Z_0} = \frac{1}{10}e^{j\frac{\pi}{4}},$$

respectively. Finally, the load current I_L is derived by

$$I_L = I_+ + I_- = \frac{1}{5}e^{j\frac{\pi}{4}}.$$

(b) When $Z_L = \infty$, from Eq. (5.117), the reflection coefficient is given by

$$\Gamma = \frac{Z_L - Z_0}{Z_L + Z_0} = 1.$$

From Eq. (5.112), the backward voltage is given by

$$V_- = \Gamma V_+ = 5e^{j\frac{\pi}{4}}.$$

Hence the forward current and the backward current are given by

$$I_+ = \frac{V_+}{Z_0} = \frac{1}{10}e^{j\frac{\pi}{4}},$$

$$I_- = -\frac{V_-}{Z_0} = -\frac{1}{10}e^{j\frac{\pi}{4}},$$

respectively. Finally, the load current is derived by

$$I_L = I_+ + I_- = 0.$$

Alternatively, since $Z_L \to \infty$, we can simply obtain I_L by

$$I_L = \frac{V_L}{Z_L} = 0.$$

∎

Reflection is an important phenomenon occurring in TX lines. It makes the analysis more complicated but also more interesting. Generally speaking, the reflection may bring two undesirable effects:

1. The power cannot be transmitted efficiently to the load.
2. The reflected signal may interfere the transmitted signal.

Therefore, when designing a high-frequency circuit, we usually adopt **impedance matching circuit** so that $Z_L = Z_0$ in order to avoid reflection. But in practice, perfect impedance matching is impossible. Thus we had better accept the following concept:

> Reflection generally occurs in TX lines so that forward waves and backward waves usually coexist.

With this concept in mind, when encountering a problem of TX lines, we naturally take reflection into consideration and adopt a proper approach to deal with the problem.

5.5   TX Line Circuit Components

In our daily life, most objects are made for a specific purpose. For example, a candle is for lighting and a bottle is for storing liquid. However, if we use our imagination, one object may have different applications. For example, a candle can also be used for decoration and bottles may be used for green buildings. Therefore, an object may be originally made for a specific purpose, but actually have different applications. It depends on how we use it.

Similarly, a TX line is originally made for conveying signal, but it also has different applications. As we have studied in the previous sections, because the voltage and the current of a TX line have a specific relationship, i.e., characteristic impedance, someone comes up with an idea that a TX line may be used as a circuit component. It turns out that by simple design, a segment of TX line can be used as a resistor, an inductor, or a capacitor in high-frequency circuits.

In this section, we start from a perspective of circuit design and see how a segment of TX line can be used as a component in various high-frequency circuits.

A Characteristic Resistance

As we have discussed in Sect. 5.3, the characteristic impedance of a TX line is given by

$$Z_0 = \sqrt{\frac{R + j\omega L}{G + j\omega C}}, \tag{5.125}$$

where ω is the frequency and (R, L, C, G) are the equivalent circuit parameters for a TX line. If a TX line is lossless and thus $R = 0$ and $G = 0$, Eq. (5.125) becomes

$$Z_0 = \sqrt{\frac{j\omega L}{j\omega C}} = \sqrt{\frac{L}{C}} = R_0, \tag{5.126}$$

where R_0 is a **positive real number** and called **characteristic resistance**.

In practice $R \neq 0$ and $G \neq 0$, but because $R << \omega L$ and $G << \omega C$ when the frequency is high, from Eq. (5.125), it is easily found that a characteristic impedance Z_0, which is a complex number, approximates the characteristic resistance R_0, which is a real number given in Eq. (5.126). In electrical engineering, TX lines having $R_0 = 50\Omega$ or $R_0 = 75\Omega$ are quite popular. For example, high-frequency circuits and test instruments usually adopt TX lines having $R_0 = 50\Omega$. Therefore, in the following we take R_0 as the characteristic resistance of TX lines under consideration.

When studying the characteristic resistance of a TX line, people usually have a question in mind: what's the difference between the characteristic resistance R_0 and a real resistor? Before we go further, it is worthwhile for readers to think about this question.

As shown in Fig. 5.26, we have a real resistor having resistance $R = 50\Omega$. Let V and I denote the voltage and the current across the resistor, respectively. We then have the following relationship

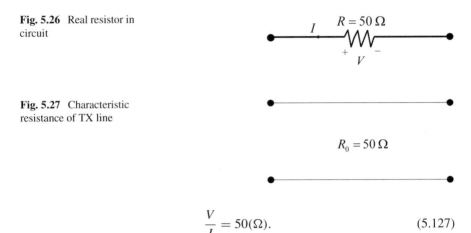

Fig. 5.26 Real resistor in circuit

Fig. 5.27 Characteristic resistance of TX line

$$\frac{V}{I} = 50(\Omega). \tag{5.127}$$

For example, when applying 10 V across the resistor, we have 0.2 A current going through the resistor and the resistor consumes power.

On the other hand, suppose we have a TX line having characteristic resistance $R_0 = 50\Omega$ as shown in Fig. 5.27. In this case, for an arbitrary point on the TX line, the ratio of forward voltage $V^+(z)$ to the forward current $I^+(z)$ is a constant given by

$$\frac{V^+(z)}{I^+(z)} = R_0. \tag{5.128}$$

Similarly, the ratio of backward voltage $V^-(z)$ to the backward current $I^-(z)$ is also a constant given by

$$\frac{V^-(z)}{I^-(z)} = -R_0. \tag{5.129}$$

Notice that the characteristic resistance R_0 is simply a ratio, but not a real resistor, because no current goes through R_0. Hence, no power is consumed in this case. In a word, characteristic resistance R_0 **simply stands for a ratio of voltage to current in a TX line**. It does not physically exist such as a real resistor.

B Input Impedance

In Fig. 5.28, we have a TX line having the length L, the characteristic resistance R_0, and the load impedance Z_L. Suppose $Z_L \neq R_0$. In this case, the reflection occurs so that a forward wave and a backward wave simultaneously exist in the TX line. From Sect. 5.4, for an arbitrary point z on the TX line, where $0 \leq z \leq L$, we have the forward voltage and the backward voltage given by

Fig. 5.28 A TX line with a load

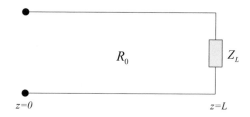

$$V^+(z) = V_a e^{-\gamma z}, \tag{5.130}$$

$$V^-(z) = V_b e^{\gamma z}, \tag{5.131}$$

where V_a and V_b are the forward voltage and the backward voltage at $z = 0$, respectively, and γ is the propagation constant.

Let V_+ and V_- represent the forward voltage and the backward voltage at $z = L$, respectively. From Eqs. (5.130) and (5.131), we obtain

$$V_+ = V_a e^{-\gamma L}, \tag{5.132}$$

$$V_- = V_b e^{\gamma L}. \tag{5.133}$$

According to the definition of reflection coefficient, we have

$$\Gamma = \frac{V_-}{V_+}. \tag{5.134}$$

Besides, from Eq. (5.117) of Sect. 5.4, we have

$$\Gamma = \frac{Z_L - R_0}{Z_L + R_0}. \tag{5.135}$$

As shown in Fig. 5.29, we consider the point at $z = L - d$. The distance between this point and the load is d. From Eqs. (5.130)–(5.133), the forward voltage and the

Fig. 5.29 A point ($z = L - d$) away from the load

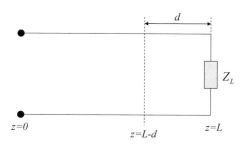

backward voltage at this point are given by

$$V^+(z) = V_a e^{-\gamma(L-d)} = V_a e^{-\gamma L} \cdot e^{\gamma d} = V_+ \cdot e^{\gamma d}, \qquad (5.136)$$

$$V^-(z) = V_b e^{\gamma(L-d)} = V_b e^{\gamma L} \cdot e^{-\gamma d} = V_- \cdot e^{-\gamma d}, \qquad (5.137)$$

respectively. Let $V(z)$ be the measured voltage at $z = L - d$. Hence $V(z)$ shall be the sum of $V^+(z)$ and $V^-(z)$. We have

$$V(z) = V_+ e^{\gamma d} + V_- e^{-\gamma d}. \qquad (5.138)$$

Equation (5.138) shows that we can use V_+, V_- and d to express the measured voltage at $z = L - d$.

In addition, let $I^+(z)$ and $I^-(z)$ be the forward current and the backward current at $z = L - d$, given by

$$I^+(z) = \frac{V^+(z)}{R_0} = \frac{V_+ e^{\gamma d}}{R_0}, \qquad (5.139)$$

$$I^-(z) = -\frac{V^-(z)}{R_0} = -\frac{V_- e^{-\gamma d}}{R_0}, \qquad (5.140)$$

respectively. Hence the measured current at $z = L - d$ is given by

$$I(z) = I^+(z) + I^-(z) = \frac{1}{R_0} \cdot \left(V_+ e^{\gamma d} - V_- e^{-\gamma d}\right). \qquad (5.141)$$

From Eqs. (5.138) and (5.141), the voltage and the current at $z = L - d$ are both functions of V_+ and V_-. Besides, they depend on d, but not on L.

After obtaining $V(z)$ and $I(z)$, we are ready to define an **input impedance** at $z = L - d$ as shown in Fig. 5.30. It is defined as

Fig. 5.30 Input impedance at a distance from the load (I)

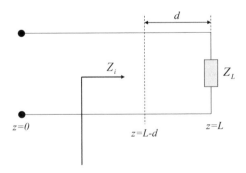

Fig. 5.31 Input impedance
at a distance from the load
(II)

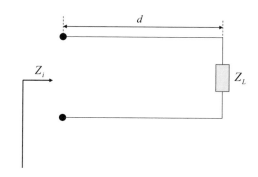

$$Z_i = \frac{V(z)}{I(z)}. \qquad (5.142)$$

Obviously, this definition is similar to that in electronic circuits, which is the ratio of input voltage to the input current. From Eqs. (5.138) and (5.141), we obtain

$$Z_i = R_0 \cdot \frac{V_+ e^{\gamma d} + V_- e^{-\gamma d}}{V_+ e^{\gamma d} - V_- e^{-\gamma d}}. \qquad (5.143)$$

From Eq. (5.134), we can rewrite Eq. (5.143) as

$$Z_i = R_0 \cdot \frac{e^{\gamma d} + \Gamma e^{-\gamma d}}{e^{\gamma d} - \Gamma e^{-\gamma d}}. \qquad (5.144)$$

Hence the input impedance Z_i depends on the reflection coefficient Γ and d. We can utilize this property and treat a segment of TX line as a circuit component. For example, we have a TX line having the length d in Fig. 5.31. The input impedance can be determined by Eq. (5.144). For a given characteristic resistance R_0, we can select proper Γ and d so that a circuit component having the desired input impedance is achieved.

In Eq. (5.144), the propagation constant is given by $\gamma = \alpha + j\beta$, where α is the attenuation constant and β is the phase constant. For a TX line, the propagation loss is usually very limited and α can be neglected. Hence we have

$$\gamma = j\beta. \qquad (5.145)$$

Then Eq. (5.144) can be rewritten as

$$Z_i = R_0 \cdot \frac{e^{j\beta d} + \Gamma e^{-j\beta d}}{e^{j\beta d} - \Gamma e^{-j\beta d}}, \qquad (5.146)$$

where β is given by

$$\beta = \frac{2\pi}{\lambda}. \tag{5.147}$$

Note that λ is the wavelength of EM wave.

Example 5.14

In Fig. 5.31, suppose we have a segment of TX line having characteristic resistance $R_0 = 50\Omega$, load impedance $Z_L = 100\Omega$, and $d = \lambda/6$. Please derive the input impedance Z_i.

Solution

From Eq. (5.135), the reflection coefficient is given by

$$\Gamma = \frac{Z_L - R_0}{Z_L + R_0} = \frac{100 - 50}{100 + 50} = \frac{1}{3}.$$

Next, from Eq. (5.146), we have

$$Z_i = R_0 \cdot \frac{e^{j\beta d} + \Gamma e^{-j\beta d}}{e^{j\beta d} - \Gamma e^{-j\beta d}} = R_0 \cdot \frac{1 + \Gamma e^{-j2\beta d}}{1 - \Gamma e^{-j2\beta d}},$$

where $R_0 = 50\Omega$ and $\Gamma = \frac{1}{3}$. Besides,

$$\beta d = \frac{2\pi}{\lambda} \cdot \frac{\lambda}{6} = \frac{\pi}{3},$$

and thus

$$e^{-j2\beta d} = e^{-j\frac{2\pi}{3}} = \cos\frac{2\pi}{3} - j\sin\frac{2\pi}{3} = -\frac{1}{2} - j\frac{\sqrt{3}}{2}.$$

Finally, we insert these values into the formula and get

$$Z_i = 50 \cdot \frac{1 - \frac{1}{3}\left(\frac{1}{2} + j\frac{\sqrt{3}}{2}\right)}{1 + \frac{1}{3}\left(\frac{1}{2} + j\frac{\sqrt{3}}{2}\right)} = 50 \cdot \frac{5 - j\sqrt{3}}{7 + j\sqrt{3}}.$$

■

C TX Line as Circuit Component

TX lines can be used as circuit components at high frequencies because ordinary lumped elements may be unavailable. In the following, we introduce three cases which are widely used when designing high-frequency circuits.

Case 1: $Z_L = 0$ (Short-circuit termination)

In Fig. 5.32, we connect two end points of a TX line with a conducting line and thus $Z_L = 0$. In this case, the termination is "**short**" in circuit perspective and the reflection coefficient is given by

$$\Gamma = \frac{Z_L - R_0}{Z_L + R_0} = -1. \tag{5.148}$$

Inserting $\Gamma = -1$ into Eq. (5.146), the input impedance of this TX line is given by

$$Z_i = R_0 \cdot \frac{e^{j\beta d} - e^{-j\beta d}}{e^{j\beta d} + e^{-j\beta d}}. \tag{5.149}$$

From **Euler's formula**, for a real number x, we have

$$e^{jx} = \cos x + j \sin x, \tag{5.150}$$

$$e^{-jx} = \cos x - j \sin x. \tag{5.151}$$

Hence

$$e^{jx} + e^{-jx} = 2 \cos x, \tag{5.152}$$

$$e^{jx} - e^{-jx} = j2 \sin x. \tag{5.153}$$

Fig. 5.32 Input impedance of short-circuit termination

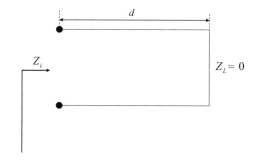

Therefore, Eq. (5.149) can be rewritten as

$$Z_i = R_0 \cdot \frac{j2 \sin \beta d}{2 \cos \beta d} = j R_0 \tan \beta d. \tag{5.154}$$

Obviously, the input impedance Z_i changes with d.

In circuitry, we learn that the impedance of an inductor is given by

$$Z = j\omega L, \tag{5.155}$$

where L is the associated inductance. Also, the impedance of a capacitor is given by

$$Z = \frac{1}{j\omega C} = -j\frac{1}{\omega C}, \tag{5.156}$$

where C is the associated capacitance. From Eq. (5.156), if we select a segment of TX line having proper length d as shown in Fig. 5.32, it can serve as the desired inductor or capacitor. In other words, it can be used as a circuit component.

When the length is smaller than a quarter wavelength, i.e., $d < \frac{\lambda}{4}$, then $\frac{2\pi d}{\lambda} < \frac{\pi}{2}$. Thus we have

$$\tan \frac{2\pi d}{\lambda} > 0. \tag{5.157}$$

Because $\beta d = \frac{2\pi d}{\lambda}$, we can rewrite Eq. (5.154) as

$$Z_i = j R_0 \tan \beta d = j B, \tag{5.158}$$

where

$$B = R_0 \tan \beta d \quad \text{(positive number)}. \tag{5.159}$$

From Eqs. (5.158) and (5.155), when $d < \frac{\lambda}{4}$, this segment of TX line has the impedance similar to what an inductor has. Therefore, it can serve as an inductor at high-frequency circuits.

On the other hand, when the length is greater than a quarter wavelength and smaller than half a wavelength, i.e., $\frac{\lambda}{4} < d < \frac{\lambda}{2}$, since $\frac{\pi}{2} < \frac{2\pi d}{\lambda} < \pi$, we have

$$\tan \beta d < 0. \tag{5.160}$$

Hence

$$Z_i = j R_0 \tan \beta d = -j B', \tag{5.161}$$

where

Fig. 5.33 Input impedance of open-circuit termination

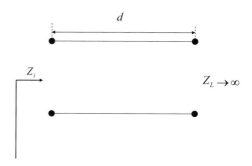

$$B' = R_0 \cdot |\tan \beta d| \quad \text{(positive number)}. \tag{5.162}$$

In this case, this segment of TX line has the impedance similar to what a capacitor has. Therefore, it can serve as a capacitor at high-frequency circuits.

In above, we can select a segment of TX line with short-circuit termination so that a desired inductor or capacitor is achieved. When designing a high-frequency circuit, we can easily generate a segment of TX line having short-circuit termination in a printed circuit board (PCB). It is a very practical and effective technique.

Case 2: $Z_L \rightarrow \infty$ (Open-circuit termination)

In Fig. 5.33, we remove the load and leave it open ($Z_L \rightarrow \infty$). In this case, the termination is "**open**" in circuit perspective and the reflection coefficient is given by

$$\Gamma = \frac{Z_L - R_0}{Z_L + R_0} = 1. \tag{5.163}$$

Inserting $\Gamma = 1$ into Eq. (5.146), the input impedance of this segment of TX line is given by

$$Z_i = R_0 \cdot \frac{e^{j\beta d} + e^{-j\beta d}}{e^{j\beta d} - e^{-j\beta d}} = R_0 \frac{\cos \beta d}{j \sin \beta d}$$
$$= -j R_0 \cdot \cot \beta d. \tag{5.164}$$

Obviously, the input impedance Z_i depends on the length d.

When the length is smaller than a quarter wavelength, i.e., $d < \frac{\lambda}{4}$, we have $\beta d = \frac{2\pi d}{\lambda} < \frac{\pi}{2}$. Hence

$$\cot \beta d > 0. \tag{5.165}$$

From Eq. (5.164), we get

$$Z_i = -j R_0 \cot \beta d = -jM \tag{5.166}$$

where

$$M = R_0 \cot \beta d \quad \text{(positive number)} \tag{5.167}$$

Hence when $d < \frac{\lambda}{4}$, this segment of TX line has the impedance similar to what a capacitor has. Therefore, it can serve as a capacitor at high-frequency circuits.

In addition, when the length is greater than a quarter wavelength and smaller than half a wavelength, i.e., $\frac{\lambda}{4} < d < \frac{\lambda}{2}$, we have $\frac{\pi}{2} < \frac{2\pi d}{\lambda} < \pi$. Thus

$$\cot \beta d < 0. \tag{5.168}$$

From Eq. (5.164), we get

$$Z_i = -j R_0 \cot \beta d = j M', \tag{5.169}$$

where

$$M' = -R_0 \cot \beta d \quad \text{(positive number)}. \tag{5.170}$$

Hence when $\frac{\lambda}{4} < d < \frac{\lambda}{2}$, this segment of TX line has the impedance similar to what an inductor has. Therefore, it can serve as an inductor at high-frequency circuits. In above, we can select a segment of TX line with open-circuit termination so that a desired capacitor or inductor is achieved.

From Eqs. (5.154) and (5.164), we can find an interesting relationship as follows. Suppose we have a segment of TX line having the length d. Let Z_{SC} be the input impedance of the **short-circuit termination** case and Z_{OC} be that of the **open-circuit termination** case. From Eqs. (5.154) and (5.164), we get

$$Z_{SC} \cdot Z_{OC} = (j R_0 \tan \beta d) \cdot (-j R_0 \cot \beta d) = R_0^2. \tag{5.171}$$

Hence

$$R_0 = \sqrt{Z_{OC} Z_{SC}}. \tag{5.172}$$

Note that the relationship in Eq. (5.172) does not depend on the length d. It tells us an effective way to measure the characteristic resistance of a TX line: we take a TX line with arbitrary length, and then we simply "open" the termination to measure Z_{OC} and "short" the termination to measure Z_{SC}, and finally calculate R_0 by using Eq. (5.172).

Case 3: $d = \frac{\lambda}{4}$ (Quarter-wavelength TX line)

In Fig. 5.34, suppose the length of a TX line is equal to the quarter wavelength, i.e., $d = \frac{\lambda}{4}$. This kind of TX line is called **quarter-wavelength TX line** (abbreviated as $\lambda/4$ TX line). It has a very special impedance feature. First, when $d = \frac{\lambda}{4}$, we have

Fig. 5.34 Input impedance
of quarter wavelength TX
line

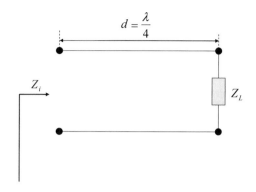

$$\beta d = \frac{2\pi d}{\lambda} = \frac{\pi}{2}. \tag{5.173}$$

Because

$$e^{j\frac{\pi}{2}} = j, \tag{5.174}$$

$$e^{-j\frac{\pi}{2}} = -j. \tag{5.175}$$

From Eq. (5.146), the input impedance is given by

$$Z_i = R_0 \cdot \frac{e^{j\frac{\pi}{2}} + \Gamma e^{-j\frac{\pi}{2}}}{e^{j\frac{\pi}{2}} - \Gamma e^{-j\frac{\pi}{2}}} = R_0 \cdot \frac{1 - \Gamma}{1 + \Gamma}. \tag{5.176}$$

Because the reflection coefficient is given by

$$\Gamma = \frac{Z_L - R_0}{Z_L + R_0}, \tag{5.177}$$

we insert Eqs. (5.177) into (5.176) and obtain

$$Z_i = \frac{R_0^2}{Z_L}. \tag{5.178}$$

From Eq. (5.178), the input impedance Z_i is inversely proportional to the load impedance Z_L. It implies that a $\lambda/4$ TX line can be used to **invert the impedance**. For example, suppose $R_0 = 50\Omega$ and $Z_L = 100\Omega$. From Eq. (5.178), the input impedance is given by

$$Z_i = \frac{(50)^2}{100} = 25\Omega. \tag{5.179}$$

If we let $Z_L = 1 K\Omega$, the input impedance becomes

$$Z_i = \frac{(50)^2}{1000} = 2.5\Omega. \tag{5.180}$$

Hence the greater the load impedance Z_L, the smaller the input impedance Z_i.

Furthermore, we know that when $b > 0$, the load impedance $Z_L = jb$ can be used as an inductor. From Eq. (5.178), we have

$$Z_i = \frac{R_0^2}{jb} = -j\frac{R_0^2}{b}. \tag{5.181}$$

Since $-R_0^2/b < 0$, the input impedance Z_i is capacitive. On the other hand, if the load impedance $Z_L = -jb$ is capacitive, from Eq. (5.178), we have

$$Z_i = \frac{R_0^2}{-jb} = j\frac{R_0^2}{b}. \tag{5.182}$$

Since $R_0^2/b > 0$, the input impedance Z_i is inductive. As a $\lambda/4$ TX line can be used to invert the impedance, it is called **impedance inverter** or **impedance transformer**.

In above, we have introduced three kinds of TX lines which are widely used as circuit components in high-frequency circuits. For a segment of TX line with a short-circuit termination ($Z_L = 0$) or an open-circuit termination ($Z_L \to \infty$), we can obtain a desired capacitor or an inductor with a proper length d. We can also invert the impedance by using a $\lambda/4$ TX line.

Example 5.15

In Fig. 5.31, suppose we have a TX line having characteristic resistance $R_0 = 50\Omega$ and $d = \frac{\lambda}{3}$. Derive the input impedance Z_i when (a). $Z_L = 0$ (b). $Z_L \to \infty$.

Solution

(a)   When $Z_L = 0$, from Eq. (5.154) we get

$$Z_i = jR_0 \tan \beta d = jR_0 \tan \frac{2\pi}{3} = j(50) \cdot \left(-\sqrt{3}\right)$$
$$= -j50\sqrt{3}.$$

(b)   When $Z_L \to \infty$, from Eq. (5.164) we get

$$Z_i = -jR_0 \cot \beta d = -jR_0 \cot \frac{2\pi}{3} = -j(50) \cdot \left(-\frac{1}{\sqrt{3}}\right)$$

$$= j\frac{50}{\sqrt{3}}.$$

∎

Example 5.16

In Fig. 5.31, suppose we have a TX line having $R_0 = 50\Omega$ and $d = \frac{\lambda}{4}$. Derive Z_i when (a) $Z_L \to \infty$, (b) $Z_L = 0$, (c) $Z_L = j5$, (d) $Z_L = 3 + j4$.

Solution

Obviously, it is a $\lambda/4$ TX line. From Eq. (5.178), we have

$$Z_i = \frac{R_0^2}{Z_L} = \frac{2500}{Z_L}$$

and hence

(a) $Z_i = \frac{2500}{Z_L} = 0,$

(b) $Z_i = \frac{2500}{Z_L} \to \infty,$

(c) $Z_i = \frac{2500}{Z_L} = \frac{2500}{j5} = -j500,$

(d) $Z_i = \frac{2500}{Z_L} = \frac{2500}{3+j4} = 300 - j400.$

∎

5.6 Standing Wave

Sometimes when we walk along seashore, we may find sea wave seems "stand still" and each point simply moves upward and downward. This is called a **standing wave**, which is actually formed by sea waves having the same wavelength but moving in opposite directions.

A standing wave occurs not only in the sea, but also in a TX line. In Sect. 5.4, we have learned that when impedance does not match ($Z_L \neq R_0$), reflection occurs and a forward wave and a backward wave coexist. The superposition of a forward wave and a backward wave results in a standing wave in a TX line. In fact, the behaviors of an EM wave in a TX line are closely related to the resultant standing wave. Hence, a number of critical parameters of a transmitted EM wave can be derived through the associated standing wave. In this section, we introduce the cause of a standing wave and utilize it to derive important parameters of an EM wave in a TX line.

A Voltage of Standing Wave

We have learned that the reflection coefficient of a TX line depends on the load impedance Z_L and the characteristic resistance R_0, given by

$$\Gamma = \frac{Z_L - R_0}{Z_L + R_0}. \tag{5.183}$$

When Z_L is given, we can readily derive the reflection coefficient from Eq. (5.183). However, what if Z_L is unknown, how do we derive Γ? In the following, we will learn how to derive Z_L and Γ by using the measured voltage of a standing wave.

In Fig. 5.35, we have a TX line and the load impedance Z_L is at $z = L$. Note that $Z_L \neq R_0$. Suppose V_+ and V_- denote the forward voltage and the backward voltage at $z = L$, respectively. Then the load voltage V_L at $z = L$ is given by

$$V_L = V_+ + V_-, \tag{5.184}$$

and the reflection coefficient is defined as

$$\Gamma = \frac{V_-}{V_+}. \tag{5.185}$$

Consider a point at $z = L - d$ and from the results in Sect. 5.5, the voltage at this point is given by

$$V(z) = V_+ e^{\gamma d} + V_- e^{-\gamma d}, \tag{5.186}$$

where γ is the propagation constant. Equation (5.186) tells us that the observed wave at any point consists of a forward wave (toward $+z$) and a backward wave (toward $-z$). The situation is similar to a sea wave encountering the bank so that a standing wave occurs. In the following, we derive the voltage of a standing wave.

For simplicity, we consider a lossless TX line having the attenuation constant $\alpha = 0$. In this case, the propagation constant is given by

Fig. 5.35 Voltage at a distance from the load

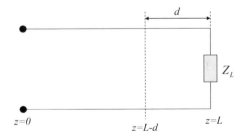

$$\gamma = j\beta, \tag{5.187}$$

where β is the phase constant. By Eqs. (5.187) and (5.185), we can rewrite Eq. (5.186) as

$$V(z) = V_+ e^{j\beta d} + V_- e^{-j\beta d} = V_+ e^{j\beta d}(1 + \Gamma e^{-j2\beta d}). \tag{5.188}$$

Note that $V(z)$ is a complex number and can be represented by

$$V(z) = A(z) \cdot e^{j\phi(z)}, \tag{5.189}$$

where

$$A(z) = |V(z)|. \tag{5.190}$$

We call $A(z)$ the **magnitude** and $\phi(z)$ the **phase** of $V(z)$.

Because $V(z)$ is the voltage phasor at $z = L - d$, the time-varying voltage $V(z, t)$ is given by

$$V(z, t) = \text{Re}\{V(z) \cdot e^{j\omega t}\} = A(z)\cos[\omega t + \phi(z)]. \tag{5.191}$$

Equation (5.191) shows that the magnitude $A(z)$ depends on z. When we try to measure the voltage along the TX line, we will get a different $A(z)$ at a different position. It means that the **measured voltage changes with the position**.

Next, we derive $A(z)$ at $z = L - d$. From Eqs. (5.190) and (5.188), because $|e^{j\beta d}| = 1$, we have

$$A(z) = |V_+| \cdot \left|1 + \Gamma e^{-j2\beta d}\right|. \tag{5.192}$$

Let the reflection coefficient Γ be represented by

$$\Gamma = |\Gamma| \cdot e^{j\theta}, \tag{5.193}$$

where $|\Gamma|$ is the magnitude, θ is the phase of Γ and $-\pi \leq \theta < \pi$. Inserting Eqs. (5.193) into (5.192), we have

$$A(z) = |V_+| \cdot \left|1 + |\Gamma| \cdot e^{-j(2\beta d - \theta)}\right|. \tag{5.194}$$

For a specific Z_L, $|\Gamma|$ is a constant and $A(z)$ is a function of the position parameter d. As will become clear later, we find that $A(z)$ varies regularly with the position. This regular variation is closely related to important properties of a TX line, which will be introduced in the next subsection.

B   Maximum Voltage and Minimum Voltage

In Eq. (5.194), $A(z)$ at $z = L - d$ depends on d and Γ, where d is the distance between the measured point and the load. Let m be an integer. When

$$2\beta d - \theta = m \cdot 2\pi, \tag{5.195}$$

because

$$e^{-j2m\pi} = 1. \tag{5.196}$$

From Eq. (5.194), the magnitude $A(z)$ reaches maximum in this case and given by

$$A(z) = V_{\max} = |V_+| \cdot (1 + |\Gamma|), \tag{5.197}$$

where $V_{\max}$ denotes the **maximum voltage** of a standing wave.

In Eq. (5.195), the phase constant β is given by

$$\beta = \frac{2\pi}{\lambda}, \tag{5.198}$$

where λ is the wavelength. From Eqs. (5.195) and (5.198), the maximum voltage occurs at

$$d = d_{\max} = m \cdot \frac{\lambda}{2} + \frac{\theta}{4\pi} \cdot \lambda, \tag{5.199}$$

where $m = 0, 1, 2\ldots$. Equation (5.199) tells us that when the distance d increases, **a maximum voltage occurs repeatedly with a period of** $\lambda/2$.

Equation (5.199) also tells us that the first $V_{\max}$ depends on θ, where $-\pi \leq \theta < \pi$. In Fig. 5.35, d is the distance between the measured point and the load, and thus $d \geq 0$. If $0 \leq \theta < \pi$, because $d \geq 0$ in Eq. (5.199), the first maximum voltage occurs when $m = 0$ and the position is given by

$$d = d_{\max,1} = \frac{\theta}{4\pi} \cdot \lambda. \tag{5.200}$$

On the other hand, if $-\pi \leq \theta < 0$, the first maximum voltage occurs when $m = 1$ and the position is given by

$$d = d_{\max,1} = \frac{\lambda}{2} + \frac{\theta}{4\pi} \cdot \lambda. \tag{5.201}$$

Hence, the position of the first maximum voltage ($d_{\max,1}$) depends on λ and θ.

Once $d_{\text{max},1}$ is determined, the maximum voltage occurs repeatedly every $\lambda/2$. Hence the other maximums occur at

$d_{\text{max},2} = d_{\text{max},1} + \frac{\lambda}{2}$ (the 2nd maximum voltage),
$d_{\text{max},3} = d_{\text{max},2} + \frac{\lambda}{2} = d_{\text{max},1} + 2 \cdot \frac{\lambda}{2}$ (the 3rd maximum voltage),

$\cdots$

$d_{\text{max},k} = d_{\text{max},k-1} + \frac{\lambda}{2} = d_{\text{max},1} + (k-1) \cdot \frac{\lambda}{2}$ (the k-th maximum voltage).

Figure 5.36 shows that when the position of the first maximum voltage $d_{\text{max},1}$ is determined, the positions of the remaining maximum voltages are readily determined with a period of $\lambda/2$ distance. This is an important property of a standing wave.

Next, we investigate the minimum voltages and the associated positions of a standing wave. In Eq. (5.194), when

$$2\beta d - \theta = (2m+1)\pi, \qquad (5.202)$$

because

$$e^{-j(2m+1)\pi} = -1, \qquad (5.203)$$

the minimum $A(z)$ in Eq. (5.194) occurs and given by

$$A(z) = V_{\text{min}} = |V_+| \cdot (1 - |\Gamma|), \qquad (5.204)$$

where V_{min} denotes the **minimum voltage** of a standing wave. Notice that from Eqs. (5.197) and (5.204), V_{max} increases with $|\Gamma|$, and V_{min} decreases with $|\Gamma|$.

Because $\beta = 2\pi/\lambda$, from Eq. (5.202), the minimum voltages occur when

$$d = d_{\text{min}} = m \cdot \frac{\lambda}{2} + (\frac{\pi + \theta}{4\pi}) \cdot \lambda, \qquad (5.205)$$

where $m = 0, 1, 2\ldots$. Equation (5.205) tells us that when the distance d increases, **a minimum voltage occurs repeatedly with a period of $\lambda/2$ distance**. Because

Fig. 5.36 Positions of maximum voltages of standing wave

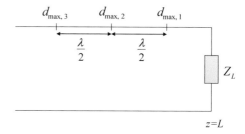

$-\pi \leq \theta < \pi$ and $d \geq 0$, in Eq. (5.205) the first minimum voltage occurs when $m = 0$ and the position is given by

$$d = d_{\min,1} = (\frac{\pi + \theta}{4\pi}) \cdot \lambda. \qquad (5.206)$$

Because the first minimum voltage always corresponds to $m = 0$, the relationship between $d_{\min,1}$ and θ is relatively easy comparing with the case of maximum voltage. This feature will be used later in deriving critical parameters of TX line.

Once $d_{\min,1}$ is determined, the minimum voltage occurs repeatedly every $\lambda/2$. Hence the other minimums occur at

$d_{\min,2} = d_{\min,1} + \frac{\lambda}{2}$ (the 2nd minimum voltage),
$d_{\min,3} = d_{\min,2} + \frac{\lambda}{2} = d_{\min,1} + 2 \cdot \frac{\lambda}{2}$ (the 3rd minimum voltage),
$\cdots$
$d_{\min,k} = d_{\min,k-1} + \frac{\lambda}{2} = d_{\min,1} + (k - 1) \cdot \frac{\lambda}{2}$ (the k-th minimum voltage).

Figure 5.37 shows that when the position of the first minimum voltage $d_{\min,1}$ is determined, the positions of the remaining minimum voltages are readily determined with a period of $\lambda/2$ distance.

Figure 5.38 shows $A(z)$ of a standing wave along a TX line, where the distance between the neighboring maximum voltages and that of the neighboring minimum voltages are both $\lambda/2$. Besides, the distance between a maximum voltage and a neighboring minimum voltage is $\lambda/4$.

Example 5.17

Suppose a TX line has $R_0 = 50\Omega$, $Z_L = j50\Omega$, and $\lambda = 4m$. Please derive (a) the position of the first maximum voltage, (b) the positions of the second and the third maximum voltage.

Solution

Fig. 5.37 Positions of minimum voltages of standing wave

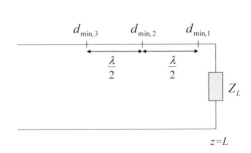

$d_{\min,3}$ $d_{\min,2}$ $d_{\min,1}$

$\frac{\lambda}{2}$ $\frac{\lambda}{2}$

Z_L

$z=L$

Fig. 5.38 Schematic plot of standing wave along a TX line

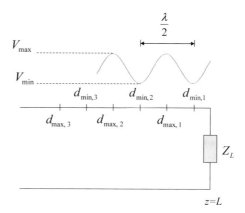

(a) First, from Eq. (5.183), the reflection coefficient is given by

$$\Gamma = \frac{Z_L - R_0}{Z_L + R_0} = \frac{j50 - 50}{j50 + 50} = \frac{50\sqrt{2} \cdot e^{j\frac{3\pi}{4}}}{50\sqrt{2} \cdot e^{j\frac{\pi}{4}}} = e^{j\frac{\pi}{2}}.$$

Hence the phase of Γ is $\theta = \frac{\pi}{2}$. Because $0 \le \theta < \pi$, from Eq. (5.200), the first maximum voltage occurs at

$$d_{max,1} = \frac{\theta}{4\pi} \cdot \lambda = \frac{\lambda}{8} = 0.5(m).$$

(b) Because a maximum voltage occurs every $\lambda/2 = 2m$, the second and the third maximum voltages occur at

$$d_{max,2} = d_{max,1} + \frac{\lambda}{2} = 2.5(m),$$

$$d_{max,3} = d_{max,2} + \frac{\lambda}{2} = 4.5(m).$$

∎

Example 5.18

Suppose a TX line has $R_0 = 50\Omega$, $Z_L = 100\Omega$, and $\lambda = 12m$. Please derive (a) the position of the first minimum voltage, (b) the positions of the second and the third minimum voltage.

Solution

(a) First, from Eq. (5.183), the reflection coefficient is given by

$$\Gamma = \frac{Z_L - R_0}{Z_L + R_0} = \frac{50}{100 + 50} = \frac{1}{3}.$$

Hence the phase of Γ is $\theta = 0°$. From Eq. (5.206), the position of the first minimum voltage is given by

$$d_{\min,1} = (\frac{\pi + \theta}{4\pi}) \cdot \lambda = \frac{\lambda}{4} = 3(m).$$

(b) Because a minimum voltage occurs every $\lambda/2 = 6m$, the second and the third minimum voltage occur at

$$d_{\min,2} = d_{\min,1} + \frac{\lambda}{2} = 9(m),$$

$$d_{\min,3} = d_{\min,1} + \lambda = 15(m).$$

■

C Derivation of Critical Parameters

Now, suppose you have a TX line having the characteristic resistance R_0. Given a voltage meter, is it possible to derive the wavelength λ, load impedance Z_L, and the reflection coefficient Γ?

The answer is "YES"! If we utilize the characteristics of a standing wave properly, we can get critical parameters of a TX line as below:

Step 1:

Starting from the load, we use the voltage meter to find the position of the first minimum voltage ($d_{\min,1}$) and that of the second minimum voltage ($d_{\min,2}$).

Step 2:

Because the distance between two neighboring minimum voltages is $\lambda/2$, we have

$$d_{\min,2} - d_{\min,1} = \frac{\lambda}{2} \quad \Rightarrow \quad \lambda = 2(d_{\min,2} - d_{\min,1}). \tag{5.207}$$

Hence the wavelength can be derived by the measured $d_{min,1}$ and $d_{min,2}$.

Step 3:

From Eq. (5.206), the distance between the first minimum voltage and the load is given by

$$d_{min,1} = (\frac{\pi + \theta}{4\pi}) \cdot \lambda, \tag{5.208}$$

where θ is the phase of the reflection coefficient Γ. Because λ is obtained in Step 2, we can derive the phase given by

$$\theta = (\frac{4d_{min,1}}{\lambda} - 1) \cdot \pi. \tag{5.209}$$

Step 4:

Next, we can measure the maximum voltage (V_{max}) and the minimum voltage (V_{min}) with the voltage meter. From Eqs. (5.197) and (5.204), the ratio of V_{max} to V_{min} is given by

$$\frac{V_{max}}{V_{min}} = \frac{1 + |\Gamma|}{1 - |\Gamma|}. \tag{5.210}$$

From Eq. (5.210), we get

$$|\Gamma| = \frac{V_{max} - V_{min}}{V_{max} + V_{min}}. \tag{5.211}$$

Hence $|\Gamma|$ can be derived by the measured V_{max} and V_{min}. Besides, since the magnitude $|\Gamma|$ and the phase θ (obtained in Step 3) are derived, the reflection coefficient is obtained by $\Gamma = |\Gamma|e^{j\theta}$.

Step 5:

From Eq. (5.183), if Γ is given, the load impedance Z_L can be derived by

$$Z_L = R_0 \cdot \frac{1 + \Gamma}{1 - \Gamma}. \tag{5.212}$$

Using Γ obtained in Step 4, we finally get the load impedance Z_L.

The above procedure might be a little tedious, but the underlining principles are fundamental. When we well understand the principles of reflection (standing wave, maximum voltage, minimum voltage… etc.), we can utilize our understanding to further derive unknown but important parameters such as λ, Γ and Z_L by simply a voltage meter!

Example 5.19

Suppose a TX line has the characteristic resistance $R_0 = 50\Omega$. Using a voltage meter, we learn that the first minimum voltage occurs at $d = 3m$ and the second minimum voltage occurs at $d = 7m$. Besides, the measured $V_{max} = 6V$ and $V_{min} = 4V$. Please derive (a) wavelength λ, (b) reflection coefficient Γ, and (c) load impedance Z_L.

Solution

(a) Because the distance between neighboring minimum voltages is $\frac{\lambda}{2}$, we have

$$\frac{\lambda}{2} = 7 - 3 \quad \Rightarrow \quad \lambda = 8(m).$$

(b) From Eq. (5.206), we get the phase θ of Γ given by

$$\theta = (\frac{4d_{min,1}}{\lambda} - 1) \cdot \pi = \frac{\pi}{2}.$$

Besides, from Eq. (5.211), we have

$$|\Gamma| = \frac{V_{max} - V_{min}}{V_{max} + V_{min}} = \frac{1}{5}.$$

Hence the reflection coefficient is given by

$$\Gamma = |\Gamma| \cdot e^{j\theta} = \frac{1}{5}e^{j\frac{\pi}{2}} = \frac{j}{5}.$$

Note that $e^{j\frac{\pi}{2}} = j$.

(c) Finally, from Eq. (5.212), the load impedance is given by

$$Z_L = R_0 \cdot \frac{1 + \Gamma}{1 - \Gamma} = 50 \cdot \frac{1 + \frac{j}{5}}{1 - \frac{j}{5}} = 50 \cdot \frac{5 + j}{5 - j}.$$

∎

D   Voltage Standing Wave Ratio (VSWR)

In many applications of standing waves, we have an important parameter called **Voltage Standing Wave Ratio (VSWR)**. Suppose a standing wave has a maximum voltage V_{max} and a minimum voltage V_{min}, and their ratio S is given by

$$S = \frac{V_{max}}{V_{min}}. \tag{5.213}$$

Then S is the VSWR of this standing wave. Because V_{max} and V_{min} are non-negative and $V_{max} \geq V_{min}$, the VSWR must be positive and satisfy

$$S \geq 1. \tag{5.214}$$

Obviously, the larger the VSWR, the larger the ratio of V_{max} to V_{min}. Besides, VSWR is closely related to the reflection coefficient Γ. Once S is obtained, the magnitude $|\Gamma|$, a critical parameter when designing a high-frequency circuit, can be readily derived.

From Eqs. (5.211) and (5.213), we get

$$|\Gamma| = \frac{S - 1}{S + 1}. \tag{5.215}$$

Hence $|\Gamma|$ can be readily derived when S is obtained. From Eq. (5.215), we find when S decreases, $|\Gamma|$ decreases, too. When S reaches the minimum, i.e., $S = 1$, then $|\Gamma| = 0$. It means that no reflection occurs.

Besides, from Eq. (5.215), we can express S as

$$S = \frac{1 + |\Gamma|}{1 - |\Gamma|}. \tag{5.216}$$

Hence if Γ is given, we can obtain S easily.

When designing a high-frequency circuit, we always prefer to reduce reflection so that a signal can be transmitted with less power loss. Therefore, S is an effective index showing the magnitude of reflection. When S approximates 1, reflection decreases to the minimum. It means that the signal can be transmitted with the least power loss.

Example 5.20

Suppose a TX line has characteristic resistance $R_0 = 50\Omega$. Please derive the VSWR in the following cases: (a) $Z_L = 50\Omega$ (b) $Z_L = 50 + j50(\Omega)$ (c) $Z_L = 0$ (d) $Z_L \to \infty$.

Solution

(a) When $Z_L = 50\Omega$, the reflection coefficient is given by

$$\Gamma = \frac{Z_L - R_0}{Z_L + R_0} = 0.$$

From Eq. (5.216), we get the VSWR given by

$$S = \frac{1 + |\Gamma|}{1 - |\Gamma|} = 1.$$

(b) When $Z_L = 50 + j50(\Omega)$, the reflection coefficient is given by

$$\Gamma = \frac{Z_L - R_0}{Z_L + R_0} = \frac{j50}{100 + j50} = \frac{j}{2 + j}.$$

Hence $|\Gamma| = \frac{1}{\sqrt{5}}$.
From Eq. (5.216), we get the VSWR given by

$$S = \frac{1 + |\Gamma|}{1 - |\Gamma|} = \frac{\sqrt{5} + 1}{\sqrt{5} - 1} = 2.62.$$

(c) When $Z_L = 0$, the reflection coefficient is given by

$$\Gamma = \frac{Z_L - R_0}{Z_L + R_0} = -1.$$

From Eq. (5.216), we get the VSWR given by

$$S = \frac{1 + |\Gamma|}{1 - |\Gamma|} \quad \rightarrow \quad \infty.$$

(d) When $Z_L \rightarrow \infty$, the reflection coefficient is given by

$$\Gamma = \frac{Z_L - R_0}{Z_L + R_0} = 1.$$

From Eq. (5.216), we get the VSWR given by

$$S = \frac{1 + |\Gamma|}{1 - |\Gamma|} \quad \rightarrow \quad \infty.$$

∎

From Example 5.20, we learn when the load impedance matches the characteristic resistance of a TX line, i.e., $Z_L = R_0$, no reflection occurs and $S = 1$. On the other hand, when the load impedance deviates from the characteristic resistance, for example, $Z_L = 0$ or $Z_L \rightarrow \infty$, the maximal reflection is achieved ($|\Gamma| = 1$) and thus $S \rightarrow \infty$.

Summary

5.1: We learn three major types of TX lines: parallel plate, two-wire, and coaxial TX lines. Besides, forward waves and backward waves are introduced.

5.2: We learn TX line equations which deal with voltage and current so that the principles and techniques of circuitry can be applied, and the analysis of a TX line can be greatly simplified.

5.3: We learn the characteristic impedance Z_0 of a TX line, which is the ratio of the voltage to the current of a forward wave. Besides, the negative of characteristic impedance, i.e., $-Z_0$, is the ratio of the voltage to the current of a backward wave.

5.4: We learn the reflection which occurs when the characteristic impedance of a TX line is not equal to the load impedance, i.e., $Z_L \neq Z_0$. The reflection coefficient is introduced to evaluate how much of an EM wave is reflected when impedance mismatch occurs.

5.5: We learn how to use a segment of TX line as a circuit component, for example, capacitor or inductor, in various high-frequency circuits.

5.6: We learn the standing wave which is closely related to the behaviors of EM waves in a TX line. Critical parameters of a transmitted EM wave can be derived through the associated standing wave.

Exercises

1. If we want to guide an EM wave to propagate for a long distance, which one of the following is a better choice?

    (a) Coaxial TX line
    (b) Two-wire TX line.

    Why?

2. Assume that the interface between a dielectric medium and a perfect conductor is the xy-plane, and the dielectric medium has $\epsilon_1 = 2 \, \epsilon_0$. When an EM wave is incident on the interface, it will be totally reflected and there is no field inside the perfect conductor. Let $\vec{E}$ be the E-field of the dielectric medium at the interface, then $\vec{E} = \vec{E}_i + \vec{E}_r$, where $\vec{E}_i$ is the incident E-field and $\vec{E}_r$ is the reflected

E-field. If $\vec{E}_i = 2\hat{x} - 3\hat{y} + 7\hat{z}$ and the surface charge density is ρ_s, please derive $\vec{E}_r$.

(Hint: Boundary conditions of a perfect conductor in Sect. 4.6.)

3. In Exercise 2, if $\vec{H}$ is the H-field of the dielectric medium at the interface, then $\vec{H} = \vec{H}_i + \vec{H}_r$, where $\vec{H}_i$ is the incident H-field and $\vec{H}_r$ is the reflected H-field. Assume that $\vec{H}_i = 4\hat{x} + 3\hat{y} - 9\hat{z}$, the surface current density is $\vec{J}_s$. Please derive $\vec{H}_r$.

4. Assume a parallel plate TX line has $\in = 4 \in_0$, $f = 100MHz$, and $\vec{E} = E_y(z) \cdot \hat{y}$.

    (a). If a forward wave has $E_y = 5 \cdot e^{j\frac{\pi}{3}}$ at $z = 0$, please derive its phasor at $z = 50m$.

    (b). If a forward wave has $E_y = 5 \cdot e^{j\frac{\pi}{3}}$ at $z = 10m$, please derive its phasor at $z = 50m$.

    (c). In (a), please derive real E-field at $z = 50m$ and $t = 20ns$ (Note: $1ns = 10^{-9}$ sec).

    (Hint: Example 5.1).

5. Assume a parallel plate TX line has $\in = 4 \in_0$, $f = 100MHz$, and $\vec{E} = E_y(z) \cdot \hat{y}$.

    (a) If a backward wave has $E_y = 10 \cdot e^{-j\frac{\pi}{6}}$ at $z = 0$, please derive its phasor at $z = 50m$.

    (b) If a backward wave has $E_y = 10 \cdot e^{-j\frac{\pi}{6}}$ at $z = 25m$, please derive its phasor at $z = 50m$.

    (c) In (b), please derive its real E-field at $z = 50m$ and $t = 5ns$.

6. A parallel plate TX line has length $L = 5m$, $\in = 4 \in_0$, and $V(z, t) = 8 \cdot \cos(\omega t - \beta z + \frac{\pi}{4})$. When $t = 0$, derive its voltages at $z = 0$ and $z = L$ for the following frequencies:

    (a) $f = 10$ KHz.
    (b) $f = 20$ MHz.
    (c) $f = 1$ GHz.

    (Hint: Example 5.3).

7. In Exercise 6, derive the voltage difference between $z = 0$ and $z = L$ when $t = 0$.

8. A parallel plate TX line has length $L = 10m$, $\in = 4 \in_0$, and $V(z, t) = 6 \cdot \cos(\omega t - \beta z + \frac{\pi}{3})$. Please roughly plot the voltage along the line at $t = 0$ when (a). $f = 1$ kHz, (b). $f = 10$ MHz, (c). $f = 100$ MHz.

9. A TX line has the following parameters at $f = 100$ MHz: $R = 0.2(\Omega/m)$, $L = 3(\mu H/m)$, $C = 10(pF/m)$, and $G = 2(\mu S/m)$. Please derive the attenuation constant α, phase constant β, and propagation constant γ.

    (Hint: Example 5.6).

10. A forward wave is propagating along a TX line, where $\alpha = 0.2(m^{-1})$ and $\beta = \frac{\pi}{3}(m^{-1})$.

(a) If the voltage phasor at $z = 0$ is $V_a = 7$, please derive the real voltage $V(z, t)$ at $z = 5m$.

(b) If the voltage phasor at $z = 0$ is $V_a = 7 \cdot e^{-j\frac{\pi}{3}}$, please derive the real voltage $V(z, t)$ at $z = 5m$.

(c) If the voltage phasor at $z = 3m$ is $V^+ = 2$, please derive real voltage $V(z, t)$ at $z = 10m$.

(d) If the voltage phasor at $z = 3m$ is $V^+ = 2 \cdot e^{-j\frac{\pi}{3}}$, please derive real voltage $V(z, t)$ at $z = 15m$.

(Hint: Example 5.4).

11. A backward wave is propagating along a TX line, where $\alpha = 0.2(m^{-1})$ and $\beta = \frac{\pi}{6}(m^{-1})$.

(a) If the voltage phasor at $z = 0$ is $V_b = 4$, please derive the real voltage $V(z, t)$ at $z = 8m$.

(b) If the voltage phasor at $z = 0$ is $V_b = 4 \cdot e^{-j\frac{\pi}{3}}$, please derive the real voltage $V(z, t)$ at $z = 8m$.

(c) If the voltage phasor at $z = 3m$ is $V^- = 2$, please derive the real voltage $V(z, t)$ at $z = 10m$.

(d) If the voltage phasor at $z = 3m$ is $V^- = 2 \cdot e^{-j\frac{\pi}{3}}$, please derive the real voltage $V(z, t)$ at $z = -5m$.

12. Assume a forward wave and a backward wave coexist in a TX line, where the voltage phasors at $z = 0$ are $V_a = V_b = 10$, $\alpha = 0.2(m^{-1})$, and $\beta = \frac{\pi}{3}(m^{-1})$. At $t = 0$ and $z = 5m$, please derive

(a) forward voltage.

(b) backward voltage.

(c) measured voltage.

13. Repeat Exercise 12 for $V_a = 5 \cdot e^{-j\frac{\pi}{4}}$ and $V_b = 3 \cdot e^{j\frac{\pi}{6}}$.

14. A TX line has the following parameters at $f = 100\,\text{MHz}$: $R = 0.1(\Omega/m)$, $L = 2(\mu H/m)$, $C = 100(pF/m)$, and $G = 2(\mu S/m)$. Please derive γ and Z_0.

(Hint: Example 5.7).

15. A TX line has the following parameters at $f = 100\,\text{MHz}$: $Z_0 = 50\Omega$, $\alpha = 0.02(m^{-1})$, and $\beta = \frac{\pi}{5}(m^{-1})$. Assume $\beta \approx \omega\sqrt{LC}$, please derive R, L, C, G of the line.

(Hint: Example 5.8).

16. Assume a TX line has the following parameters: $Z_0 = 50\Omega$, $\alpha = 0.01(m^{-1})$, and $\beta = \frac{\pi}{6}(m^{-1})$. If the voltage phasor of a forward wave at $z = 0$ is $V_a = 10 \cdot e^{j\frac{\pi}{3}}$, please derive the real voltage and current at $z = 18m$.

(Hint: Example 5.9).

17. In Exercise 16, if the current phasor at $z = 0$ is $I_a = 2 \cdot e^{-j\frac{2\pi}{3}}$, please derive the real voltage and current at $z = 24m$.

18. Assume a TX line has the following parameters: $Z_0 = 75\Omega$, $\alpha = 0.01(m^{-1})$, and $\beta = \frac{\pi}{3}(m^{-1})$. If the voltage phasor of a backward wave at $z = 0$ is $V_b = 9 \cdot e^{j\frac{\pi}{4}}$, please derive the real voltage and current at $z = 10m$.

19. In Exercise 18, if the current phasor at $z = 0$ is $I_b = 0.1 \cdot e^{-j\frac{2\pi}{3}}$, please derive the real voltage and current at $z = 24m$.

20. A TX line has the following parameters at $f = 200\,MHz$: $Z_0 = 50\Omega$, $\alpha = 0.02(m^{-1})$, and $\beta = \frac{\pi}{5}(m^{-1})$. If a forward wave and a backward wave coexist along the line, where forward voltage and backward voltage at $z = 0$ are $V^+(t) = 10\cos 2\pi f t$ and $V^-(t) = 8\cos 2\pi f t$. Please derive

   (a) the measured voltage and current at $z = 0$,
   (b) the measured voltage and current at $z = 12m$,
   (c) the measured voltage and current at $z = -20m$.

   (Hint: Example 5.10).

21. Explain physically and mathematically why reflection is necessary if $Z_L \neq Z_0$.

   (Hint: Sect. 5.4).

22. A lossless TX line has $Z_0 = 50\Omega$, $Z_L = 30\Omega$, length $L = 10m$ and propagation constant $\gamma = j\beta = j\frac{\pi}{3}$. Assume a forward wave propagates from $z = 0$ to $z = L$, where the forward voltage at $z = 0$ is $V_a = 4 \cdot e^{-j\frac{4\pi}{3}}$. Please derive the following phasors at $z = L$:

   (a) forward voltage and current.
   (b) backward voltage and current.
   (c) load voltage and current.

   (Hint: Example 5.11).

23. A TX line having $Z_0 = 50\Omega$ is terminated with a load $Z_L = 70\Omega$, and the measured voltage at the load is $V_L(t) = 6\cos \omega t$. Please derive

   (a) the reflection coefficient,
   (b) the forward voltage and the backward voltage at the load,
   (c) the forward current and the backward current at the load,
   (d) the load current.

24. Repeat Exercise 23 for $Z_L = 100 + j50(\Omega)$.

25. A lossless TX line has $Z_0 = 50\Omega$, length $L = 20m$, and propagation constant $\gamma = j\beta = j\frac{\pi}{3}$. Suppose a forward wave propagates from $z = 0$ to $z = L$, where the forward voltage at $z = 0$ is $V_a = 7 \cdot e^{j\frac{\pi}{4}}$. Please derive the forward voltage, backward voltage, and load voltage at $z = L$ for the following loads:

   (a) $Z_L = 0$.
   (b) $Z_L = 50\Omega$.
   (c) $Z_L \to \infty$.

(Hint: Example 5.13).

26. A lossless TX line has $R_0 = 75\Omega$ and $\beta = \frac{\pi}{3}(m^{-1})$. The input signal is at $z = 0$ and the load is at $z = 10m$. If $Z_L = 25\Omega$ and the measured load voltage is $V_L = 5$(phasor), please derive

    (a) the forward voltage V_+ and the backward voltage V_- at $z = 10m$,
    (b) the forward voltage and the backward voltage at $z = 0$,
    (c) the forward current and the backward current at $z = 0$,
    (d) the input impedance at $z = 0$.

(Hint: Example 5.14).

27. Repeat Exercise 26 for $Z_L = j75\Omega$.

28. A lossy TX line has $R_0 = 75\Omega$, $\alpha = 0.02(m^{-1})$, and $\beta = \frac{\pi}{4}(m^{-1})$. The input signal is at $z = 0$ and the load is at $z = 10m$. If the measured load voltage is $V_L = 5$ (phasor) and $Z_L = 50\Omega$. Please derive

    (a) the forward voltage V_+ and the backward voltage V_- at $z = 10m$,
    (b) the forward voltage and the backward voltage at $z = 3m$,
    (c) the forward current and the backward current at $z = 3m$,
    (d) the input impedance at $z = 3m$.

29. A lossless TX line has $R_0 = 50\Omega$ and $\beta = 0.2\pi(m^{-1})$.

    (a) Please design an open-circuited TX line circuit to get input impedance $Z_i = -j20(\Omega)$.
    (b) Design a short-circuited TX line circuit to get $Z_i = -j20(\Omega)$.
    (c) Design a $\lambda/4$ TX line circuit to get $Z_i = 10(\Omega)$.

(Hint: Refer to Sect. 5.5C).

30. A lossless transmission having $R_0 = 50\Omega$ and $\beta = \frac{4\pi}{3}(m^{-1})$ is terminated with a load $Z_L = 100\Omega$. If the measured load voltage phasor is $V_L = 4$. Please derive

    (a) the reflection coefficient,
    (b) the forward voltage V_+ and backward voltage V_- at the load,
    (c) the maximum voltage (V_{max}) and minimum voltage (V_{min}) of standing wave,
    (d) the locations where the second and the third maximum and minimum voltages occur,
    (e) VSWR.

(Hint: Examples 5.17 and 5.20).

31. Repeat Exercise 30 for $Z_L = 100 + j50$.

32. A lossless transmission having $R_0 = 50\Omega$ and $\beta = \frac{\pi}{2}(m^{-1})$ is terminated with a load $Z_L = 30\Omega$. If the forward voltage phasor at the load is $V_+ = 6$, please derive

   (a) the reflection coefficient,
   (b) the load voltage,
   (c) the maximum voltage (V_{max}) and minimum voltage (V_{min}) of the standing wave,
   (d) the locations where the first and the second maximum and minimum voltages occur,
   (e) VSWR.

33. Repeat Exercise 32 for $Z_L = 50 - j50(\Omega)$.
34. A TX line has $R_0 = 50\Omega$, the first minimum voltage of standing wave occurs at $d = 2m$, and the first maximum voltage occurs at $d = 5m$. If $V_{max} = 5$ and $V_{min} = 1$, please derive

   (a) wavelength λ,
   (b) reflection coefficient Γ,
   (c) load Z_L.

   (Hint: Example 5.19).
35. A TX line has $R_0 = 50\Omega$, the first maximum voltage of standing wave occurs at $d = 0.5m$, and the first minimum voltage occurs at $d = 2m$. If the VSWR is $S = 4$, please derive

   (a) wavelength λ,
   (b) reflection coefficient Γ,
   (c) load Z_L.

36. Suppose a TX line has $V_{min} = 2V$. Please derive V_{max} and $|\Gamma|$ in the following cases:

   (a) $S = 1$,
   (b) $S = 1.5$,
   (c) $S = 3$.

37. Suppose a TX line has $R_0 = 75\Omega$. Please derive VSWR in the following cases:

   (a) $Z_L = 75\Omega$,
   (b) $Z_L = 50 + j100\Omega$,
   (c) $Z_L = 40 - j60\Omega$.

Chapter 6
Smith Chart and Matching Circuit Design

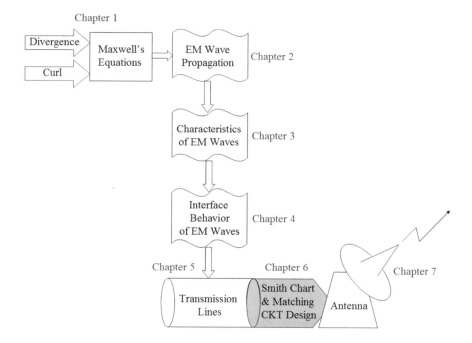

© Springer Nature Switzerland AG 2020

M.-S. Kao and C.-F. Chang, *Understanding Electromagnetic Waves*,

https://doi.org/10.1007/978-3-030-45708-2_6

Abstract Why is the design of a 10 GHz amplifier much more complicated than that of a 10 kHz amplifier? It is a question for many EE students. This chapter answers it as follows.

A. Motivation and build-up of Smith chart

Smith chart is the most fundamental and important tool in high frequency circuit design. We introduce the motivation of Smith chart in a heuristic way, and show how to build it step by step, so that readers can have an insightful understanding of Smith chart.

B. Applications of Smith chart

We provide abundant examples of Smith chart with details in various applications to help readers learn how to apply Smith chart in different cases.

C. Impedance matching design

Impedance matching is indispensable in high frequency circuit design, which is widely used to avoid reflection and achieve maximum power transfer. We introduce three useful impedance matching methods to help readers catch the core idea and techniques of impedance matching.

D. High frequency amplifier design

In order to help readers well understand high frequency amplifier design, we start from the design of low frequency amplifiers in kHz range, and then medium frequency amplifiers in MHz range, and finally get into the design of high frequency amplifiers in GHz range. Important concepts of high frequency amplifier design such as S-parameters, maximum power transfer theorem, and input/output matching circuits, are also introduced. The process will build a solid background for advanced study in high frequency circuit design.

Keywords Smith chart · Normalized load impedance · Reflection coefficient · r-circle · x-curve · Load impedance · Input impedance · Load admittance · Impedance matching circuit · Lumped element · Distributed element · Quarter wavelength matching circuit · S-parameters · Maximum power transfer theorem

6.1   Principles of Smith Chart

A successful painting can make a painter famous all over the world. Similarly, a useful chart can make an electrical engineer known globally in his professional field. This chart is invented by an American engineer Phillip H. Smith (1905–1987) and hence called **Smith chart**. The motivation of Smith chart is simply to reduce the computational complexity with the aid of a graph. But the applications of Smith chart are finally far beyond the original motivation and become an important tool when designing a high-frequency circuit.

This chapter introduces the principles and applications of Smith chart. The concept of impedance matching is also given with the emphasis on high-frequency circuit design. An example of impedance matching of a BJT amplifier between a signal source and a load is finally given. Readers who are interested in circuit design should invest more time and effort into this chapter.

A Motivation of Smith Chart

Suppose R_0 is the characteristic resistance of a TX line and Z_L is the load impedance. Then the reflection coefficient is given by

$$\Gamma = \frac{Z_L - R_0}{Z_L + R_0}. \tag{6.1}$$

In Eq. (6.1), Z_L is a complex number and can be expressed by

$$Z_L = R_L + jX_L, \tag{6.2}$$

where R_L and X_L are real part and imaginary part of Z_L, respectively. Inserting Eq. (6.2) into Eq. (6.1), we have

$$\Gamma = \frac{(R_L - R_0) + jX_L}{(R_L + R_0) + jX_L} = \frac{\left[(R_L - R_0) + jX_L\right] \cdot \left[(R_L + R_0) - jX_L\right]}{\left[(R_L + R_0) + jX_L\right] \cdot \left[(R_L + R_0) - jX_L\right]} = \Gamma_r + j\Gamma_i, \tag{6.3}$$

where Γ_r and Γ_i are real part and imaginary part of Γ, respectively. From Eq. (6.3), we have

$$\Gamma_r = \frac{R_L^2 + X_L^2 - R_0^2}{(R_L + R_0)^2 + X_L^2}, \tag{6.4}$$

$$\Gamma_i = \frac{2R_0 X_L}{(R_L + R_0)^2 + X_L^2}. \tag{6.5}$$

In 1930s, electronic computers have not been invented yet. Hence hand calculations of Eqs. (6.4) and (6.5) are heavy work and time-consuming. In addition, it is very easy to make an error in calculation. Therefore, an idea comes to Mr. Smith's mind:

Is it possible to have a graphic tool so that given a load impedance (Z_L), we can skip hand calculation and derive the corresponding reflection coefficient (Γ) simply with the aid of this graph?

This is a good idea since no electronic computers are available. A graphic tool may save a lot of time and avoid potential errors of hand calculations. The remaining problem is: how do we get this tool?

First, from Eq. (6.1), equivalently, we get

$$\Gamma = \frac{\frac{Z_L}{R_0} - 1}{\frac{Z_L}{R_0} + 1} = \frac{u - 1}{u + 1}, \tag{6.6}$$

where

$$u = \frac{Z_L}{R_0}. \tag{6.7}$$

In Eq. (6.7), u is the ratio of Z_L to R_0, called **normalized load impedance**.

From Eq. (6.6), Γ solely depends on u. Generally, u is a complex number and can be expressed by

$$u = r + jx, \tag{6.8}$$

where r is the real part and x is the imaginary part of u.

In Eq. (6.6), we can regard Γ as a function of u. A critical breakthrough adopted by Mr. Smith is to reversely regard u as a function of Γ. Readers may ask:

Why is it a critical breakthrough to regard u as a function of Γ?

The reason is simply because the absolute value of a reflection coefficient must be smaller or equal to 1. That is

$$|\Gamma| \leq 1. \tag{6.9}$$

It follows that

$$|\Gamma|^2 = \Gamma_r^2 + \Gamma_i^2 \leq 1 \Rightarrow -1 \leq \Gamma_r \leq 1, -1 \leq \Gamma_i \leq 1 \tag{6.10}$$

Hence in a two-dimensional complex plane, **both Γ_r and Γ_i fall within the unit circle whose radius is one**.

Because $u = r + jx$ and $\Gamma = \Gamma_r + j\Gamma_i$, when we regard u as a function of Γ, we can express (r, x) as a function of (Γ_r, Γ_i) in a complex plane. Because Γ_r and Γ_i are bounded in the unit circle, the chart is limited to the unit circle in complex plane. Otherwise, if we regard Γ as a function of u, because $0 \leq r < \infty$ and $-\infty < x < \infty$, the chart will cover half of the complex plane, which is difficult to realize. Hence reversely regarding u as a function of Γ is the first critical breakthrough to achieve Smith chart.

In the following, we try to express (r, x) as a function of (Γ_r, Γ_i). From Eq. (6.6), we get

$$u = \frac{1 + \Gamma}{1 - \Gamma}. \tag{6.11}$$

Because $\Gamma = \Gamma_r + j\Gamma_i$ and $u = r + jx$, we can rewrite Eq. (6.11) as

Fig. 6.1 Illustrating Γ-plane

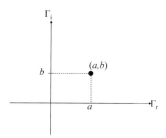

$$r + jx = \frac{(1 + \Gamma_r) + j\Gamma_i}{(1 - \Gamma_r) - j\Gamma_i}. \tag{6.12}$$

On the right side of Eq. (6.12), we multiply both numerator and denominator by $(1 - \Gamma_r) + j\Gamma_i$ and then we obtain

$$r + jx = \frac{(1 - \Gamma_r^2 - \Gamma_i^2) + j2\Gamma_i}{(1 - \Gamma_r)^2 + \Gamma_i^2}. \tag{6.13}$$

Because the real part and the imaginary part in both sides must be equal, we have

$$r = \frac{1 - \Gamma_r^2 - \Gamma_i^2}{(1 - \Gamma_r)^2 + \Gamma_i^2}, \tag{6.14}$$

$$x = \frac{2\Gamma_i}{(1 - \Gamma_r)^2 + \Gamma_i^2}. \tag{6.15}$$

In Eqs. (6.14) and (6.15), r and x are functions of Γ_r and Γ_i. By using these two equations, we can build up Smith chart.

B Γ-Plane

Before further introducing Smith chart, we shall introduce a complex plane called Γ-plane.

As discussed previously, a reflection coefficient Γ is a complex number and can be expressed by $\Gamma = \Gamma_r + j\Gamma_i$. Mathematically, we can geometrically represent Γ by a two-dimensional Cartesian coordinate system with the real part along the x-axis (Γ_r-axis) and the imaginary part along y-axis (Γ_i-axis). This complex plane is called the Γ-**plane**, and a point (a, b) in Γ-plane represents $\Gamma = a + jb$ as shown in Fig. 6.1. For example, the point $(0.2, 0.5)$ represents the reflection coefficient $\Gamma = 0.2 + j0.5$ and the point $(0.3, -0.7)$ represents the reflection coefficient $\Gamma = 0.3 - j0.7$.

On the other hand, a reflection coefficient Γ can be expressed by a polar coordinate system on Γ-plane and given by

Fig. 6.2 Representation in
polar coordinate system on
Γ-plane

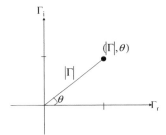

$$\Gamma = |\Gamma|e^{j\theta}, \tag{6.16}$$

where $|\Gamma|$ is the magnitude and θ is the phase. As shown in Fig. 6.2, $\Gamma = |\Gamma|e^{j\theta}$ corresponds to a point $(|\Gamma|, \theta)$ in a polar coordinate, where $|\Gamma|$ is the distance between the point and the origin O, and θ is the phase angle. For example, the point $(0.6, 30°)$ represents the reflection coefficient $\Gamma = 0.6 \cdot e^{j30°}$ and the point $(0.4, -130°)$ represents the reflection coefficient $\Gamma = 0.4 \cdot e^{-j130°}$. Note that the range of phase angle θ on Γ-plane is $-180° \leq \theta \leq 180°$. For the upper plane, it corresponds to $0° \leq \theta \leq 180°$, and $-180° \leq \theta \leq 0°$ for the lower plane.

In the following, we show the relationship between (r, x) and (Γ_r, Γ_i) on Γ-plane and based on this, we finally get a graphic tool to precisely describe their relationship—Smith chart.

C r-Circles

From Eq. (6.14), we can rewrite the relationship between r and (Γ_r, Γ_i) given by

$$r \cdot [(1 - \Gamma_r)^2 + \Gamma_i^2] = 1 - \Gamma_r^2 - \Gamma_i^2$$
$$\Rightarrow (1 + r)\Gamma_r^2 - 2r\Gamma_r + (1 + r)\Gamma_i^2 = 1 - r. \tag{6.17}$$

From Eq. (6.17), equivalently, we can get

$$\left(\Gamma_r - \frac{r}{1 + r}\right)^2 + \Gamma_i^2 = \left(\frac{1}{1 + r}\right)^2. \tag{6.18}$$

From Geometry, Eq. (6.18) forms a circle on Γ-plane. The circle has the center at $(\frac{r}{1+r}, 0)$ and the radius is given by

$$h = \frac{1}{1 + r}. \tag{6.19}$$

Fig. 6.3 Illustrating
r-circles on Γ-plane

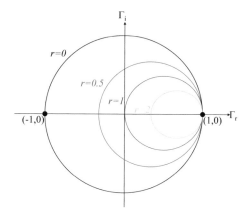

Therefore, a specific r yields a specific circle on Γ-plane. The center of the circle lies on Γ_r-axis and the radius h decreases when r increases ($0 \leq r < \infty$). Four example circles are illustrated in Fig. 6.3:

- When $r = 0$, the center is at the origin and the radius $h = 1$. Hence it corresponds to the unit circle.
- When $r = 0.5$, the center is at ($\frac{1}{3}$, 0) and the radius $h = \frac{2}{3}$.
- When $r = 1$, the center is at ($\frac{1}{2}$, 0) and the radius $h = \frac{1}{2}$.
- When $r = 2$, the center is at ($\frac{2}{3}$, 0) and the radius $h = \frac{1}{3}$.

Hence different values of r yield different circles. These circles are called r-circles. It is important to mention that all the r-circles must pass through the point $(\Gamma_r, \Gamma_i) = (1, 0)$. The reason is because the center of each circle is at ($\frac{r}{1+r}$, 0) and the corresponding radius is $h = \frac{1}{1+r}$. Therefore

$$\frac{r}{1+r} + h = 1. \tag{6.20}$$

Hence all the r-circles must pass through the point $(1, 0)$. In fact, the point $(\Gamma_r, \Gamma_i) = (1, 0)$ is an important point of Smith chart which will become clear later. Notice that when r keeps growing up and approaches infinity, i.e., $r \to \infty$, the center approaches $(1, 0)$ and the radius approaches zero. It is shown in Fig. 6.4.

D x-Curves

Next, we explore the relationship between x and (Γ_r, Γ_i). Notice that the value of x might be positive or negative because $-\infty < x < \infty$. From Eq. (6.15), we can rewrite the relationship between x and (Γ_r, Γ_i) given by

$$x \cdot [(1 - \Gamma_r)^2 + \Gamma_i^2] = 2\Gamma_i$$

Fig. 6.4 r-circles on
Γ-plane

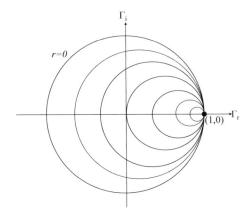

$$\Rightarrow (1 - \Gamma_r)^2 + \Gamma_i^2 - \frac{2}{x} \cdot \Gamma_i = 0. \tag{6.21}$$

From Eq. (6.21), equivalently, we get

$$(\Gamma_r - 1)^2 + \left(\Gamma_i - \frac{1}{x}\right)^2 = \left(\frac{1}{x}\right)^2. \tag{6.22}$$

Similar to Eq. (6.18), Eq. (6.22) forms a circle on Γ-plane. The circle has the center at $(1, \frac{1}{x})$ and the radius h is given by

$$h = \frac{1}{|x|}. \tag{6.23}$$

Obviously, for a specific x, the center of the circle lies on the line of $\Gamma_r = 1$ and the radius decreases when $|x|$ increases. Besides, because the center is at $(1, \frac{1}{x})$ and the corresponding radius is $\frac{1}{|x|}$, the circle must pass through the point **(1, 0)**.

In Fig. 6.5, we have examples of $x = \pm 0.5$, $x = \pm 1$, and $x = \pm 2$.

- When $x = 0.5$, the center of the corresponding circle is at $(1, 2)$ and the radius is $h = 2$.
- When $x = -0.5$, the center of the corresponding circle is at $(1, -2)$ and the radius is $h = 2$.
- When $x = 1$, the center of the corresponding circle is at $(1, 1)$ and the radius is $h = 1$.
- When $x = -1$, the center of the corresponding circle is at $(1, -1)$ and the radius is $h = 1$.
- When $x = 2$, the center of the corresponding circle is at $(1, 1/2)$ and the radius is $h = 1/2$.
- When $x = -2$, the center of the corresponding circle is at $(1, -1/2)$ and the radius is $h = 1/2$.

Fig. 6.5 Illustrating circles corresponding to x on Γ-plane

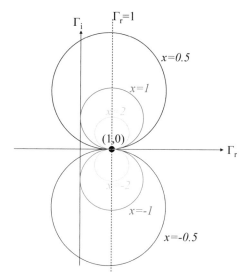

Fig. 6.6 Circles corresponding to x on Γ-plane

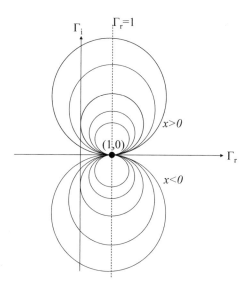

All the examples follow the rules we just discussed. A more general case is plotted in Fig. 6.6. **When $x > 0$, the corresponding circles are in the upper plane; when $x < 0$, the corresponding circles are in the lower plane**. A radius increases when $|x|$ decreases as shown in Eq. (6.23). When x approaches zero, i.e., $x \to 0$, the corresponding radius approaches infinity, i.e., $h \to \infty$.

Fig. 6.7 *x*-curves within
unit circle

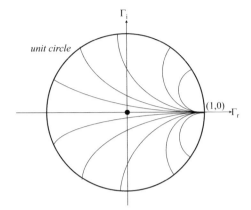

Fig. 6.8 Combination of
r-circles and *x*-curves within
unit circle

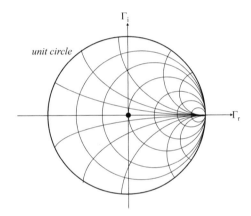

   Finally, because a reflection coefficient must satisfy the condition of $|\Gamma| \leq 1$, we
only consider those points falling within the unit circle. Hence we truncate those
x-circles in Fig. 6.6 with the unit circle and get the results as shown in Fig. 6.7. Note
that the remaining part of a truncated circle actually forms a curve in Fig. 6.7 and we
call it an *x*-**curve**. For a specific *x*, we have a corresponding *x*-curve. **When $x > 0$,
the curve is on the upper plane; when $x < 0$, the curve is on the lower plane.
When $x = 0$, the curve becomes Γ_r-axis.** Hence we have many *x*-curves in Fig. 6.7
showing the relationship between *x* and (Γ_r, Γ_i) within the unit circle.

E   Smith Chart

As discussed in the previous subsections, Fig. 6.4 reveals the relationship between
r and reflection coefficient (Γ_r, Γ_i) and Fig. 6.7 reveals the relationship between *x*
and (Γ_r, Γ_i). If we combine these two figures together, we get Fig. 6.8, which shows

the fundamental structure of Smith chart. In Fig. 6.8, for a specific value of r, we have a corresponding r-circle, and for a specific value of x, we have a corresponding x-curve. A Smith chart is a graph of r-circles and x-curves in Γ-plane. A typical Smith chart is simply a refined version of Fig. 6.8, which is shown in Fig. 6.9. The resolution of r-circles and x-curves is improved as shown in Fig. 6.9.

In Fig. 6.9, the center of Smith chart is the origin, the horizontal axis is Γ_r-axis and the vertical axis is Γ_i-axis. Please notice that a typical Smith chart does not symbolize the origin nor the Γ_r-axis and the Γ_i-axis. **A tip for readers to learn Smith chart effectively is to establish the origin, the Γ_r-axis and the Γ_i-axis in your own mind.** Therefore, you can easily link $\Gamma = (\Gamma_r, \Gamma_i)$ with r-circle as well as x-curve given in Eqs. (6.18) and (6.22) through Smith chart.

Besides, a smart way to get familiar with Smith chart is to plot it step by step. First, we can start from r-circles. For a specific value of r, we can draw a corresponding circle. For example, in Fig. 6.9, we draw the unit circle when $r = 0$. Then we draw another circle with radius $h = 1/1.1$ when $r = 0.1$ and so on. For a typical Smith

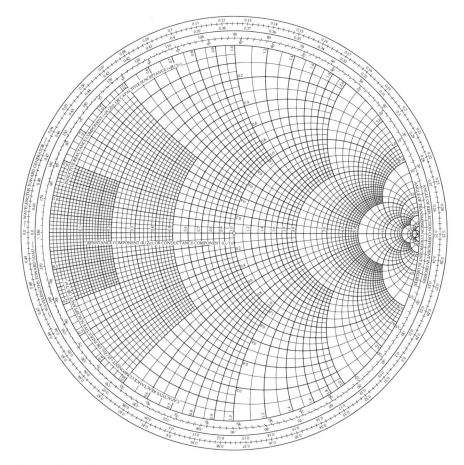

Fig. 6.9 Smith chart

chart, we need to plot circles of $r = 0, 0.1, 0.2, 0.3 \ldots$ and finally as $r \to \infty$, we have the point $(\Gamma_r, \Gamma_i) = (1, 0)$.

Next, we plot x-curves. For a specific value of x, we can draw a corresponding circle, and then we truncate it as an x-curve within the unit circle. When $x > 0$, we draw each x-curve in the upper Γ-plane. When $x < 0$, we draw each x-curve in the lower Γ-plane. Notice that a typical **Smith chart does not include a negative sign "−" for x-curves in the lower plane**. It is regarded as background knowledge in using Smith chart.

Finally, if you can patiently draw each r-circle as well as x-curve step by step, you probably will find Smith chart is simple and start to enjoy it!

When using Smith chart, we shall keep in mind that both rectangular (Cartesian) coordinate and polar coordinate can be applied in Γ-plane according to the best use. For example, a point (Γ_r, Γ_i) in Cartesian coordinate corresponds to the reflection coefficient $\Gamma = \Gamma_r + j\Gamma_i$. Hence the point $(0.3, -0.5)$ in Smith chart corresponds to $\Gamma = 0.3 - j0.5$. In addition, a point $(|\Gamma|, \theta)$ in polar coordinate system corresponds to the reflection coefficient $\Gamma = |\Gamma|e^{j\theta}$, where $|\Gamma|$ is the distance between the point and the origin, and θ is the azimuth angle from Γ_r-axis. In order to help users find θ efficiently, it is labeled with the degrees around the unit circle as shown in Fig. 6.9. The range of θ is $-180° \leq \theta \leq 180°$. For example, a point $(\Gamma_r, \Gamma_i) = (1, 0)$ can be represented as $\Gamma = |\Gamma|e^{j\theta} = 1^{j0°}$ in the polar coordinate and the phase angle is $\theta = 0°$. Similarly, the point $(0.2, 0.2)$ can be represented as $\Gamma = \sqrt{0.08}e^{j45°}$ and the phase angle is $\theta = 45°$. Note that beside the scale of phase angle, there are other scales around the circumference of the unit circle. We will explain their meanings later.

Finally, as we observe Fig. 6.9 carefully, we will find that each r-circle and each x-curve have only one intersection point. That means, for a normalized load impedance $u = r + jx$, we have only one corresponding reflection coefficient in Smith chart. For example, let A be the intersection point of $r = r_1$-circle and $x = x_1$-curve. Then the associated normalized load impedance of the point A is $u = r_1 + jx_1$. If A is represented in Cartesian coordinate by (a, b), then the corresponding reflection coefficient is $\Gamma = a + jb$. Therefore, $u = r_1 + jx_1$ corresponds to $\Gamma = a + jb$. Namely, given $u = r_1 + jx_1$, we can utilize Smith chart to get the corresponding reflection coefficient $\Gamma = a + jb$ immediately! It is exactly the original idea of Mr. Smith.

Example 6.1

Suppose we have reflection coefficients

(a) $\Gamma = 0.4 + j0.8$,
(b) $\Gamma = -0.3 + j0.5$.

   Please identify the corresponding points P and Q, respectively, in Smith chart.

Solution

(a) When we have $\Gamma = 0.4 + j0.8$, the corresponding point is expressed with Cartesian coordinate in Γ-plane and $(\Gamma_r, \Gamma_i) = (0.4, 0.8)$. Suppose in Fig. 6.10, the radius of the unit circle has the length of l. We start from the origin and horizontally shift right a distance of $0.4\ l$. Then we move vertically up $0.8\ l$ and reach the corresponding point P.

(b) When $\Gamma = -0.3 + j0.5$, the corresponding point in Γ-plane is $(\Gamma_r, \Gamma_i) = (-0.3, 0.5)$. Suppose in Fig. 6.10, the radius of the unit circle has the length of l. We start from the origin and horizontally shift left a distance of $0.3\ l$. Then we move vertically up $0.5\ l$ and reach the corresponding point Q.

■

Example 6.2

Suppose we have reflection coefficients

(a) $\Gamma = 0.5 \cdot e^{j20°}$,
(b) $\Gamma = 0.35 \cdot e^{-j110°}$.

   Please identify the corresponding points P and Q, respectively, in Smith chart.

Solution

(a) When we have $\Gamma = 0.5 \cdot e^{j20°}$, the corresponding point is expressed with polar coordinate in Γ-plane, and the associated magnitude is $|\Gamma| = 0.5$ and the phase angle is 20°. Suppose in Fig. 6.11, the radius of the unit circle has the length of l. First, we find the point labeled $\theta = 20°$ on the circumference of the unit circle in upper Γ-plane. Then we draw a line between this point and the origin. On this line, the desired point P is $0.5\ l$ away from the origin.

(b) When we have $\Gamma = 0.35 \cdot e^{-j110°}$, the associated magnitude is $|\Gamma| = 0.35$ and the phase angle is $-110°$ in Γ-plane. Suppose in Fig. 6.11, the radius of the unit circle has the length of l. First, we find the point labeled $\theta = -110°$ on the circumference of the unit circle in lower Γ-plane (negative sign of the phase angle indicates the lower Γ-plane). Then we draw a line between this point and the origin. On this line, the desired point Q is $0.35\ l$ away from the origin.

■

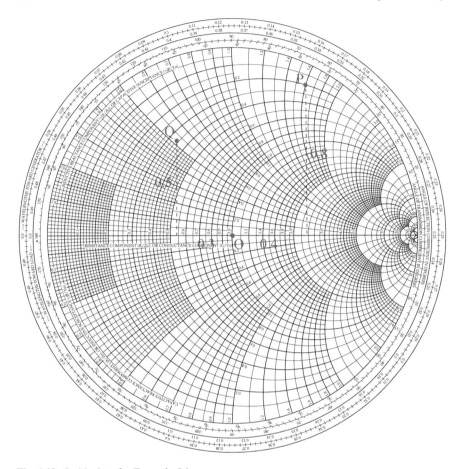

Fig. 6.10 Smith chart for Example 6.1

Example 6.3

Suppose we have the normalized load impedance

(a) $u = 0.6 + j2$,
(b) $u = 1 - j1.7$.

Please identify the corresponding points P and Q, respectively, in Smith chart.

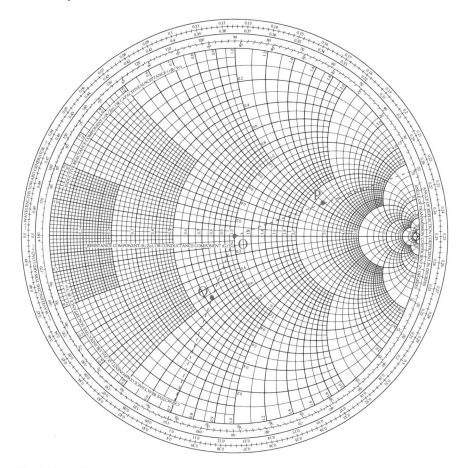

Fig. 6.11 Smith chart for Example 6.2

Solution

(a) When we have the normalized load impedance $u = 0.6 + j2$, we find the $r = 0.6$ circle and the $x = 2$ curve in upper plane of Smith chart as shown in Fig. 6.12. The intersection of $r = 0.6$ circle and $x = 2$ curve is the desired point P.

(b) When we have the normalized load impedance $u = 1 - j1.7$, we find the $r = 1$ circle and the $x = -1.7$ curve in lower plane of Smith chart as shown in Fig. 6.12. (Note: Smith chart does not explicitly label the negative sign "−" of x.) The intersection of $r = 1$ circle and $x = -1.7$ curve is the desired point Q.

■

In Example 6.3, once the desired point P or Q is identified in Smith chart, we can immediately get the corresponding reflection coefficient geometrically with simple calculations. It will become clear in the next section.

6.2   Applications of Smith Chart

In the previous section, we introduce Smith chart from the mathematical formula and illustrate how to build Smith chart step by step. In this section, we simply focus on the applications of Smith chart —how we use it. In principle, Smith chart is a graphic

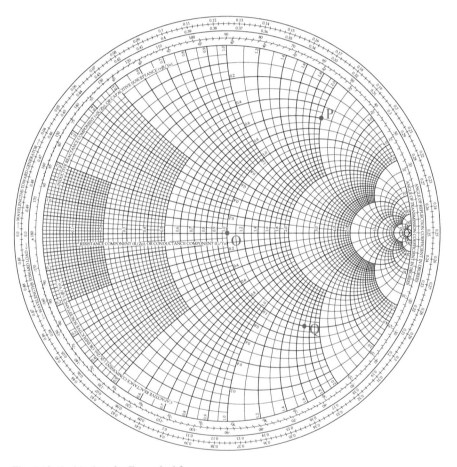

Fig. 6.12 Smith chart for Example 6.3

tool describing the relationship between a normalized load impedance $u = r + jx$ and the corresponding reflection coefficient $\Gamma = \Gamma_r + j\Gamma_i$. It is an important tool when designing high-frequency circuits and can be used to resolve lots of practical problems. In the following, we first overview the complex operation in Γ-plane and then give a number of popular applications of Smith chart.

A Complex Operation in Γ-Plane

Smith chart is established in Γ-plane with the horizontal Γ_r-axis and the vertical Γ_i-axis. In Γ-plane, each point represents a complex reflection coefficient Γ given by

$$\Gamma = \Gamma_r + j\Gamma_i. \tag{6.24}$$

Hence we can assign two numbers in Cartesian coordinate system—(Γ_r, Γ_i) to represent a reflection coefficient Γ. For example, the point $(0.2, 0.4)$ represents $\Gamma = 0.2 + j0.4$, and the point $(-0.3, 0.7)$ represents $\Gamma = -0.3 + j0.7$.

On the other hand, Γ can be equivalently expressed as

$$\Gamma = |\Gamma|e^{j\theta}. \tag{6.25}$$

In this case we apply polar coordinate and assign two numbers—$(|\Gamma|, \theta)$ to represent the reflection coefficient Γ. For example, the point $(0.4, 70°)$ represents $\Gamma = 0.4 \cdot e^{j70°}$, and the point $(0.5, -20°)$ represents $\Gamma = 0.5 \cdot e^{-j20°}$.

In Fig. 6.13, suppose $\Gamma = \Gamma_r + j\Gamma_i$. Then we have

$$|\Gamma| = \sqrt{\Gamma_r^2 + \Gamma_i^2}, \tag{6.26}$$

$$\theta = \tan^{-1}\frac{\Gamma_i}{\Gamma_r}, \tag{6.27}$$

Fig. 6.13 Representation in polar coordinate system on Γ-plane

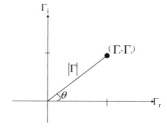

where $-180° \leq \theta \leq 180°$. Equations (6.26) and (6.27) show how we convert a point from Cartesian coordinate to polar coordinate. On the other hand, given $\Gamma = |\Gamma|e^{j\theta}$ and then we have

$$\Gamma_r = |\Gamma| \cdot \cos\theta, \tag{6.28}$$

$$\Gamma_i = |\Gamma| \cdot \sin\theta. \tag{6.29}$$

Equations (6.28) and (6.29) show how we convert a point from polar coordinate to Cartesian coordinate. Furthermore, when we have another reflection coefficient given by

$$\Gamma' = \Gamma \cdot e^{-j\phi}, \tag{6.30}$$

we then have

$$\Gamma' = |\Gamma|e^{j\theta} \cdot e^{-j\phi} = |\Gamma| \cdot e^{j(\theta-\phi)}. \tag{6.31}$$

Because Γ' can be represented by $|\Gamma'| \cdot e^{j\theta'}$, comparing it with Eq. (6.31), we have

$$\left|\Gamma'\right| = |\Gamma|, \tag{6.32}$$

$$\theta' = \theta - \phi. \tag{6.33}$$

Fig. 6.14 Phase rotation on Γ-plane

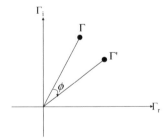

The physical meaning of Eqs. (6.31)–(6.33) is shown in Fig. 6.14. In Fig. 6.14, suppose $\Gamma = |\Gamma| \cdot e^{j\theta}$ is represented by a point in Γ-plane. Using the same magnitude $|\Gamma|$ and clockwise rotating a phase angle ϕ, we get the point which represents $\Gamma' = \Gamma \cdot e^{-j\phi}$.

The above operations in Γ-plane are actually well known in complex domain. They are fundamental when using Smith chart in various applications.

B Reflection Coefficient

According to the original idea of Mr. Smith, given a load impedance Z_L, we would like to get the associated reflection coefficient geometrically without complicated calculation. This is achieved by Smith chart and the procedure is provided in the following:

Step 1: We derive the normalized load impedance $u = \frac{Z_L}{R_0} = r_1 + jx_1$.

Step 2: As shown in Fig. 6.15, we find the r_1-circle and the x_1-curve in Smith chart. They have an intersection and we designate it as point P.

Step 3: Suppose the corresponding reflection coefficient of point P is $\Gamma = |\Gamma|e^{j\theta}$. We draw a line from the origin O through point P to the circumference of the unit circle. Then we read the scale of phase angle and get θ.

Step 4: Let l be the radius of the unit circle, i.e., $l = 1$ (unit length). By comparing l with the length of line $\overline{OP}$, we get the magnitude $|\Gamma|$. For example, if the length of l is 6 cm, then 6 cm is equivalent to 1 unit length. Suppose the length of line $\overline{OP}$ is 3 cm. Then $|\Gamma| = 3\,\text{cm}/6\,\text{cm} = 0.5$.

Step 5: After deriving θ and $|\Gamma|$, we can immediately get the corresponding reflection coefficient $\Gamma = |\Gamma|e^{j\theta}$.

Example 6.4
Suppose a TX line has the characteristic resistance $R_0 = 50\,\Omega$ and the load impedance is $Z_L = 30 + j60\,(\Omega)$. Please derive the reflection coefficient Γ.

Solution

Step 1: First, we derive the normalized load impedance by $u = \frac{Z_L}{R_0} = 0.6 + j1.2$.
Step 2: As shown in Fig. 6.16, we find the intersection of $r = 0.6$-circle and $x = 1.2$-curve. We designate it as point P.

Step 3: We draw a line from the origin O through point P to the circumference of the unit circle. Then we read phase angle and get $\theta = 71.7°$.

Step 4: Suppose the measured radius of the unit circle is l. Since the length of line $\overline{OP}$ is 0.63 l, we get the magnitude of the reflection coefficient given by $|\Gamma| = 0.63$.

Step 5: The reflection coefficient is given by $\Gamma = 0.63 \cdot e^{j71.7°}$.

∎

Example 6.5

Suppose a TX line has the characteristic resistance $R_0 = 50\ \Omega$ and the load impedance is $Z_L = 20 - j100\ (\Omega)$. Please derive the reflection coefficient Γ.

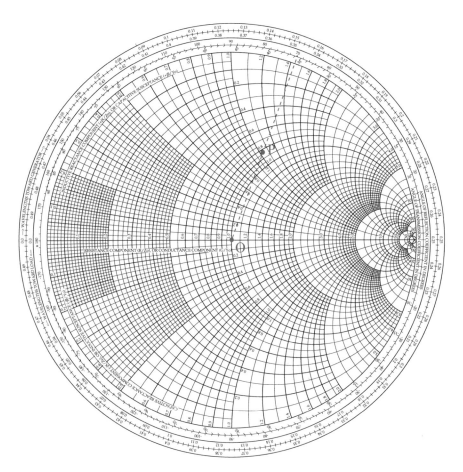

Fig. 6.15 Illustrating Smith chart calculation of reflection coefficient

Solution

Step 1: First we derive the normalized load impedance by $u = \frac{Z_L}{R_0} = 0.4 - j2$.

Step 2: As shown in Fig. 6.16, we find the intersection of $r = 0.4$-circle and $x = -2$-curve (in lower Γ-plane). We designate it as point Q.

Step 3: We draw a line from the origin O through point Q to the circumference of the unit circle. Then we read the phase angle and get $\theta = -51.7°$.

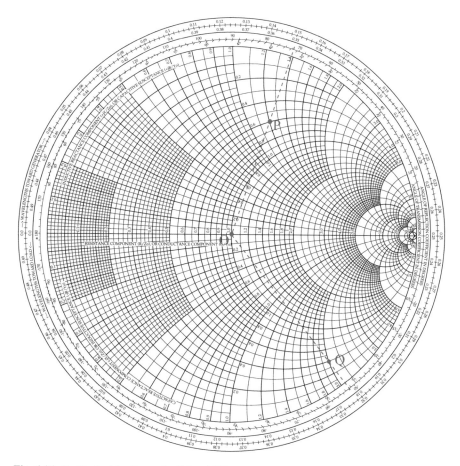

Fig. 6.16 Smith chart for Examples 6.4 and 6.5

Step 4: Suppose the measured radius of the unit circle is l. Since the length of line $\overline{OQ}$ is $0.85\,l$, we get the magnitude of the reflection coefficient given by $|\Gamma| = 0.85$.

Step 5: The reflection coefficient is given by $\Gamma = 0.85 \cdot e^{-j51.7°}$.

■

From the above two examples, we can use Smith chart as a graphic tool to derive the reflection coefficient Γ from a given Z_L. Comparing with hand calculation, it saves a lot of time and avoids potential computational errors.

C   Load Impedance

From the previous section, using Smith chart, we can derive the reflection coefficient Γ from a given load impedance Z_L. On the other hand, we can derive the load impedance Z_L from a given reflection coefficient. The procedure is provided in the following:

Step 1: Given a reflection coefficient Γ, we can find the associated point P in Γ-plane of Smith chart.

Step 2: From the r-circle and the x-curve closest to the point P, we get the best estimate of r and x so that their intersection is point P.

Step 3: From the normalized load impedance $u = r + jx$, we immediately obtain the corresponding load impedance given by $Z_L = R_0 \cdot u = R_0 \cdot (r + jx)$.

Example 6.6
Suppose a TX line has the characteristic resistance $R_0 = 50\,\Omega$ and reflection coefficient $\Gamma = 0.7e^{j128°}$. Please derive the corresponding load impedance Z_L.

Solution

Step 1: Suppose the radius of the unit circle in Fig. 6.17 is l. From the given reflection coefficient $\Gamma = 0.7e^{j128°}$, the length of the magnitude is given by $|\Gamma| = 0.7\,l$. Next, from the scale of phase angle along the circumference, we find the point of $\theta = 128°$. We draw a line between this point and the origin. Then the point which has a distance of $0.7\,l$ away from the origin along this line is point P.

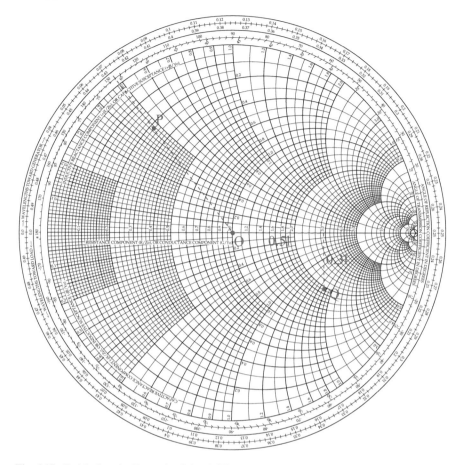

Fig. 6.17 Smith chart for Examples 6.6 and 6.7

Step 2: In Smith chart, we can find the best estimate of r (circle) and x (curve) so that the intersection is point P. Hence we get the normalized load impedance $r + jx$, where $r = 0.22$ and $x = 0.47$.

Step 3: From the normalized load impedance, we immediately obtain the corresponding load impedance given by $Z_L = R_0 \cdot (r + jx) = 11 + j23.5$ (Ω).

∎

Example 6.7
Suppose a TX line has the characteristic resistance $R_0 = 50\,\Omega$ and reflection coefficient $\Gamma = 0.5 - j0.3$. Please derive the corresponding load impedance Z_L.

Solution

Step 1: Suppose the radius of the unit circle in Fig. 6.17 is l. The reflection coefficient is represented by Cartesian coordinate and given by $(\Gamma_r, \Gamma_i) = (0.5, -0.3)$. Hence from the origin, we horizontally shift right $0.5\,l$ and vertically shift down $0.3\,l$. Then we reach the desired point Q.

Step 2: In Smith chart, we can find the best estimate of r (circle) and x (curve) so that the intersection is point Q. Hence we get the normalized load impedance $r + jx$, where $r = 1.95$ and $x = -1.75$. Note that negative sign is not displayed in Smith chart. We need to put a negative sign for x in the lower Γ-plane.

Step 3: From the normalized load impedance, we immediately obtain the corresponding load impedance given by $Z_L = R_0 \cdot (r + jx) = 97.5 - j87.5$ (Ω).

∎

D   Input Impedance of a Segment of Lossless TX Line

In this subsection, we introduce another application of Smith chart. Suppose we have a lossless TX line having the propagation constant $\gamma = j\beta$, where $\beta = 2\pi/\lambda$. Now, we want to derive the input impedance Z_i looking toward the load at a distance d as shown in Fig. 6.18. In other words, we want to derive the **input impedance of a segment of TX line**.

First, the relationship between the reflection coefficient Γ and the normalized load impedance u is given by

$$u = \frac{1 + \Gamma}{1 - \Gamma}. \tag{6.34}$$

From the result in Sect. 5.5, the input impedance of the TX line is given by

$$Z_i = R_0 \cdot \frac{e^{j\beta d} + \Gamma e^{-j\beta d}}{e^{j\beta d} - \Gamma e^{-j\beta d}} = R_0 \cdot \frac{1 + \Gamma e^{-j2\beta d}}{1 - \Gamma e^{-j2\beta d}}. \tag{6.35}$$

Fig. 6.18 Input impedance of a segment of TX line

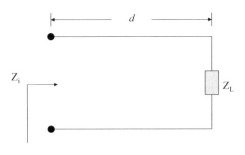

We define the normalized input impedance as

$$u_i = \frac{Z_i}{R_0}. \tag{6.36}$$

Let

$$\Gamma' = \Gamma \cdot e^{-j2\beta d}, \tag{6.37}$$

Then Eq. (6.35) can be rewritten as

$$u_i = \frac{1 + \Gamma'}{1 - \Gamma'}. \tag{6.38}$$

Obviously, Eq. (6.38) has an identical form as Eq. (6.34). Hence we can get u_i from a given Γ' using Smith chart, as we get u from a given Γ. First, we describe how we derive Γ' from Γ in Smith chart. Suppose

$$\Gamma = |\Gamma| \cdot e^{j\theta}, \tag{6.39}$$

and $\phi = 2\beta d$.

From Eqs. (6.37) and (6.39), Γ' is given by

$$\Gamma' = \Gamma \cdot e^{-j\phi} = |\Gamma| \cdot e^{j(\theta - \phi)}. \tag{6.40}$$

From Eq. (6.40), we have $|\Gamma'| = |\Gamma|$. It means Γ' and Γ have an identical magnitude, but different phase angles. According to Eq. (6.40), starting from the phase angle θ, we can clockwise rotate an angle ϕ and then attain the phase angle of Γ', i.e., $\theta' = \theta - \phi$.

Since the propagation constant is given by $\beta = 2\pi/\lambda$, when $d = 0.01\lambda$, the phase angle is given by

$$\phi = 2\beta d = 2 \cdot \frac{2\pi}{\lambda} \cdot (0.01\lambda) = 0.04\pi = 7.2°. \quad (\pi = 180°) \tag{6.41}$$

Similarly, when $d = 0.02\lambda$, the phase angle is $\phi = 14.4°$. When $d = 0.25\lambda$, the phase angle $\phi = 180°$. When $d = 0.5\lambda$, the phase angle $\phi = 360°$, which is exactly a complete cycle.

In Smith chart, in order to help users get Γ' from Γ, the outmost scale of the unit circle designates the relationship between d and $\phi = 2\beta d$. The unit is a wavelength. As illustrated in Fig. 6.19, we can see the words "wavelength toward generator" and the numbers 0.04, 0.05, ..., 0.48, 0.49. It shows the relationship between d and ϕ for every 0.01 wavelength. As shown in Fig. 6.19, we start from $d = 0$ at $(\Gamma_r, \Gamma_i) = (-1, 0)$, i.e., the leftmost point of the Γ_r-axis, and rotate clockwise. We reach $d = 0.125\lambda$ at $(\Gamma_r, \Gamma_i) = (0, 1)$. We continue to rotate clockwise to $d = 0.25\lambda$ at

$(\Gamma_r, \Gamma_i) = (1, 0)$. Similarly, we go on to $d = 0.375\lambda$ at $(0, -1)$ and finally we come back to $(-1, 0)$ which represents $d = 0.5\lambda$.

In the following examples, we will learn how useful the outmost wavelength scale is so that we can easily derive Γ' from Γ in Smith chart.

Example 6.8

Suppose a TX line has the reflection coefficient $\Gamma = 0.4 \cdot e^{j49.5°}$. If the length of this TX line is $d = 0.23\lambda$. Please derive the reflection coefficient Γ' of the segment of TX line.

Solution

Step 1: Similar to what we did in Example 6.6, we can find the point P representing Γ in Smith chart as shown in Fig. 6.19.

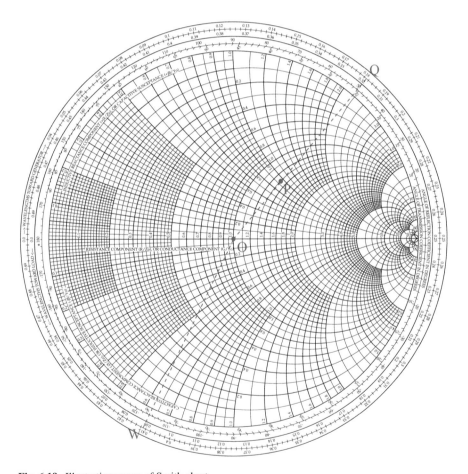

Fig. 6.19 Illustrating usage of Smith chart

Step 2: We draw a line between the origin O and point P, and extend the line $\overline{OP}$ to the outmost scale of the unit circle. The intersection is point Q and we can read the scale as 0.181 (λ).

Step 3: Because the length of the TX line is $d = 0.23\lambda$ and $0.181 + 0.23 = 0.411(\lambda)$, we go clockwise to the scale of $0.411(\lambda)$ and reach the point W at the outmost scale of the unit circle.

Step 4: We connect the origin O and the point W, and then we can read the corresponding phase angle given by $\theta' = -116°$. According to our discussions, θ' is the phase angle of Γ' and hence the reflection coefficient is given by $\Gamma' = |\Gamma| \cdot e^{-j116°} = 0.4 \cdot e^{-j116°}$.

∎

Example 6.9

Suppose a TX line has the reflection coefficient $\Gamma = 0.3 \cdot e^{-j129.5°}$. If the length of this TX line is $d = 0.2\lambda$. Please derive the reflection coefficient Γ' of the section of TX line.

Solution

Step 1: Similar to what we did in Example 6.6, we can find the point P representing Γ in Smith chart as shown in Fig. 6.20.

Step 2: We draw a line between the origin O and the point P, and extend the line $\overline{OP}$ to the outmost scale of the unit circle. The intersection is point Q and we can read the scale as $0.43(\lambda)$.

Step 3: Because the length of the TX line is $d = 0.2\lambda$, we calculate $0.43 + 0.2 = 0.63(\lambda) = 0.13 (\lambda)$ (Note that **every half wavelength, 0.5λ, is reset**). Hence we go clockwise to the scale of $0.13(\lambda)$ and reach the point W.

Step 4: We connect the origin O and point W, and then we can read the corresponding phase angle given by $\theta' = 86°$. According to the previous discussions, θ' is the phase angle of Γ' and hence the reflection coefficient is given by $\Gamma' = |\Gamma| \cdot e^{j86°} = 0.3 \cdot e^{j86°}$.

∎

From Examples 6.8 and 6.9, given Γ, we simply do some "rotation" in Smith chart and get the reflection coefficient Γ' of a segment of TX line. Similarly, given a load impedance Z_L, we can derive the input impedance Z_i of a segment of TX line having a specific length d using Smith chart.

Step 1: Given a load impedance Z_L, we can derive normalized impedance by $u = Z_L/R_0 = r + jx$.

Step 2: Similar to what we did in Example 6.4, from u, we can find the corresponding point P in Smith chart. Then we get the associated reflection coefficient Γ.

Step 3: Similar to what we did in Example 6.8, we can derive Γ' from Γ.

Step 4: Once Γ' is obtained, we can derive the corresponding normalized input impedance $u_i = r' + jx'$. It is similar to what we did in Example 6.6.

Step 5: Finally, the input impedance is given by $Z_i = R_0 \cdot u_i = R_0 \cdot (r' + jx')$.

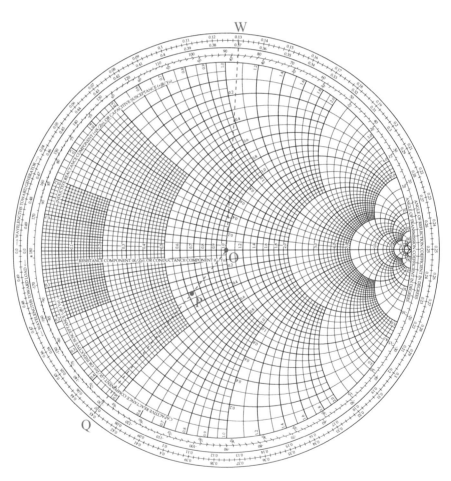

Fig. 6.20 Smith chart for Example 6.9

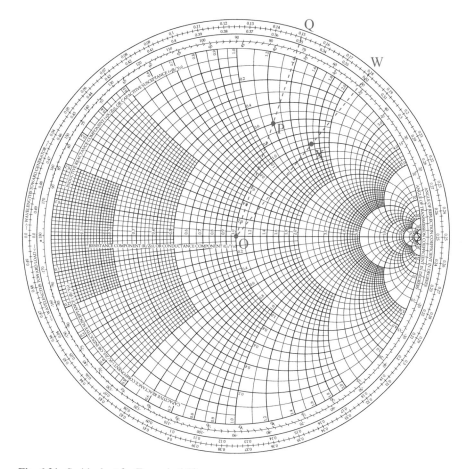

Fig. 6.21 Smith chart for Example 6.10

Example 6.10

In Fig. 6.18, suppose a TX line has the characteristic resistance $R_0 = 50\,\Omega$ and the load impedance $Z_L = 30 + j60\,(\Omega)$. Please derive the input impedance Z_i of a segment of TX line having $d = 0.03\lambda$.

Solution

Step 1: First, we derive the normalized load impedance given by $u = \frac{Z_L}{R_0} = 0.6 + j1.2$.

Step 2: In Fig. 6.21, we find the intersection of $r = 0.6$-circle and $x = 1.2$-curve and designate it as point P.

Step 3: We draw a line between the origin O and point P, and extend the line $\overline{OP}$ to the outmost scale of the unit circle. The intersection is point Q and we can read the scale as $0.15(\lambda)$.

Step 4: Because the length of the TX line is $d = 0.03\lambda$ and $0.15 + 0.03 = 0.18$ (λ), we go clockwise to the scale of $0.18(\lambda)$ and reach the point W along the outmost scale of the unit circle.

Step 5: We connect the origin O and point W. Along $\overline{OW}$, we can designate a point X so that $\overline{OX} = \overline{OP}$. Then the point X represents Γ'. (Because $|\Gamma| = \overline{OP}$ and $|\Gamma'| = \overline{OX}$, we have $|\Gamma| = |\Gamma'|$.)

Step 6: From Smith chart, we find that point X is the intersection of $r' = 1$-circle and $x' = 1.65$-curve. Hence the normalized input impedance is given by $u_i = 1 + j1.65$.

Step 7: Finally, the input impedance is given by $Z_i = R_0 \cdot u_i = 50 + j82.5$ (Ω). ∎

Example 6.11
In Fig. 6.18, suppose a TX line has the characteristic resistance $R_0 = 50\,\Omega$ and load impedance $Z_L = 10 - j150$ (Ω). Please derive the input impedance Z_i of a segment of TX line having $d = 0.27\lambda$.

Solution

Step 1: First, we derive the normalized load impedance given by $u = \frac{Z_L}{R_0} = 0.2 - j3$.

Step 2: In Fig. 6.22, we find the intersection of $r = 0.2$-circle and $x = -3$-curve and designate it as point P.

Step 3: We draw a line between the origin O and point P, and extend the line $\overline{OP}$ to the outmost scale of the unit circle. The intersection is point Q and we can read the scale as $0.301(\lambda)$.

Step 4: Because the length of the TX line is $d = 0.27\lambda$, we calculate $0.301 + 0.27 = 0.571(\lambda) = 0.071$ (λ). Note that every half wavelength (0.5λ) is reset. Hence we go clockwise to the scale of $0.071(\lambda)$ and reach the point W along the outmost scale of the unit circle.

Step 5: We connect the origin O and point W. Along $\overline{OW}$, we designate a point X so that $\overline{OX} = \overline{OP}$. Then point X represents Γ'.

Step 6: From Smith chart, we find that point X is the intersection of $r' = 0.02$-circle and $x' = 0.48$-curve. Hence the normalized input impedance is given by $u_i = 0.02 + j0.48$.

Step 7: Finally, the input impedance is given by $Z_i = R_0 \cdot u_i = 1 + j24$ (Ω). ∎

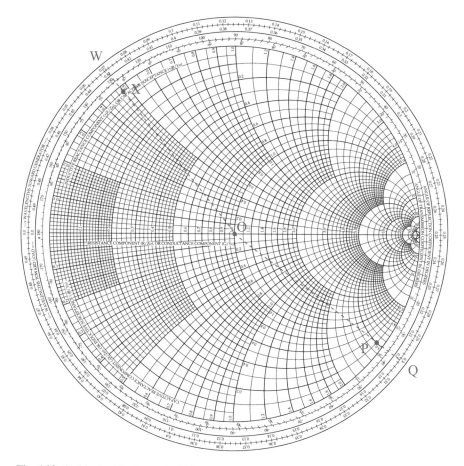

Fig. 6.22 Smith chart for Example 6.11

E   Input Impedance of a Segment of Lossy TX Line

As discussed previously, for a lossy TX line, the propagation constant includes a real part and is expressed by $\gamma = \alpha + j\beta$, where $\alpha > 0$. From Eq. (5.144) of Sect. 5.5, the input impedance is given by

$$Z_i = R_0 \cdot \frac{e^{\gamma d} + \Gamma e^{-\gamma d}}{e^{\gamma d} - \Gamma e^{-\gamma d}} = R_0 \cdot \frac{1 + \Gamma e^{-2\gamma d}}{1 - \Gamma e^{-2\gamma d}}. \tag{6.42}$$

Let

$$\Gamma^* = \Gamma e^{-2\gamma d}. \tag{6.43}$$

Then from Eq. (6.42), the normalized input impedance is given by

$$u_i = \frac{Z_i}{R_0} = \frac{1 + \Gamma^*}{1 - \Gamma^*}. \tag{6.44}$$

Notice that Eq. (6.44) is similar to Eq. (6.38). Hence we can apply similar procedure to derive the input impedance of a segment of lossy TX line using Smith chart. The only difference is the conversion between Γ' and Γ^*.

Because $\gamma = \alpha + j\beta$, Eq. (6.43) can be rewritten as

$$\Gamma^* = \Gamma e^{-j2\beta d} \cdot e^{-2\alpha d} = \Gamma' \cdot e^{-2\alpha d}, \tag{6.45}$$

where Γ' is the reflection coefficient of a lossless TX line. Since $\alpha > 0$ and $d > 0$, the last term in Eq. (6.45) is a positive real number smaller than 1, i.e., $0 < e^{-2\alpha d} < 1$. Besides, from Eq. (6.45), once Γ' is obtained, we can simply multiply it with $e^{-2\alpha d}$ to get Γ^*. Examples are provided in the following.

Example 6.12
In Fig. 6.18, suppose a TX line has the characteristic resistance $R_0 = 50\,\Omega$, wavelength $\lambda = 10$ cm, attenuation constant $\alpha = 0.2/\text{cm}$, and load impedance $Z_L = 10 - j150\,(\Omega)$. Please derive the input impedance Z_i of a segment of TX line having $d = 1$ cm.

Solution

Step 1: First, we derive the normalized load impedance given by $u = \frac{Z_L}{R_0} = 0.2 - j3$.

Step 2: As shown in Fig. 6.23, we find the intersection of $r = 0.2$-circle and $x = -3$-curve and designate it as point P.

Step 3: We draw a line between the origin O and point P, and extend the line $\overline{OP}$ to the outmost layer of the unit circle. The intersection is point Q and we can read the scale as $0.301\,(\lambda)$.

Step 4: Because the length of the TX line is $d = 1$ cm $= 0.1\lambda$, we calculate $0.301 + 0.1 = 0.401(\lambda)$. Hence we go clockwise to the scale of $0.401(\lambda)$ and reach the point W.

Step 5: Because $d = 1$ cm and $\alpha = 0.2/\text{cm}$, we have the degradation factor $e^{-2\alpha d} = e^{-0.4} = 0.67$. We connect the origin O and point W. Along $\overline{OW}$, we designate a point X so that $\overline{OX} = (0.67) \cdot \overline{OP}$. (Because $|\Gamma'| = |\overline{OP}|$ and $|\Gamma^*| = \overline{OX}$, we have $|\Gamma^*| = |\Gamma'| \cdot e^{-2\alpha d}$.) Then the point X represents Γ^*.

Step 6: From Smith chart, we find that point X is the intersection of $r = 0.32$-circle and $x = -0.66$-curve (in lower Γ-plane). Hence the normalized input impedance is given by $u_i = 0.32 - j0.66$.

Step 7: Finally, the input impedance is given by $Z_i = R_0 \cdot u_i = 16 - j33\,(\Omega)$. ∎

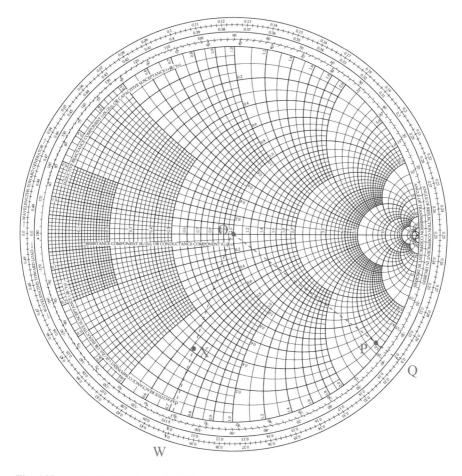

Fig. 6.23 Smith chart for Example 6.12

Example 6.13

In Fig. 6.18, a segment of TX line has the length $d = 3$ cm, characteristic resistance $R_0 = 50\,\Omega$, wavelength $\lambda = 15$ cm, and attenuation constant $\alpha = 0.1$/cm. Suppose the measured input impedance is given by $Z_i = 70 - j30$. Please derive the load impedance Z_L.

Solution

In this example, we shall notice that it is different from Example 6.12. In Example 6.12, we derived an input impedance Z_i from a given load impedance Z_L. In this example, we need to derive a load impedance Z_L from a given input impedance Z_i.

Step 1: First, we derive the normalized input impedance, the length d in the unit of wavelength, and the conversion factor $e^{2\alpha d}$ given by

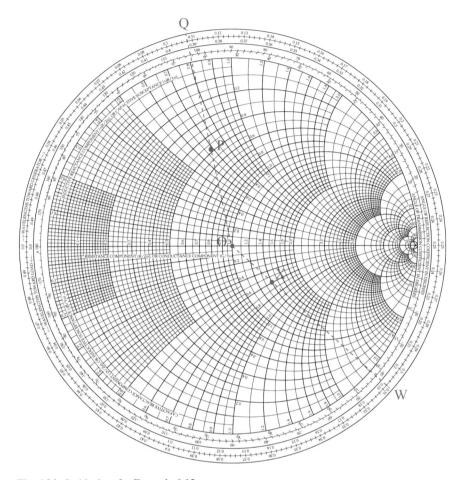

Fig. 6.24 Smith chart for Example 6.13

$$u_i = \frac{Z_i}{R_0} = 1.4 - j0.6,$$

$$d = 3/15 = 0.2(\lambda),$$

$$e^{2\alpha d} = e^{0.6} = 1.82,$$

respectively.

Step 2: As shown in Fig. 6.24, we find the intersection of $r = 1.4$-circle and $x = -0.6$-curve and designate it as point X, which represents Γ^* in Smith chart.

Step 3: We draw a line between the origin O and point X, and extend the line $\overline{OX}$ to the outmost scale of the unit circle. The intersection is point W and we can read the scale as $0.308(\lambda)$.

Step 4: Because the length of the TX line is $d = 0.2\lambda$, we calculate $0.308 - 0.2 = 0.108$ (λ). Hence we rotate counter-clockwise and reach the point Q.

Step 5: We connect the origin O and point Q. Along $\overline{OQ}$, we designate a point P so that $\overline{OP} = e^{2\alpha d} \cdot \overline{OX} = (1.82) \cdot \overline{OX}$ (Because $|\Gamma^*| = \overline{OX}$ and $|\Gamma| = \overline{OP}$, we have $|\Gamma| = e^{2\alpha d} \cdot |\Gamma^*|$ from Eq. (6.45).) Hence point P represents Γ in Smith chart.

Step 6: From Smith chart, we find that point P is the intersection of $r = 0.48$-circle and $x = 0.68$-curve. Hence the normalized load impedance is given by $u = 0.48 + j0.68$.

Step 7: Finally, the load impedance is given by $Z_L = R_0 \cdot u = 24 + j34$ (Ω).

■

F Load Admittance

Suppose that a normalized load impedance u corresponding to a reflection coefficient Γ is given by

$$u = \frac{1 + \Gamma}{1 - \Gamma}. \tag{6.46}$$

Let u' be another normalized load impedance having the corresponding reflection coefficient $-\Gamma$. Then we have

$$u' = \frac{1 + (-\Gamma)}{1 - (-\Gamma)} = \frac{1 - \Gamma}{1 + \Gamma} = \frac{1}{u} \tag{6.47}$$

Therefore, the following relationship holds:

If u corresponds to Γ, then $u' = \frac{1}{u}$ corresponds to $-\Gamma$.

In Smith chart, Γ and $-\Gamma$ are symmetric with respect to the origin. Therefore, if u is obtained, we can easily get $u' = \frac{1}{u}$ by using Smith chart. The procedure is shown in the following.

Step 1: First, we find the point P corresponding to $u = r + jx$ in Smith chart. Then P represents the reflection coefficient Γ.

Step 2: Next, we find the point Q which is symmetric to P with respect to the origin O.

Step 3: Finally, because Q represents the reflection coefficient $-\Gamma$, we can immediately derive the associated normalized load impedance $u' = r' + jx'$ in Smith chart.

Example 6.14

Suppose a TX line has a normalized load impedance $u = 0.2 + j0.5$. Please derive $u' = \frac{1}{u}$.

Solution

Step 1: In Fig. 6.25, we can find the point P which is the intersection of $r = 0.2$-circle and $x = 0.5$-curve in Smith chart.

Step 2: We then find the point Q symmetric to point P with respect to the origin O as shown in Fig. 6.25.

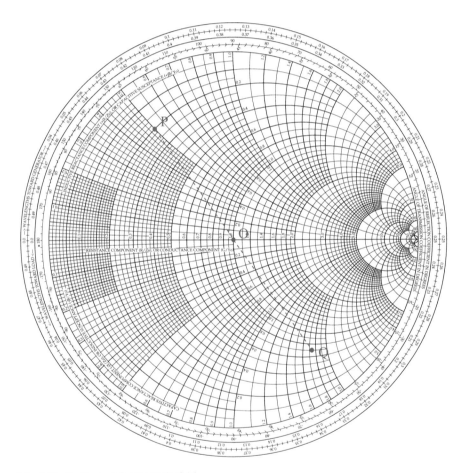

Fig. 6.25 Smith chart for Example 6.14

Step 3: In Smith chart, we can immediately find that the point Q is the intersection of $r' = 0.7$-circle and $x' = -1.72$-curve. Then we have $u' = \frac{1}{u} = 0.7 - j1.72$.

∎

From Example 6.14, if u is given, we can immediately derive $u' = \frac{1}{u}$ by using Smith chart without any calculations.

Now, let Z_L be a load impedance. The corresponding **load admittance** is defined as

$$Y_L = \frac{1}{Z_L}. \tag{6.48}$$

In addition, let R_0 be the characteristic resistance. Then the corresponding **characteristic admittance** is defined as

$$Y_0 = \frac{1}{R_0}. \tag{6.49}$$

From the above definitions, we can define a **normalized load admittance** as

$$y = \frac{Y_L}{Y_0} \tag{6.50}$$

From Eqs. (6.48)–(6.50), the relationship between y and u is given by

$$y = \frac{R_0}{Z_L} = \frac{1}{u}. \tag{6.51}$$

Note that in circuitry, an admittance is the reciprocal of the associated impedance. Hence the result in Eq. (6.51) is consistent with what we learned in circuitry. Besides, because we can derive $1/u$ from u easily by using Smith chart, we can derive y easily when u is given as in Example 6.14. It follows that when a load impedance Z_L is given, the corresponding load admittance Y_L can be easily derived. Two examples are provided in the following.

Example 6.15
Suppose a TX line has the characteristic resistance $R_0 = 50 \, \Omega$ and load impedance $Z_L = 30 + j60 \, (\Omega)$. Please derive the load admittance Y_L.

Solution

Step 1: First, we derive the normalized load impedance given by $u = \frac{Z_L}{R_0} = 0.6 + j1.2$.

Step 2: In Fig. 6.26, we can find the point P which is the intersection of $r = 0.6$-circle and $x = 1.2$-curve in Smith chart.

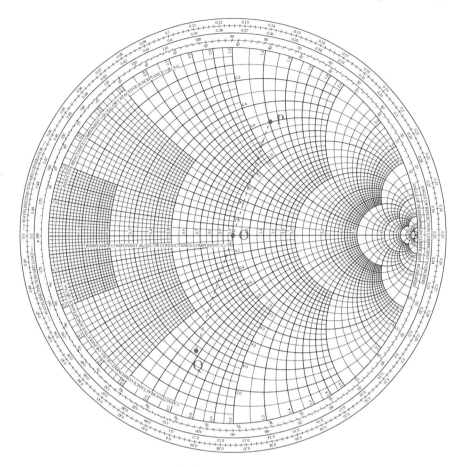

Fig. 6.26 Smith chart for Example 6.15

Step 3: Then we can find the point Q symmetric to the point P with respect to the origin O.

Step 4: In Smith chart, we can immediately find that the point Q is the intersection of $r' = 0.33$-circle and $x' = -0.67$-curve. Hence $y = \frac{1}{u} = 0.33 - j0.67$.

Step 5: Finally, $Y_L = y \cdot Y_0 = \frac{y}{R_0} = 0.0066 - j0.0134$ (Ω^{-1}).   ∎

Example 6.16

Suppose a TX line has the characteristic resistance $R_0 = 50\,\Omega$ and load impedance $Z_L = 10 - j20$ (Ω). Please derive the load admittance Y_L.

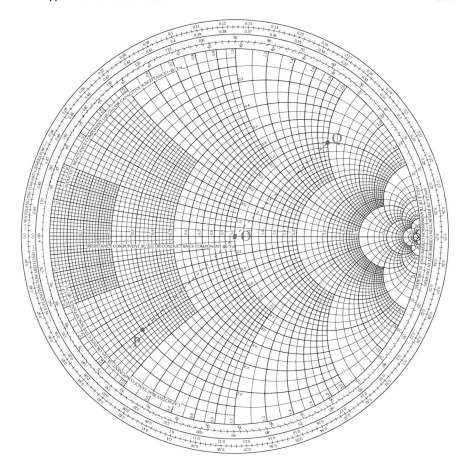

Fig. 6.27 Smith chart for Example 6.16

Solution

Step 1: First, we derive the normalized load impedance given by $u = \frac{Z_L}{R_0} = 0.2 - j0.4$.

Step 2: In Fig. 6.27, we find the point P which is the intersection of $r = 0.2$-circle and $x = -0.4$-curve in Smith chart.

Step 3: Then we can find the point Q symmetric to the point P with respect to the origin O.

Step 4: In Smith chart, we immediately find that the point Q is the intersection of $r' = 1$-circle and $x' = 2$-curve. Hence $y = \frac{1}{u} = 1 + j2$.

Step 5: Finally, $Y_L = yY_0 = \frac{y}{R_0} = 0.02 + j0.04 \; (\Omega^{-1})$.

■

In the above, for a given u, we can derive y by using Smith chart. Therefore, a normalized impedance u and the associated admittance y are actually one-to-one mapping. It implies that **each point in Smith chart corresponds to a unique y. Hence Smith chart can also be used as an admittance chart, and we can derive the reflection coefficient Γ for a given y.** The derivation is similar to what we do for a given u.

G Summary

In this section, we have introduced a number of popular applications of Smith chart, including

- Given a load impedance Z_L, we derive the reflection coefficient Γ.
- Given a reflection coefficient Γ, we derive the load impedance Z_L.
- Given a load impedance Z_L, we derive the input impedance Z_i of a lossless TX line.
- Given a load impedance Z_L, we derive the input impedance Z_i of a lossy TX line.
- Given a load impedance Z_L, we derive the associated load admittance Y_L.

Once we do more practice and get familiar with Smith chart, the above applications will become quite easy. Besides, Smith chart has many other applications. For example, when designing a high-frequency circuit, we usually need to attain **impedance matching** between different circuit blocks in order to avoid reflection or achieve maximum power transfer. As will be learned in the following sections, Smith chart is an important tool to achieve this goal.

6.3 Impedance Matching Design

In the previous sections, we learn how to use a Smith chart to derive the load impedance and the input impedance of a segment of TX line. In high-frequency circuit design, impedance matching is an important practice and can be implemented to reduce reflection and achieve the maximal power transfer. An example is shown in Fig. 6.28, where we use a TX line to deliver signals from a power amplifier to an antenna, which will radiate the signal to the space. Note that a power amplifier and an antenna are important elements in wireless communication systems.

Let R_0 be the characteristic resistance of the transmission line and Z_L be the input impedance of the antenna. Assume $Z_L \neq R_0$. In this case, reflection occurs so that the amplifier cannot effectively deliver signal power to the antenna.

In order to resolve the above issue, as shown in Fig. 6.29, we design an impedance matching circuit between the power amplifier and the antenna. Let Z_i be the input impedance of the matching circuit. When $Z_i = R_0$, there is no reflection occurring at the input of the matching circuit. Hence, all the signal power from the amplifier can be delivered to the matching circuit. As the matching circuit is composed of lossless components, e.g., capacitors, inductors, or TX lines, almost all the signal power will be delivered to the antenna afterwards. In the following, we will introduce the essential idea of matching circuits and the way to implement them.

Because matching circuits can be implemented with lumped elements or with distributed elements, we first introduce these two kinds of elements. Let l be the physical length of a circuit component and λ be the operating wavelength. Depending on the relative size between l and λ, a circuit component can be regarded as a lumped element or as a distributed element given below:

1. Lumped element

When the physical size of a circuit component is much smaller than the operating wavelength, i.e., $l \ll \lambda$, the component is regarded as a **lumped element**. In this case, we find the voltage and the current of the circuit component are almost independent of position. Hence, we take the voltage and the current as functions of time, i.e., $V = V(t)$ and $I = I(t)$. In short, a lumped element is a circuit component whose voltage and current are functions of time only.

In low-frequency circuits, because of the long wavelength, almost all the circuit components are regarded as the lumped elements. In high-frequency circuits, there are many specially designed inductors, capacitors, and resistors, their physical length is normally much smaller than the operating wavelength, i.e., $l \ll \lambda$ still holds even though the frequency is high. Hence, they are regarded as lumped elements too. When applying these circuit components in high-frequency circuits, we only consider time variation of their voltages and currents, and completely ignore their spatial dependence.

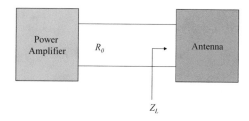

Fig. 6.28 Signal transmission from power amplifier to antenna with TX line

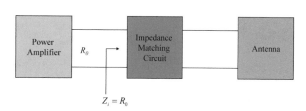

Fig. 6.29 Illustration of Impedance matching circuit

Fig. 6.30 Impedance
matching design using
lumped elements (I)

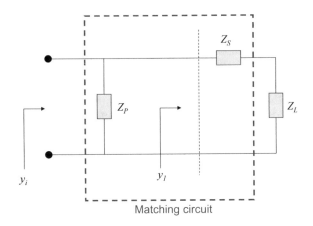

Matching circuit

2. Distributed element

When the physical size of a circuit component is comparable with the operating wave-length, i.e., l is comparable with λ, a circuit component is regarded as a **distributed element**. In this case, the voltage and the current of a circuit component not only depend on time, but also depend on position. A TX line is a good example of distributed element, because its length is usually comparable with the operating wavelength. Therefore, the voltage and the current along a TX line are functions of time and position, i.e., $V = V(t, z)$ and $I = I(t, z)$, where z denotes position. Because of the spatial dependency of voltages and currents, the design and analysis of distributed circuits is much more complicated than that of lumped circuits.

In high-frequency circuits, both lumped elements and distributed elements are widely used to achieve desired functions. Because the design of matching circuits using lumped elements is relatively easy, in the following we first introduce this design to catch the core idea of matching circuits.

A Matching Circuit with Lumped Elements

Figure 6.30 shows the matching circuit to be designed, where a lumped element with the impedance Z_S is in series with the load, and another lumped element with the impedance Z_P is in parallel with the load. Let

$$Z_S = jS, \tag{6.52}$$

$$Z_P = jP, \tag{6.53}$$

where S and P are real numbers. Note that because capacitors or inductors are used, both Z_S and Z_P are pure imaginary numbers. Ideally, these two components do not consume any power.

Assume Z_i is the input impedance of the matching circuit. Then we have the normalized input impedance given by

$$z_i = \frac{Z_i}{R_0}. \tag{6.54}$$

Assume the load impedance is given by

$$Z_L = R_L + jX_L. \tag{6.55}$$

Then the normalized load impedance is given by

$$u = \frac{Z_L}{R_0}. \tag{6.56}$$

As will become clear later, we prefer admittance, instead of impedance, in designing the matching circuit. From Eqs. (6.54) and (6.56), we get the normalized input admittance of the matching circuit and the normalized load admittance given by

$$y_i = \frac{1}{z_i} = \frac{R_0}{Z_i}, \tag{6.57}$$

$$y_L = \frac{1}{u} = \frac{R_0}{Z_L}, \tag{6.58}$$

respectively. These two parameters are very critical for matching circuit design. Note that when the impedance is matched, i.e., $Z_i = R_0$, we get

$$y_i = 1. \tag{6.59}$$

Hence, the goal of the matching circuit is to transfer y_L to y_i, and $y_i = 1$.

As shown in Fig. 6.30, assume y_1 is the normalized admittance of the series connection of Z_S and Z_L and given by

$$y_1 = \frac{R_0}{Z_S + Z_L}. \tag{6.60}$$

The core idea of the matching circuit is to achieve the goal $y_i = 1$ in two steps:

1. First, use the series component Z_S to transfer y_L to y_1, where $y_1 = 1 + jb$ and b is a real number.
2. Next, use the parallel component Z_P to transfer $y_1 = 1 + jb$ to y_i by canceling the imaginary term "jb" in y_1. Hence, $y_i = 1$ as desired.

If the above idea is caught, readers will find the design of matching circuits is actually very easy. Now, we describe the way to implement the matching circuit in two steps.

Step 1: Transferring y_L to y_1, and $y_1 = 1 + jb$.

From Eqs. (6.52), (6.55) and (6.60), we get

$$y_1 = \frac{R_0}{R_L + j(S + X_L)} = \frac{R_0 R_L - jR_0(S + X_L)}{R_L^2 + (S + X_L)^2}. \tag{6.61}$$

When the condition $y_1 = 1 + jb$ is met, we get the following results.

$$1 = \frac{R_0 R_L}{R_L^2 + (S + X_L)^2}, \tag{6.62}$$

$$b = -\frac{R_0(S + X_L)}{R_L^2 + (S + X_L)^2}. \tag{6.63}$$

When Eqs. (6.62) and (6.63) are satisfied, we successfully transfer y_L to $y_1 = 1 + jb$.

In Eq. (6.62), it is easy to see that if $R_L > R_0$, the equality cannot hold. It means $y_1 = 1 + jb$ cannot be achieved under this condition. Hence, this matching circuit is applicable only for the case of $R_L < R_0$.

In the following, we assume that $R_L < R_0$. From Eq. (6.62), it is easy to get

$$S = -X_L \pm \sqrt{R_0 R_L - R_L^2}. \tag{6.64}$$

Note that S has two possible solutions, both are valid ones since S can be any real number.

In addition, when Eq. (6.62) is satisfied, we get

$$R_L^2 + (S + X_L)^2 = R_0 R_L. \tag{6.65}$$

Hence, inserting Eq. (6.65) into Eq. (6.63), we obtain b as

$$b = -\frac{S + X_L}{R_L}. \tag{6.66}$$

From Eq. (6.66), when S is obtained, the corresponding b is obtained too.

Step 2: Transferring y_1 to y_i, and $y_i = 1$.

In Fig. 6.30, the normalized admittance of the parallel component is given by

$$y_P = \frac{R_0}{Z_P} = \frac{R_0}{jP}. \tag{6.67}$$

Because Z_P is in parallel with Z_S and Z_L, we get

$$y_i = y_1 + y_P. \tag{6.68}$$

We design y_P so that

$$y_P = -jb. \tag{6.69}$$

Hence y_i is given by

$$y_i = (1 + jb) + (-jb) = 1. \tag{6.70}$$

Therefore, by using y_P to cancel the term "jb" in y_1, we achieve the goal $y_i = 1$. Finally, from Eqs. (6.67), (6.69) and (6.66), we get

$$P = \frac{R_0}{jy_P} = \frac{R_0}{b} = -\frac{R_0 R_L}{S + X_L}. \tag{6.71}$$

After understanding the idea behind the above two steps, it is easy to design S and P by using Eq. (6.64) and Eq. (6.71), respectively. Then, we can use reactive components, i.e., inductors and capacitors, to implement the matching circuit. The impedance of an inductor is given by

$$Z = j\omega L, \tag{6.72}$$

where ω is the frequency and L is the inductance. If S is a positive number, we can use an inductor for the series component. The corresponding inductance is calculated by

$$j\omega L = jS \Rightarrow L = \frac{S}{\omega}. \tag{6.73}$$

On the other hand, the impedance of a capacitor is given by

$$Z = \frac{1}{j\omega C} = \frac{-j}{\omega C}, \tag{6.74}$$

where C is the capacitance. If S is a negative number, we can use a capacitor for the series component. The corresponding capacitance is calculated by

$$\frac{-j}{\omega C} = -j|S| \Rightarrow C = \frac{1}{\omega |S|}. \tag{6.75}$$

Hence, depending on the sign of S, we can select an inductor or a capacitor to implement the series component. The same idea can be applied to the design of the parallel component too.

Example 6.17
For the matching circuit design using lumped elements as shown in Fig. 6.30, assume $R_0 = 50\,\Omega$, $Z_L = 30 + j80\,(\Omega)$, and the operation frequency is $f = 1$ GHz. Please design the matching circuit.

Solution
First, we take the negative sign in Eq. (6.64), then we get S as

$$S = -X_L - \sqrt{R_0 R_L - R_L^2}.$$

Because $R_L = 30$ and $X_L = 80$, we get

$$S = -\left(80 + 10\sqrt{6}\right)$$

From Eq. (6.71) we get

$$P = -\frac{R_0 R_L}{S + X_L} = 25\sqrt{6}.$$

Next, because S is a negative number, we use a capacitor to implement the series component. From Eq. (6.75), the associated capacitance is given by

$$C = \frac{1}{2\pi f |S|} = 1.52 \times 10^{-12}(\text{F}).$$

As P is a positive number, we use an inductor to implement the parallel component. Hence from Eq. (6.72), we have

$$Z_P = jP = j\omega L.$$

Then the associated inductance is given by

$$L = \frac{P}{2\pi f} = 9.75 \times 10^{-9}(\text{H}).$$

∎

Finally, as mentioned above, the matching circuit shown in Fig. 6.30 works only for the case of $R_L < R_0$. If $R_L > R_0$, we can use another matching circuit shown in Fig. 6.31 to get $Z_i = R_0$. The design idea of this matching circuit is similar to

Fig. 6.31 Impedance matching design using lumped elements (II)

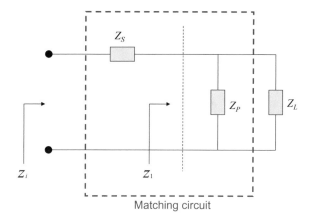

Matching circuit

the previous one, except that we first use a reactive component in parallel with the load, and then use another reactive component in series with the load. The operation principle of this circuit is described below:

Step 1: Using the parallel component with the impedance Z_P to transfer the normalized load impedance u to be a normalized impedance z_1, where $z_1 = 1 + jh$ and h is a real number.

Step 2: Using the series component with the impedance Z_S to cancel the term "jh" in z_1. Then we get the normalized input impedance $z_i = 1$, i.e., $Z_i = R_0$.

Because the design and implementation of this matching circuit are similar to those of the previous matching circuit, we omit them and leave it to readers to work out by themselves.

B Matching Circuit with Distributed Elements (TX Lines)

After learning matching circuits by using lumped elements, i.e., $l << \lambda$, now we learn the design by using distributed elements, i.e., the physical size l of components, is comparable to the operating wavelength λ. Because TX lines can be easily implemented on the printed circuit board (PCB), they are the best candidates for this purpose. As discussed in Sect. 5.5, when a load is connected to a TX line, its input impedance varies with the length of the line. In addition, a segment of TX line can be used as a circuit component. These two features are readily applicable to impedance matching circuits.

In the following, we present a matching circuit using TX lines, called **single-stub matching circuit**. The design of this matching circuit is given below:

Step 1: First, for a given load impedance, we use Eq. (6.58) to obtain the normalized load admittance y_L.

Fig. 6.32 Impedance
matching using distributed
elements (I)

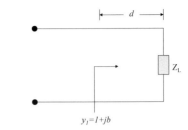

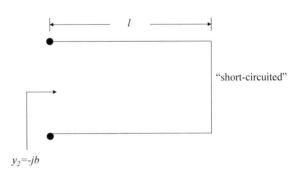

$y_1 = 1 + jb$

Fig. 6.33 Impedance
matching using distributed
elements (II)

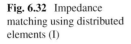

"short-circuited"

$y_2 = -jb$

Step 2: As shown in Fig. 6.32, we deliberately select a point along the TX line whose
distance from the load is d so that the normalized input admittance at this point is

$$y_1 = 1 + jb, \tag{6.76}$$

where b is a real number. Notice that in Eq. (6.76), **the real part of y_1 must be 1**
and b can be an arbitrary number.

Step 3: As shown in Fig. 6.33, we take another TX line whose end points are shorted
and its length is l. We deliberately choose l so that the normalized input admittance
of this line is given by

$$y_2 = -jb, \tag{6.77}$$

where b is the number we got in Step 2.

Step 4: Finally, we parallelly connect the shorted TX line to the TX line obtained in
Step 2, the result is shown in Fig. 6.34. Suppose y_i is the normalized input admittance
of this circuit. Then we have

$$y_i = y_1 + y_2. \tag{6.78}$$

Thus we obtain

$$y_i = (1 + jb) + (-jb) = 1. \tag{6.79}$$

Fig. 6.34 Impedance
matching using distributed
elements (III)

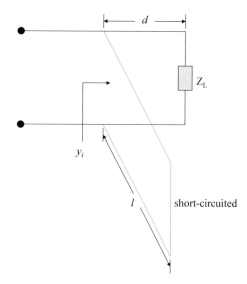

Hence, the impedance is matched.

Readers may already find the design idea of this circuit is the same as what we
have in Fig. 6.30, except that we use two segments of TX lines, instead of two
lumped elements, to implement the circuit. In the following, we present an example
to implement the above idea with Smith chart.

Example 6.18

For the matching circuit design using distributed elements as shown in Fig. 6.34,
suppose we have a TX line with the characteristic resistance $R_0 = 50\,\Omega$, wavelength
$\lambda = 10$cm, and the load impedance $Z_L = 30 - j100$ (Ω). Please design an impedance
matching circuit so that $Z_i = R_0$.

Solution

Step 1: First, from Eq. (6.58), we have the normalized load admittance given by

$$y_L = \frac{R_0}{Z_L} = 0.14 + j0.46.$$

In Fig. 6.35, we treat Smith chart as an admittance chart and we can find a point
P corresponds to y_L. In this chart, the use of y_L is the same as the way we use the
normalized load impedance u in Sect. 6.2.

Step 2: In Fig. 6.35, we consider a circle having the center at the origin O and the
radius being the length of line $\overline{OP}$. Considering clockwise rotation, the intersection
of this circle and the $y = 1$-circle is the point Q. From the scales in Smith chart,
point Q corresponds to the normalized admittance

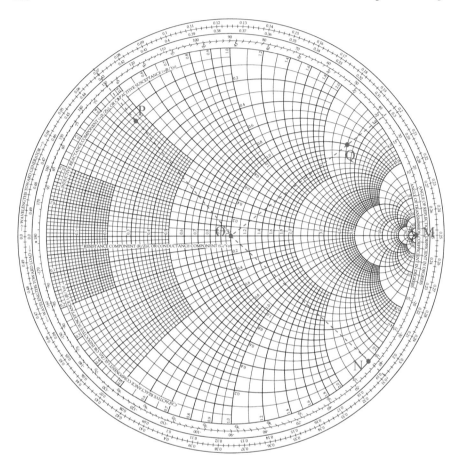

Fig. 6.35 Smith chart for Example 6.18

$$y_1 = 1 + j2.6.$$

From the wavelength scale around the unit circle, the corresponding wavelength of line $\overline{OP}$ is 0.07λ and that of $\overline{OQ}$ is 0.198λ. Hence we obtain the distance

$$d = 0.198 - 0.07 = 0.128(\lambda).$$

At the position which is $d = 0.128\lambda = 1.28$cm away from the load, the normalized input admittance is $y_1 = 1 + j2.6$.

Step 3: Next, we design a TX line having short-circuited terminal as shown in Fig. 6.33 so that the length is l and the normalized input admittance is given by

$$y_2 = -j2.6.$$

Fig. 6.36 Impedance matching design for Example 6.18

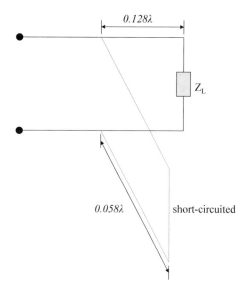

Because the terminal is short circuited, the corresponding normalized admittance is given by $y_{short} \to \infty$. In Fig. 6.35, $y_{short} \to \infty$ corresponds to the point M which is the rightmost point of the Γ_r-axis in Smith chart. Then we consider a circle having the center at the origin O and the radius is the length of line $\overline{OM} = 1$. Considering clockwise rotation, the intersection of this circle and $y = -j2.6$-curve is the point N.

From the wavelength scale around the unit circle, the corresponding wavelength of line $\overline{OM}$ is 0.25λ and that of $\overline{ON}$ is 0.308λ. Hence we obtain

$$l = 0.308 - 0.25 = 0.058(\lambda).$$

Therefore, we get $l = 0.058(\lambda) = 0.58\text{cm}$ and the normalized admittance of the TX line is $y_2 = -j2.6$.

Step 4: As shown in Fig. 6.36, we connect the TX line in Step 3 in parallel with that of the original circuit in Step 2. The resultant normalized input admittance (y_i) is given by

$$y_i = y_1 + y_2 = (1 + j2.6) + (-j2.6) = 1.$$

Hence $Z_i = R_0$ and the impedance is matched.

■

From mathematical viewpoint, when we want to transfer an arbitrary normalized load admittance $y_L = p + jq$ to $y_i = 1$, we need at least two components to adjust it. In circuitry, these two components can be two lumped elements as shown in Fig. 6.30, or two distributed elements such as two segments of TX lines as shown in Fig. 6.34, depending on which one is the best choice.

C   Quarter Wavelength Matching Circuit

Let l be the length of a TX line and λ be the operating wavelength. When $l = \lambda/4$, this line is called a quarter wavelength TX line, abbreviated as $\lambda/4$ TX line. As will be shown, a $\lambda/4$ TX line is useful for matching a resistive load.

Figure 6.37 is a matching circuit using a $\lambda/4$ TX line to transfer a resistive load R_L to R_0. Let R_A be the characteristic resistance of the $\lambda/4$ TX line. From the results of Sect. 5.5, the input impedance of the $\lambda/4$ TX line is given by

$$Z_i = \frac{R_A^2}{R_L}. \tag{6.80}$$

Because $Z_i = R_0$ is desired, we get

$$R_A = \sqrt{R_L R_0}. \tag{6.81}$$

Hence, by using a $\lambda/4$ TX line with a characteristic resistance R_A given in Eq. (6.81), we get impedance matching, i.e., $Z_i = R_0$. Note that if it is a complex load with the impedance $Z_L = R_L + jX_L$, we cannot use a $\lambda/4$ TX line to achieve impedance matching, since it is not feasible to adjust two numbers (R_L, X_L) using only one component. Therefore, a $\lambda/4$ TX line is usually used to match a resistive load, but not a complex load.

Example 6.19
Let $R_L = 200\ \Omega$ be the load resistance and $R_0 = 50\ \Omega$ be the characteristic resistance of the TX line. Please design a $\lambda/4$ TX line with the characteristic resistance R_A to achieve impedance matching, i.e., $Z_i = R_0$.

Fig. 6.37 Impedance
matching circuit using
quarter wavelength TX line

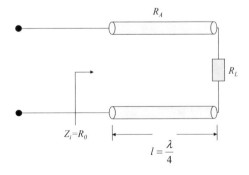

Solution
From Eq. (6.81), we get

$$R_A = \sqrt{R_L R_0} = \sqrt{200 \times 50} = 100 \ \Omega.$$

■

6.4 Briefly on High-Frequency Amplifier Design

After learning the matching circuit design in the previous section, we proceed to study its application in high-frequency amplifiers. The knowledge of high-frequency amplifiers is essential for those who are interested in high-frequency circuit design, because in electronic circuits, an amplifier is an important building block that is widely used in various applications. In practice, the signal frequency we need to amplify can be from below 1 Hz up to tens of GHz. Within such a wide signal spectrum, the considerations of amplifier design may change at different frequency ranges. In the following, we introduce design consideration of amplifiers at different frequency ranges with special emphasis on high-frequency amplifiers. Besides, we assume that readers are familiar with bipolar junction transistor (BJT) and use BJT amplifiers for illustration.

A Low-Frequency Amplifier

In the beginning, we show an npn BJT in Fig. 6.38, where the three terminals are Base (B), Emitter (E), and Collector (C). The function of BJT is described by the following formula:

Fig. 6.38 An npn BJT

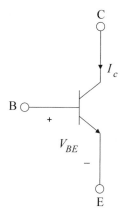

Fig. 6.39 CE amplifier
using npn BJT

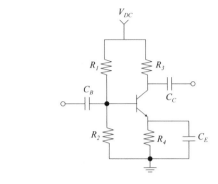

Fig. 6.40 Small-signal
equivalent model of BJT for
low operation frequency

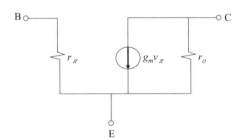

$$I_C = I_S \cdot e^{\frac{V_{BE}}{V_T}} \left(1 + \frac{V_{CE}}{V_A}\right), \qquad (6.82)$$

where I_S is the reverse saturation current, V_T is the thermal voltage, and V_A is the Early voltage. Equation (6.82) shows that the collector current (I_C) is mainly controlled by the base-emitter voltage (V_{BE}), so we can consider a BJT as a voltage-controlled current source.

At low-frequency range, e.g., $f = 1$ kHz, the design of amplifiers is simple. For example, Fig. 6.39 is a typical Common Emitter (CE) amplifier, where four resistors (R_1, R_2, R_3, R_4) are used for setting appropriate bias voltages and currents, and three capacitors (C_B, C_E, C_C) are used for DC blocking and AC bypassing purposes.

In designing low-frequency amplifiers, we usually use the small-signal equivalent model to ease circuit analysis. Figure 6.40 is the small-signal equivalent model of BJT, where r_π is the equivalent resistor between base and emitter, r_o is the equivalent resistor between collector and emitter, and $g_m v_\pi$ is a current source controlled by the base-emitter voltage v_π—the core function of BJT.

The small-signal equivalent model can help us calculate critical amplifier parameters, for example, voltage/current gain, input resistance, and output resistance. Unfortunately, this model is no longer valid when the operation frequency increases beyond around 1 MHz.

Fig. 6.41 Parasitic capacitances of BJT

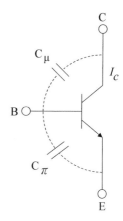

Fig. 6.42 Small-signal equivalent model of BJT for medium operation frequency

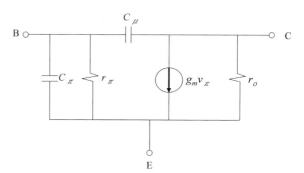

B Medium-Frequency Amplifier

When the operation frequency increases to the range of MHz, we still regard a BJT as a voltage-controlled current source, whose function is described by Eq. (6.82). However, since the frequency is high, the effect of parasitic capacitance inherent in BJT can no longer be neglected. As shown in Fig. 6.41, there are two parasitic capacitors in BJT: the base-emitter capacitor C_π and the collector-base capacitor C_μ. Hence, the small-signal equivalent model of Fig. 6.40 should be modified to include these two parasitic capacitors. Figure 6.42 shows the modified small-signal equivalent model, where C_π and C_μ are included in this model. You may have a question: why do we have to consider these two capacitors in the MHz range?

The typical parasitic capacitance of BJT is around several pico Farads (1 pF $=$ 10^{-12} F). For example, assume a BJT has $C_\mu = 5$ pF. The impedance of C_μ is given by

$$Z = \frac{1}{j\omega C_\mu},$$

where ω is the radian frequency. When the operation frequency is low, for example, $f = 10$ kHz, the magnitude of Z is given by

$$|Z| = \left| \frac{1}{j\omega C_\mu} \right| = \frac{1}{2\pi f C_\mu} = \frac{1}{2\pi \cdot 10^4 \cdot 5 \cdot 10^{-12}} = 3.18 \times 10^6 \, \Omega = 3.18 \, M\Omega.$$

Since $|Z|$ is very large, we can **regard C_μ as an open circuit** and ignore its influence. Hence in this case, we consider the small-signal equivalent model as in Fig. 6.40, instead of Fig. 6.42.

When $f = 10$ MHz, we have

$$|Z| = \left| \frac{1}{j\omega C_\mu} \right| = \frac{1}{2\pi f C_\mu} = \frac{1}{2\pi \cdot 10^7 \cdot 5 \cdot 10^{-12}} = 3.18 \times 10^3 \, \Omega = 3.18 \, k\Omega.$$

In this case, we find that the magnitude of $|Z|$ is comparable with that of the equivalent resistors of the small-signal model. Moreover, due to the Miller effect we learned in Electronics, the influence of C_μ is greatly enhanced by the amplifier gain, which leads to a significant gain degradation. Therefore, C_μ cannot be neglected. Owing to the effect of parasitic capacitance, the analysis of medium-frequency amplifiers is much more complicated than that of low-frequency amplifiers.

C   High-Frequency Amplifier

When the operation frequency keeps going up, for example, $f = 1$ GHz, because the frequency is so high and the corresponding wavelength is very small, we find that not only parasitic capacitances of BJT, even lead inductance associated with the device package may have a noticeable influence on amplifier performance. Besides, the inherent reflection along transmission lines cannot be neglected since it seriously degrades amplifier performance. Therefore, we seek another way to describe the behavior of BJT at high-frequency range, and the method of scattering parameters is introduced.

As shown in Fig. 6.43, the idea of scattering parameters is taking a BJT as a black box, whose behavior is described by the incident waves and the reflected waves of this box. That is, as shown in Fig. 6.43b, we consider a BJT as a two-port network with two incident waves (a_1, a_2) and two outgoing waves (b_1, b_2) at both ports. The relationship between the four waves (a_1, a_2, b_1, b_2) is given by

$$\begin{bmatrix} b_1 \\ b_2 \end{bmatrix} = \begin{bmatrix} S_{11} & S_{12} \\ S_{21} & S_{22} \end{bmatrix} \begin{bmatrix} a_1 \\ a_2 \end{bmatrix}, \tag{6.83}$$

where

a_1 the wave getting into port 1,

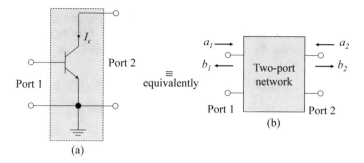

Fig. 6.43 Two-port network model of BJT for high operation frequency

a_2 the wave getting into port 2,
b_1 the wave leaving port 1,
b_2 the wave leaving port 2.

In Eq. (6.83), S_{11}, S_{12}, S_{21}, and S_{22} are called **scattering parameters** and abbreviated as **S-parameters,** and their values are determined by properties of BJT. In general, the wave mentioned above can represent voltage wave, current wave, or power wave. In high-frequency amplifier design, we normally take it as the normalized voltage wave. From Eq. (6.83), we get the following relationship:

$$b_1 = S_{11}a_1 + S_{12}a_2, \tag{6.84}$$

$$b_2 = S_{21}a_1 + S_{22}a_2, \tag{6.85}$$

where the four S-parameters are defined as

S_{11} the reflection coefficient at port-1,
S_{22} the reflection coefficient at port-2,
S_{12} the transmission coefficient from port-2 to port-1,
S_{21} the transmission coefficient from port-1 to port-2.

Readers should pay special attention to the definition of S_{12} and S_{21}, because it is easy to mistake S_{12} for "transmission coefficient from port 1 to port 2", so is S_{21}. The definition of transmission coefficient S_{ij} is "from port j to port i", but not "from port i to port j". The ordering in the definition is somewhat contrary to our intuition, so it is easy to induce an error.

Furthermore, Eq. (6.84) shows us the wave leaving port 1, i.e., b_1, consisting of two components: one is the reflected wave at port 1 ($S_{11}a_1$), the other is the transmitted wave from port 2 to port 1 ($S_{12}a_2$). Meanwhile, Eq. (6.85) shows us the wave leaving port 2, i.e., b_2, consisting of two components: one is the reflected wave at port 2 ($S_{22}a_2$), the other is the transmitted wave from port 1 to port 2 ($S_{21}a_1$).

Next, the S-parameters of a BJT depend not only on frequency, but also on bias voltage and current. For instance, a high-frequency BJT may have the following S-parameters at different frequencies for a specific bias condition:

$$f = 1\,\text{GHz}: \quad S_{11} = 0.42e^{j130°}, \ S_{21} = 5.57e^{j66°}, \ S_{12} = 0.88e^{j47°}, \ S_{22} = 0.23e^{-j59°},$$

$$f = 2\,\text{GHz}: \quad S_{11} = 0.44e^{j120°}, \ S_{21} = 2.43e^{j56°}, \ S_{12} = 0.09e^{j53°}, \ S_{22} = 0.21e^{-j90°},$$

$$f = 3\,\text{GHz}: \quad S_{11} = 0.47e^{j115°}, \ S_{21} = 1.66e^{j47°}, \ S_{12} = 0.11e^{j61°}, \ S_{22} = 0.25e^{-j87°}.$$

Besides, a high-frequency BJT may have the following S-parameters for different bias conditions when $f = 1$ GHz.

$$V_{CE} = 1\,\text{V}, \ I_C = 10\,\text{mA}: \quad S_{11} = 0.62e^{j170°}, \ S_{21} = 4.65e^{j76°}, \ S_{12} = 0.075e^{j97°}, \ S_{22} = 0.41e^{-j85°}$$

$$V_{CE} = 3\,\text{V}, \ I_C = 10\,\text{mA}: \quad S_{11} = 0.66e^{j160°}, \ S_{21} = 6.2e^{j86°}, \ S_{12} = 0.082e^{j92°}, \ S_{22} = 0.33e^{-j82°}$$

$$V_{CE} = 5\,\text{V}, \ I_C = 10\,\text{mA}: \quad S_{11} = 0.62e^{j155°}, \ S_{21} = 6.5e^{j93°}, \ S_{12} = 0.086e^{j77°}, \ S_{22} = 0.32e^{-j77°}$$

In practice, the S-parameters at different frequencies and bias conditions are provided in the datasheet. A circuit designer must find a BJT with appropriate S-parameters at the operation frequency and design the corresponding bias circuit to meet the requirements.

Example 6.20
Suppose we have a BJT properly biased in a circuit and its S-parameters at operation frequency $f = 3$ GHz is given by

$$S = \begin{bmatrix} 0.1e^{j30°} & 0.02e^{j50°} \\ 8e^{j75°} & 0.4e^{-j40°} \end{bmatrix}.$$

If the incident wave at port-1 is $a_1 = 2e^{j0°}$ and that at port-2 is $a_2 = 0.3e^{j60°}$, please derive the waves leaving the two ports, i.e., b_1 and b_2, respectively.

Solution
From Eq. (6.83), we have

$$\begin{bmatrix} b_1 \\ b_2 \end{bmatrix} = \begin{bmatrix} S_{11} & S_{12} \\ S_{21} & S_{22} \end{bmatrix} \begin{bmatrix} a_1 \\ a_2 \end{bmatrix} = \begin{bmatrix} 0.1e^{j30°} & 0.02e^{j50°} \\ 8e^{j75°} & 0.4e^{-j40°} \end{bmatrix} \begin{bmatrix} 2e^{j0°} \\ 0.3e^{j60°} \end{bmatrix}.$$

Hence

$$b_1 = 0.2e^{j30°} + 0.006e^{j110°},$$

$$b_2 = 16e^{j75°} + 0.12e^{j20°}.$$

■

Fig. 6.44 Circuit model for illustrating maximum power transfer theorem

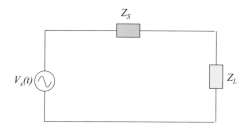

D High-Frequency Amplifier Design

Using the method of S-parameters, the design of high-frequency amplifiers becomes easier because the internal reflection of the amplifier circuit can be included in the design process. In the following, we briefly introduce the design of high-frequency amplifiers.

First, we introduce **maximum power transfer theorem**, that is very critical in high-frequency circuit design. As shown in Fig. 6.44, we have a signal source with a source impedance Z_S and a load with the impedance Z_L. In general, the source impedance is given by

$$Z_S = R_S + jX_S, \tag{6.86}$$

where R_S is the real part and X_S is the imaginary part. Similarly, Z_L can be expressed as

$$Z_L = R_L + jX_L, \tag{6.87}$$

where R_L is the real part and X_L is the imaginary part.

For a given Z_S, it can be proved that maximum power can be transferred from the source to the load, if the following condition meets:

$$Z_L = Z_S^*, \tag{6.88}$$

where Z_S^* is the complex conjugate of Z_S. That is, if $R_L = R_S$ and $X_L = -X_S$, then maximum power transfer is achieved. This theorem is useful in circuit design, because it tells us the way to achieve maximum power transfer from a source to a load, or more generally from a circuit to another circuit.

Next, we learn the design of high-frequency amplifiers. The design includes two steps.

Fig. 6.45 Design of DC bias
circuit with an inductor

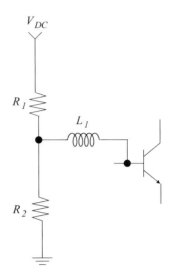

Step 1: Design of DC bias circuit

The DC bias circuit is to set appropriate voltages and currents of BJT to obtain the desired S-parameters. For instance, if the desired S-parameters of a BJT is given at $V_{CE} = 3$ V and $I_C = 10$ mA, we shall design the bias circuit to make the BJT operate at $V_{CE} = 3$ V and $I_C = 10$ mA.

The bias circuit for high-frequency amplifiers is basically the same as that of low-frequency amplifiers. First, we select the desired S-parameters from the BJT datasheet at a specific bias voltage/current. Then we use resistors to obtain the desired bias voltage/current, and use capacitors for DC blocking and AC bypassing. Besides, because the impedance of an inductor is zero at DC and approaches infinity when the frequency is sufficiently high, we may take it as a short circuit at DC, and an open circuit at high frequency. As shown in Fig. 6.45, the inductor L_1 brings the bias voltage set by resistors R_1 and R_2 to the base of BJT, while the influence of R_1 and R_2 on the input signal is removed because L_1 acts as an open circuit at high frequency. Note that the inductors and capacitors used in high-frequency amplifiers are specifically selected, so that the associated parasitic effects are minimized. Meanwhile, an automatic bias control circuit may be designed to achieve a stable bias condition.

Step 2: Design of input and output matching circuits

Because reflection may reduce the power transfer and interfere the input signal, we have to deal with reflection in high-frequency amplifiers. This is the major difference between high-frequency amplifiers and low-frequency amplifiers. Figure 6.46 shows a typical high-frequency amplifier with two matching circuits: the input matching circuit (IMC) deals with the reflection between the signal source and the BJT, and the output matching circuit (OMC) deals with the reflection between the BJT and the load. Suppose $R_0 = 50 \, \Omega$ is the characteristic resistance of the TX line. For

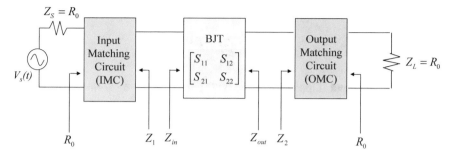

Fig. 6.46 Impedance matching design using IMC and OMC

Fig. 6.47 Impedance matching design between IMC, BJT and OMC

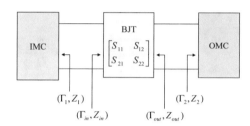

simplicity, we assume the source impedance $Z_S = 50\,\Omega$, and the load impedance $Z_L = 50\,\Omega$ too. In general, the input impedance and output impedance of a BJT are not 50 Ω.

The goal of the IMC is twofold. First, as shown in Fig. 6.46, its input impedance should match R_0, so that no reflection occurs at the IMC input. In this case, all the signal powers from the source will deliver to the IMC. Second, assume the output impedance of the IMC is Z_1, and the input impedance of the BJT is Z_{in}. From the maximum power transfer theorem, when $Z_1 = Z_{in}^*$, maximum power can be transferred from the IMC to the BJT. When the above two conditions are satisfied, we obtain maximum power transfer from the signal source to the BJT.

On the other hand, the goal of the OMC is twofold as well. First, its output impedance should match R_0, so that all the output powers of the OMC can deliver to the load with $Z_L = R_0$. Second, assume the input impedance of the OMC is Z_2, and the output impedance of the BJT is Z_{out}. From the maximum power transfer theorem, when $Z_2 = Z_{out}^*$, maximum power can be transferred from the BJT to the OMC. When the above two conditions are satisfied, we obtain maximum power transfer from the BJT to the load.

From the above descriptions of IMC and OMC, when the design goals of both matching circuits are met, we shall get maximum power transfer from the signal source to the BJT, and then from the BJT to the load. In a word, we get maximum power transfer from the signal source to the load.

In order to help readers gain more insight of high-frequency amplifier design, we present the relationship between several important parameters in the following. First, as shown in Fig. 6.47, the reflection coefficient at the BJT input, denoted as Γ_{in}, is given by

$$\Gamma_{in} = \frac{Z_{in} - R_0}{Z_{in} + R_0}, \tag{6.89}$$

where Z_{in} is the input impedance of the BJT. Let Γ_1 be the reflection coefficient at the IMC output. The relationship between Γ_1 and the output impedance of the IMC, i.e., Z_1, is given by

$$\Gamma_1 = \frac{Z_1 - R_0}{Z_1 + R_0}. \tag{6.90}$$

When $Z_1 = Z_{in}^*$, i.e., maximum power is transferred from IMC to the BJT, then we have

$$\Gamma_1 = \frac{Z_{in}^* - R_0}{Z_{in}^* + R_0} = \Gamma_{in}^*. \tag{6.91}$$

Consequently, if (Z_1, Z_{in}) is a complex conjugate pair, then (Γ_1, Γ_{in}) is a complex conjugate pair too. Hence, in order to achieve maximum power transfer, Eq. (6.91) shall be satisfied.

Next, we consider the reflection coefficient at the BJT output, denoted as Γ_{out}. As shown in Fig. 6.47, the relationship between Γ_{out} and Z_{out} is given by

$$\Gamma_{out} = \frac{Z_{out} - R_0}{Z_{out} + R_0}, \tag{6.92}$$

where Z_{out} is the output impedance of the BJT. Let Γ_2 be the reflection coefficient at the OMC input. The relationship between Γ_2 and input impedance of the OMC, i.e., Z_2, is given by

$$\Gamma_2 = \frac{Z_2 - R_0}{Z_2 + R_0}. \tag{6.93}$$

Hence, if $Z_2 = Z_{out}^*$, i.e., maximum power is transferred from the BJT to OMC, then we have

$$\Gamma_2 = \frac{Z_{out}^* - R_0}{Z_{out}^* + R_0} = \Gamma_{out}^*. \tag{6.94}$$

Consequently, if (Z_2, Z_{out}) is a complex conjugate pair, then (Γ_2, Γ_{out}) is a complex conjugate pair too. Hence, in order to achieve maximum power transfer, Eq. (6.94) shall be satisfied.

In above, we derive the relationship between Γ_{in} and Γ_1 at the BJT input as well as the relationship between Γ_{out} and Γ_2 at the BJT output. Now, we proceed to find the relationship between Γ_{in} and S-parameters. First, because of the nonzero S-parameter S_{12}, the reflection at the BJT input is actually related to the OMC. Owing to the multiple reflections between the BJT output and the OMC in Fig. 6.47, Γ_{in} is given by (the proof is given in Example 6.21)

$$\Gamma_{in} = S_{11} + \frac{S_{12}S_{21}\Gamma_2}{1 - S_{22}\Gamma_2}. \tag{6.95}$$

Equation (6.95) shows that Γ_{in} not only depends on S_{11}, but also depends on the other three S-parameters and Γ_2, i.e., the reflection coefficient at the OMC input.

On the other hand, due to the nonzero S-parameter S_{21}, the reflection coefficient Γ_{out} at the BJT output is actually related to the IMC. Similar to Eq. (6.95), Γ_{out} is given by

$$\Gamma_{out} = S_{22} + \frac{S_{12}S_{21}\Gamma_1}{1 - S_{11}\Gamma_1}. \tag{6.96}$$

Equation (6.96) shows that Γ_{out} not only depends on S_{22}, but also depends on the other three S-parameters and Γ_1.

The above results show the interplay between critical parameters in the design of high-frequency amplifiers. From Eq. (6.95), we find that Γ_{in} is a function of Γ_2, and from Eq. (6.96), we find that Γ_{out} is a function of Γ_1. Moreover, from Eq. (6.91) and Eq. (6.94), the following two conditions should be satisfied in order to achieve maximum power transfer from the source to the load, given by

$$\Gamma_{in} = \Gamma_1^* \Rightarrow S_{11} + \frac{S_{12}S_{21}\Gamma_2}{1 - S_{22}\Gamma_2} = \Gamma_1^*, \tag{6.97}$$

$$\Gamma_{out} = \Gamma_2^* \Rightarrow S_{22} + \frac{S_{12}S_{21}\Gamma_1}{1 - S_{11}\Gamma_1} = \Gamma_2^*. \tag{6.98}$$

For a given set of $(S_{11}, S_{22}, S_{12}, S_{21})$, we have to design Z_1 and Z_2 so that the corresponding Γ_1 and Γ_2 satisfy Eqs. (6.97) and (6.98) simultaneously. This is not an easy task in practice, which is the reason why high-frequency amplifier design is usually much more difficult than that of low-frequency ones.

The above is a brief introduction of impedance matching in high-frequency amplifier design. If readers are interested in this topic, you can take advanced courses on "high-frequency amplifier design", where you can learn more details about the concept and the practice.

Example 6.21

In Fig. 6.47, assume the reflection coefficient at the input of the OMC is Γ_2. Prove that Eq. (6.95), i.e., the reflection coefficient at the BJT input, is given by

$$\Gamma_{in} = S_{11} + \frac{S_{12}S_{21}\Gamma_2}{1 - S_{22}\Gamma_2}.$$

Solution

Let a_1 be the incident wave of the BJT in Fig. 6.47. Then the wave leaving the BJT input is given by

$$b_1 = S_{11}a_1 + S_{12}a_2, \tag{6.99}$$

where a_2 is the incident wave at the BJT output. Because the reflection coefficient of the OMC is Γ_2, we can express a_2 as

$$a_2 = v + v \cdot S_{22}\Gamma_2 + v \cdot (S_{22}\Gamma_2)^2 + v \cdot (S_{22}\Gamma_2)^3 + \cdots, \tag{6.100}$$

where

$$v = S_{21}a_1 \cdot \Gamma_2.$$

In the above, v denotes the wave of a_1 after passing through the BJT and reflected by OMC; the term $v \cdot S_{22}\Gamma_2$ denotes the wave v being further reflected by the BJT and the OMC; the term $v \cdot (S_{22}\Gamma_2)^2$ denotes the wave $v \cdot S_{22}\Gamma_2$ being reflected by the BJT and the OMC again; and so on. Hence, due to the multiple reflections between the BJT and the OMC, from Eq. (6.100), we have

$$\begin{aligned} a_2 &= v[1 + S_{22}\Gamma_2 + (S_{22}\Gamma_2)^2 + (S_{22}\Gamma_2)^3 + \cdots \\ &= \frac{v}{1 - S_{22}\Gamma_2} \\ &= \frac{S_{21}a_1\Gamma_2}{1 - S_{22}\Gamma_2}. \end{aligned} \tag{6.101}$$

Finally, from Eqs. (6.99) and (6.101), we obtain

$$b_1 = S_{11}a_1 + \frac{S_{12}S_{21}\Gamma_2}{1 - S_{22}\Gamma_2}a_1.$$

Hence, the reflection coefficient at the BJT input is given by

$$\Gamma_{in} = \frac{b_1}{a_1}$$

$$= S_{11} + \frac{S_{12}S_{21}\Gamma_2}{1 - S_{22}\Gamma_2}.$$

It completes the proof. Following the same logic as above, Eq. (6.96) can be proved. Readers are encouraged to do it by themselves.

■

Finally, after introducing the concept of impedance matching for high-frequency amplifier design, we are ready to learn a very important element for a wireless communication system—antenna, which is used to effectively transmit and receive EM waves in free space. This topic will be introduced in the next chapter.

Summary

6.1: We learn the principles of Smith chart and how to build it step by step.
6.2: We learn how to use Smith chart as a graphic tool in different applications.
6.3: We learn the concept of impedance matching circuit design by using lumped elements or distributed elements.
6.4: We learn the design consideration of an amplifier at different frequency ranges with the emphasis on the high-frequency impedance matching design.

Exercises

Please use Smith chart to solve the following problems. Suppose $R_0 = 50\,\Omega$ in all the problems.

1. If $Z_L = 10 + j30\,(\Omega)$, please derive the reflection coefficient Γ. (Hint: Example 6.4).
2. If $Z_L = 90 - j50\,(\Omega)$, please derive the reflection coefficient Γ.
3. If $\Gamma = 0.3e^{j30°}$, please derive the load impedance Z_L. (Hint: Example 6.6).
4. If $\Gamma = 0.5e^{-j120°}$, please derive the load impedance Z_L.
5. If $\Gamma = 0.4 + j0.7$, please derive the load impedance Z_L. (Hint: Example 6.7).
6. If $\Gamma = 0.2 - j0.3$, please derive the load impedance Z_L.
7. Suppose $\Gamma = 0.2e^{j80°}$ and $d = 0.15\lambda$. Please derive $\Gamma' = \Gamma e^{-j2\beta d}$. (Hint: Example 6.8).
8. Suppose $\Gamma = 0.7e^{-j110°}$ and $d = 0.3\lambda$. Please derive $\Gamma' = \Gamma e^{-j2\beta d}$.
9. In Fig. 6.18, if $Z_L = 10 + j20\,(\Omega)$ and $d = 0.08\lambda$, please derive the input impedance Z_i. (Hint: Example 6.10).
10. In Fig. 6.18, if $Z_L = 20 - j80\,(\Omega)$ and $d = 0.3\lambda$, please derive the input impedance Z_i.
11. In Fig. 6.18, suppose $Z_L = 50 - j50\,(\Omega)$, $d = 2\,cm$, attenuation constant $\alpha = 0.2/cm$ and wavelength $\lambda = 10\,cm$. Please derive the input impedance Z_i. (Hint: Example 6.12).

12. In Fig. 6.18, suppose $Z_L = 20 + j15\,(\Omega)$, $d = 3\,\text{cm}$, $\alpha = 0.2/\text{cm}$, and $\lambda = 10\,\text{cm}$. Please derive the input impedance Z_i.

13. In Fig. 6.18, suppose $d = 1\,\text{cm}$, $\alpha = 0.1/\text{cm}$, and $\lambda = 10\,\text{cm}$. If the measured input impedance is $Z_i = 40 - j30$, please derive the load impedance Z_L. (Hint: Example 6.13).

14. In Fig. 6.18, suppose $d = 3\,\text{cm}$, $\alpha = 0.1/\text{cm}$, and $\lambda = 10\,\text{cm}$. If the measured input impedance $Z_i = 70 + j30$, please derive the load impedance Z_L.

15. Suppose $u = 0.4 + j0.6$. Please derive $u' = 1/u$. (Hint: Example 6.14).

16. Suppose $u = 0.2 - j0.4$. Please derive $u' = 1/u$.

17. Suppose $Z_L = 15 + j20\,(\Omega)$. Please derive Y_L. (Hint: Example 6.15).

18. Suppose $Z_L = 30 - j40\,(\Omega)$. Please derive Y_L.

19. What is the meaning of lumped elements and distributed elements in a circuit? Explain the difference between them.
     (Hint: Refer to Sect. 6.3).

20. For the matching circuit design using lumped elements as shown in Fig. 6.30, suppose $R_0 = 75\,\Omega$ and $Z_L = 15 + j50\,(\Omega)$. The operation frequency is $f = 3\,\text{GHz}$. Please design the matching circuit.
     (Hint: Example 6.17).

21. For the matching circuit design using TX lines as shown in Fig. 6.34, suppose $R_0 = 50\,\Omega$ and $Z_L = 50 + j150\,(\Omega)$. Please design the matching circuit so that $Z_i = R_0$.
     (Hint: Example 6.18).

22. Suppose we have $R_0 = 50\,\Omega$ and $Z_L = 100 - j50\,(\Omega)$. Please design a matching circuit using TX lines as shown in Fig. 6.34 so that $Z_i = R_0$.
     (Hint: Example 6.18).

23. Suppose $R_L = 300\,\Omega$ is the load resistance and $R_0 = 75\,\Omega$ is the characteristic resistance of the TX line. If we want to use a $\lambda/4$ TX line with the characteristic resistance R_A to achieve impedance matching, i.e., $Z_i = R_0$, please derive R_A.
     (Hint: Example 6.19).

24. Refer to Sect. 6.4 explain why a BJT amplifier has different models in low frequency and medium frequency.

25. Please draw a two-port network and explain the meaning of S-parameters.
     (Hint: Refer to Sect. 6.4).

26. Suppose we have a BJT properly biased in a circuit and its S-parameters at operation frequency $f = 1\,\text{GHz}$ is given by

$$S = \begin{bmatrix} 0.05e^{j80°} & 0.3e^{j30°} \\ 4e^{j65°} & 0.2e^{-j15°} \end{bmatrix}.$$

If the incident voltage wave at port-1 is $a_1 = 3e^{j0°}$ and that at port-2 is $a_2 = 0.2e^{j50°}$, please derive b_1 and b_2.
     (Hint: Example 6.20).

27. Suppose we have a BJT in Fig. 6.47 and its S-parameters is given by

$$S = \begin{bmatrix} 0.1e^{j30°} & 0.04e^{j50°} \\ 7e^{j75°} & 0.4e^{-j40°} \end{bmatrix}.$$

If the reflection coefficient at the input of the OMC is $\Gamma_2 = 0.2$. Please derive the reflection coefficient Γ_{in} at the BJT input.
(Hint: Example 6.21).

28. Following the same logic as in deriving Eq. (6.95) in Example 6.21, please derive Eq. (6.96).

29. Assume the BJT in Fig. 6.47 has the same S-parameters as those given in Exercise 27. If the reflection coefficient at the BJT output is $\Gamma_{out} = 0.3e^{j30°}$, please derive the reflection coefficient at the IMC output, i.e., Γ_1.

Chapter 7
Antenna

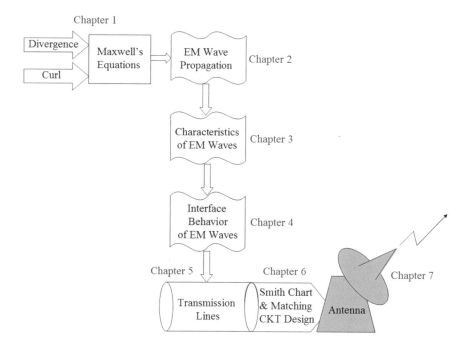

© Springer Nature Switzerland AG 2020
M.-S. Kao and C.-F. Chang, *Understanding Electromagnetic Waves*,
https://doi.org/10.1007/978-3-030-45708-2_7

Abstract Antenna is a critical component in wireless communication systems because it can effectively transmit and receive EM waves. Unfortunately, the study of antenna is typically not easy since it involves lengthy mathematics. By contrast, our approach emphasizes the physical meaning and avoid unnecessary mathematics so that readers can gain the insight of antenna principle step by step:

Step 1. Learning the radiation of a tiny conducting line.
Step 2. Learning a simple dipole antenna based on knowledge gained in Step 1.
Step 3. Learning critical parameters of general antennas.
Step 4. Learning the receiving mechanism of antenna.

In above, we use abundant examples and illustrations to show the fundamental principles of antenna in a heuristic way. For instance, we show why a transmission line cannot effectively radiate EM waves by physical intuition and simple mathematics. Then with a slight modification of the structure, it becomes an antenna which can effectively radiate EM waves.

Furthermore, we use the principle of fishing net to explain the idea of effective area of a receiving antenna, so that readers can fully understand its physical meanings. The study of this chapter will help readers understand core ideas and properties of antennas which are very important in wireless communication systems.

Keywords Antenna · Spherical coordinate · Near field · Far field · Dipole antenna · Radiation pattern · E-plane pattern · H-plane pattern · Beamwidth · Directivity · Omnidirectional antenna · Directional gain · Radiation efficiency · Reciprocity theorem · Effective area · Friis transmission formula

Briefing

Freedom, everybody pursues by nature. Not only human beings, but also EM waves like freedom! When an EM wave is confined in a limited space, it is like a prisoner and tries to escape from the space. Using this character of EM waves and some physical laws, we can confine an EM wave in a specific space, and release it (let it radiate) at a desired timing and scenario.

For EM waves, **an antenna plays a different role from a TX line**. The function of TX line is to transmit signal effectively in a specific space. For example, a parallel-plate TX line confines EM waves between two metal plates in order to convey waves to a target destination. In this case, we try to reduce radiation to the minimum and thus prevent power loss. For an antenna, we try to radiate EM waves as much as possible so that more power is transmitted and then received by users. Therefore, how to radiate EM waves effectively is the key point when designing an antenna.

In this chapter, we focus on antennas and six sections are provided. For each section, we introduce an important principle with a number of critical parameters. We start from the most fundamental concept of antennas and introduce the simplest antenna. Then we discuss important concepts of antennas step by step. Finally, we discuss the relationship between transmitting antennas and receiving antennas. These sections shall help readers understand the properties of antennas and thus build a solid background of antennas.

7.1 Introduction

In this section, we first introduce a useful coordinate system when designing an antenna. Then we discuss the difference between an antenna and a TX line in order to further explore fundamental principles of antennas.

A *Spherical Coordinate System*

When dealing with antenna problems, we find **spherical coordinate** is much more useful than rectangular (Cartesian) coordinate in many scenarios because a space-varying EM field can be well specified with spherical coordinate. We introduce spherical coordinate system in the following.

First, in a two-dimensional plane, we can use **polar coordinate** specifying the position of a point by two numbers: the radial distance of that point from the origin and its azimuth angle. As shown in Fig. 7.1, let A be a point on the xy-plane and O be the origin. Suppose the distance between O and point A is r, i.e., the length of $\overline{OA}$ is r, and ϕ is the azimuth angle between x-axis and $\overline{OA}$. Hence, given the distance r and angle ϕ, we can precisely specify the position of point A. This is the basic idea of polar coordinate system. As shown in Fig. 7.1, point A locates at (r, ϕ) in a polar coordinate system. It means we can start from the origin O along the direction specified by the azimuth angle ϕ. After walking the distance r, we reach point A.

From Fig. 7.1, we can convert point A from polar coordinate (r, ϕ) into rectangular coordinate (x_0, y_0) by

$$x_0 = r \cdot \cos \phi, \tag{7.1}$$

$$y_0 = r \cdot \sin \phi. \tag{7.2}$$

Fig. 7.1 A point in rectangular coordinate and polar coordinate

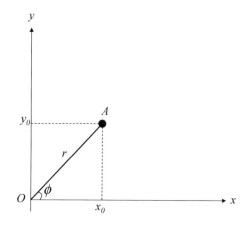

Conversely, from Fig. 7.1, we can convert point A from rectangular coordinate (x_0, y_0) into polar coordinate (r, ϕ) by

$$r = \sqrt{x_0^2 + y_0^2}, \tag{7.3}$$

$$\phi = \tan^{-1} \frac{y_0}{x_0}. \tag{7.4}$$

Using Eqs. (7.1)–(7.4), the conversion between polar coordinate and rectangular coordinate can be readily achieved.

Extending the idea of polar coordinate to three-dimensional (3-D), we get spherical coordinate. As shown in Fig. 7.2, let A be a point in three-dimensional space and O be the origin. Suppose the distance between O and point A is R, i.e., the length of $\overline{OA}$ is R, and θ is the angle between $\overline{OA}$ and z-axis. Let $\overline{OB}$ be the projection of $\overline{OA}$ on the xy-plane, and ϕ be the azimuth angle between x-axis and $\overline{OB}$. Then, given the distance R, angles θ and ϕ, we can precisely specify the position of point A. As shown in Fig. 7.2, the spherical coordinate of point A is (R, θ, ϕ). It means we can start from the origin O along the direction specified by the angle pair (θ, ϕ). After walking the distance R, we reach point A. For example, suppose point A locates at $(7, 30°, 43°)$. It means the distance between point A and the origin is 7 unit length, the angle between $\overline{OA}$ and z-axis is 30°, and the azimuth angle between x-axis and $\overline{OB}$ is 43°. When we move from the origin along the direction specified by $(\theta = 30°, \phi = 43°)$, after a distance of 7 unit length, we reach point A.

In 3-D space, it is easy to convert spherical coordinate to rectangular coordinate, and vice versa. Suppose point A locates at (R, θ, ϕ) of a spherical coordinate system. From Fig. 7.2, we have $\overline{OB} = R \cdot \cos(\pi/2 - \theta) = R \cdot \sin\theta$. Hence, we can convert the position to rectangular coordinate (x_0, y_0, z_0) by

$$x_0 = \overline{OB} \cdot \cos\phi = R \cdot \sin\theta \cdot \cos\phi, \tag{7.5}$$

Fig. 7.2 A point in spherical coordinate

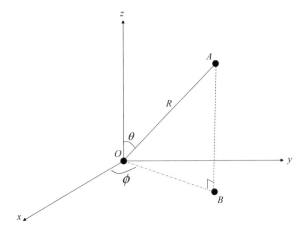

$$y_0 = \overline{OB} \cdot \sin \phi = R \cdot \sin \theta \cdot \sin \phi, \tag{7.6}$$

$$z_0 = R \cdot \cos \theta. \tag{7.7}$$

On the other hand, suppose point A locates at (x_0, y_0, z_0) of a rectangular coordinate system. From Fig. 7.2, we can convert the position to spherical coordinate (R, θ, ϕ) by

$$R = \sqrt{x_0^2 + y_0^2 + z_0^2}, \tag{7.8}$$

$$\theta = \cos^{-1} \frac{z_0}{R} = \cos^{-1} \frac{z_0}{\sqrt{x_0^2 + y_0^2 + z_0^2}}, \tag{7.9}$$

$$\phi = \tan^{-1} \frac{y_0}{x_0}. \tag{7.10}$$

Note that in spherical coordinate, the range of θ is $0 \le \theta \le \pi$, and the range of ϕ is $-\pi \le \phi \le \pi$.

Example 7.1
Suppose point A has the spherical coordinate $(4, 30°, 120°)$ and point B has the rectangular coordinate $(2, -3, 6)$. Please derive (a) the length of $\overline{AB}$ (b) the angle between $\overrightarrow{OA}$ and $\overrightarrow{OB}$.

Solution

(a) Suppose the rectangular coordinate of point A is (x_0, y_0, z_0). From Eqs. (7.5) to (7.7), we have

$$x_0 = R \cdot \sin \theta \cdot \cos \phi = 4 \cdot \sin 30° \cdot \cos 120° = -1$$
$$y_0 = R \cdot \sin \theta \cdot \sin \phi = 4 \cdot \sin 30° \cdot \sin 120° = \sqrt{3}.$$
$$z_0 = R \cdot \cos \theta = 4 \cdot \cos 30° = 2\sqrt{3}$$

Hence point A locates at $(-1, \sqrt{3}, 2\sqrt{3})$ and the length of $\overline{AB}$ is given by

$$\overline{AB} = \sqrt{(-1-2)^2 + (\sqrt{3} - (-3))^2 + (2\sqrt{3} - 6)^2} = \sqrt{69 - 18\sqrt{3}}.$$

(b) From the above, we have

$$\overrightarrow{OA} = (-1, \sqrt{3}, 2\sqrt{3})$$
$$\overrightarrow{OB} = (2, -3, 6).$$

The dot product of two vectors $\overrightarrow{OA}$ and $\overrightarrow{OB}$ is calculated by

$$\overrightarrow{OA} \cdot \overrightarrow{OB} = (-1 \times 2) + (\sqrt{3} \times -3) + (2\sqrt{3} \times 6) = 9\sqrt{3} - 2.$$

From the definition of dot product, we have

$$\overrightarrow{OA} \cdot \overrightarrow{OB} = |\overrightarrow{OA}| \cdot |\overrightarrow{OB}| \cdot \cos \angle\theta_{AB},$$

where $\angle\theta_{AB}$ is the angle between $\overrightarrow{OA}$ and $\overrightarrow{OB}$. Since

$$|\overrightarrow{OA}| = R = 4,$$
$$|\overrightarrow{OB}| = \sqrt{2^2 + (-3)^2 + 6^2} = 7.$$

Finally, we get

$$\theta_{AB} = \cos^{-1} \frac{\overrightarrow{OA} \cdot \overrightarrow{OB}}{|\overrightarrow{OA}| \cdot |\overrightarrow{OB}|} = \cos^{-1} \left(\frac{9\sqrt{3} - 2}{28} \right).$$

∎

In vector analysis, we can define associated **unit vectors** for each point in spherical coordinate system. As shown in Fig. 7.3, a point A locates at (R, θ, ϕ) in a spherical coordinate system. We can define three unit vectors $(\hat{a}_R, \hat{a}_\theta, \hat{a}_\phi)$ whose length is equal to unity, i.e., $|\hat{a}_R| = |\hat{a}_\theta| = |\hat{a}_\phi| = 1$, and they are mutually orthogonal, i.e., $\hat{a}_R \perp \hat{a}_\theta \perp \hat{a}_\phi$. These three unit vectors are specified as follows.

Fig. 7.3 Unit vectors in spherical coordinate

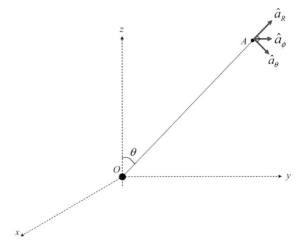

$\hat{a}_R$: the unit vector regarding R. It has the same direction as the vector $\overrightarrow{OA}$.

$\hat{a}_\theta$: the unit vector regarding θ. It locates on the plane spanned by the z-axis and the vector $\overrightarrow{OA}$, and it is perpendicular to $\hat{a}_R$.

$\hat{a}_\phi$: the unit vector regarding ϕ. It is given by $\hat{a}_\phi = \hat{a}_R \times \hat{a}_\theta$.

From above, we can specify $(\hat{a}_R, \hat{a}_\theta, \hat{a}_\phi)$ in rectangular coordinate system by the following formulas:

$$\hat{a}_R = \sin\theta\cos\phi \cdot \hat{x} + \sin\theta\sin\phi \cdot \hat{y} + \cos\theta \cdot \hat{z}, \tag{7.11}$$

$$\hat{a}_\theta = \cos\theta\cos\phi \cdot \hat{x} + \cos\theta\sin\phi \cdot \hat{y} - \sin\theta \cdot \hat{z}, \tag{7.12}$$

$$\hat{a}_\phi = -\sin\phi \cdot \hat{x} + \cos\phi \cdot \hat{y}. \tag{7.13}$$

For different points in space, we have different unit vectors $(\hat{a}_R, \hat{a}_\theta, \hat{a}_\phi)$. These three vectors are very useful in antenna design because they can effectively specify the variation of EM fields induced by an antenna.

Example 7.2

Suppose point A locates at $(4, 30°, 120°)$ in a spherical coordinate system. Please derive the associated unit vectors $(\hat{a}_R, \hat{a}_\theta, \hat{a}_\phi)$.

Solution

Because $\theta = 30°$ and $\phi = 120°$, from Eqs. (7.11) to (7.13), we get

$$\hat{a}_R = \sin\theta\cos\phi \cdot \hat{x} + \sin\theta\sin\phi \cdot \hat{y} + \cos\theta \cdot \hat{z}$$

$$= \left(\frac{1}{2}\right)\left(-\frac{1}{2}\right) \cdot \hat{x} + \left(\frac{1}{2}\right)\left(\frac{\sqrt{3}}{2}\right) \cdot \hat{y} + \left(\frac{\sqrt{3}}{2}\right) \cdot \hat{z}$$

$$= -\frac{1}{4} \cdot \hat{x} + \frac{\sqrt{3}}{4} \cdot \hat{y} + \frac{\sqrt{3}}{2} \cdot \hat{z},$$

$$\hat{a}_\theta = \cos\theta\cos\phi \cdot \hat{x} + \cos\theta\sin\phi \cdot \hat{y} - \sin\theta \cdot \hat{z}$$

$$= \left(\frac{\sqrt{3}}{2}\right)\left(-\frac{1}{2}\right) \cdot \hat{x} + \left(\frac{\sqrt{3}}{2}\right)\left(\frac{\sqrt{3}}{2}\right) \cdot \hat{y} - \left(\frac{1}{2}\right) \cdot \hat{z}$$

$$= -\frac{\sqrt{3}}{4} \cdot \hat{x} + \frac{3}{4} \cdot \hat{y} - \frac{1}{2} \cdot \hat{z},$$

$$\hat{a}_\phi = -\sin\phi \cdot \hat{x} + \cos\phi \cdot \hat{y}$$

$$= -\frac{\sqrt{3}}{2} \cdot \hat{x} - \frac{1}{2} \cdot \hat{y}.$$

Readers are encouraged to verify $|\hat{a}_R| = |\hat{a}_\theta| = |\hat{a}_\phi| = 1$ and $\hat{a}_R \perp \hat{a}_\theta \perp \hat{a}_\phi$ by themselves.

∎

B Principles of Antennas

In Sect. 2.1, we start from Maxwell's equations and with a little imagination, we can realize how an EM wave propagates outward from a conducting line as shown in Fig. 2.8. In this section, we introduce antenna—a device designed to convert electric power into radiated EM waves.

First, an antenna mainly consists of pieces of conductor. When a current flows through a conductor, from Ampere's law, we have

$$\nabla \times \vec{H} = \vec{J}, \tag{7.14}$$

where $\vec{J}$ is the current density and $\vec{H}$ is the induced magnetic field. From Eq. (7.14), when a time-varying current flows through a conductor, a magnetic field $\vec{H}$ is induced on the plane perpendicular to the direction of the current. The induced magnetic field $\vec{H}$ will generate an electric field $\vec{E}$, and then $\vec{E}$ will generate another magnetic field $\vec{H}$, and so on. Hence an EM wave propagates outward (radiates) as shown in Fig. 2.8. Now, our problem is:

How can we **effectively** radiate an EM wave?

In Fig. 7.4, we transmit a signal $V_S(t)$ by a TX line, and the TX line conveys the signal to an antenna. Then the antenna will radiate the signal. First, we explain why a TX line cannot radiate EM waves effectively. Then we explain why an antenna can radiate EM waves effectively.

Suppose we have a TX line along y-axis as shown in Fig. 7.5. It consists of two parallel conducting lines: line A and line B. From the characteristics of TX lines, for a specific position $z = z_0$, the magnitude of a current in line A is equal to that of the corresponding current in line B, but they flow in opposite directions. That is, if $I_0\hat{y}$ flows in line A, then $-I_0\hat{y}$ flows in line B.

Now, imagine that we are very far away from the TX line and look back at the TX line. In this case, because the distance is so far, we cannot actually distinguish line A

Fig. 7.4 Signal transmission from source to antenna with TX line

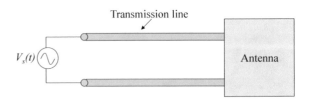

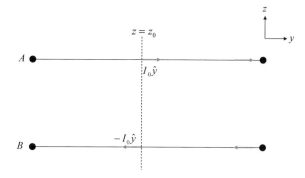

Fig. 7.5 Illustrating currents in TX line

Fig. 7.6 TX line with small bended segments at terminals

and line B, so they seem combining together as a single line, instead of two separated lines. The current of the combined line is the sum of two currents (identical magnitude with opposite directions) and the resultant current seems mutually canceled. In other words, the TX line seems like a conducting line having no current flow. Hence for us at a far distance, the TX line seems not radiating any EM waves.

In fact, we do expect very limited radiation from a TX line to prevent power loss during transmission. In this case, the signal can be conveyed to the destination efficiently. However, what we have in Fig. 7.5 is obviously not a good candidate of antenna.

Let us make a little change in Fig. 7.5 by bending a small segment of terminal section as shown in Fig. 7.6. A small segment of line A and a small segment of line B are bended to $+z$ and $-z$, respectively. After this change, although the current in line A and line B may remain the same as what we have in Fig. 7.5, it looks quite different for an observer at a far distance. Because the currents on the small bended segments of line A and line B have the same direction as shown in Fig. 7.6, the combined current is not canceled. In this case, two bended segments look like a conducting line toward $+z$ having nonzero current. In other words, the bended segments form a simple antenna that can radiate EM waves.

From the above example, it tells us that with a little modification of conducting lines, we may effectively radiate EM power which is originally preserved in a TX line.

C Radiation of a Small Conducting Line

No matter how complicated the structure of an antenna maybe, we can divide it into lots of conducting pieces, and each conducting piece has its own contribution to the overall radiation. For example, when we use a conducting line as an antenna, we can cut it into lots of tiny segments. The radiating EM field of the conducting line is actually the sum of the respective radiating field contributed by each tiny segment. Hence learning the radiating EM field generated by a tiny conducting line is a good start to realize how an antenna works.

Suppose we have a tiny conducting line as an antenna and its length is $d\ell$, where $d\ell$ is much smaller than the wavelength λ_0 of the radiated EM wave. Hence, the current throughout the line can be regarded as a constant at a specific time, given by

$$i(t) = I \cdot \cos \omega t, \tag{7.15}$$

where I is a constant and ω is the frequency. Assume the line locates at the origin and points toward $+z$ as shown in Fig. 7.7. As discussed previously, this tiny conducting line radiates EM power. Suppose we have a point P with the spherical coordinate (R, θ, ϕ) and we want to know the induced EM fields of this line at P. We can use Maxwell's equations to derive the induced EM fields. Because the derivation is somewhat complicated, in order to focus on the principles of antenna, we skip the derivation and give the result directly. The induced magnetic field at point P is given

Fig. 7.7 Radiation of a tiny conducting line

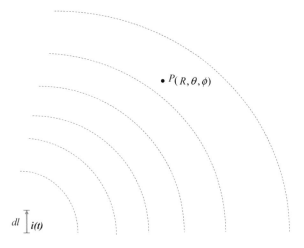

by

$$\vec{H} = H_\phi \cdot \hat{a}_\phi$$
$$= -\frac{Id\ell}{4\pi}\beta^2 \sin\theta \cdot e^{-j\beta R}\left[\frac{1}{j\beta R} + \frac{1}{(j\beta R)^2}\right] \cdot \hat{a}_\phi, \qquad (7.16)$$

where $\hat{a}_\phi$ is the unit vector in the direction of ϕ, H_ϕ is the component of $\vec{H}$ along $\hat{a}_\phi$ and the phase constant $\beta = 2\pi/\lambda_0$. Note that $\vec{H}$ only has the component along $\hat{a}_\phi$.

From Eq. (7.16), when the current $i(t)$ flows along z-axis, the induced magnetic field $\vec{H}$ at the point P is along the direction $\hat{a}_\phi$. The magnitude of $\vec{H}$ is proportional to the length $d\ell$ and decreases as the distance R increases. In addition, H_ϕ depends on $\sin\theta$. The maximal magnitude occurs at $\theta = 90°$, and the minimum occurs at $\theta = 0°$ and $\theta = 180°$ $(H_\phi = 0)$. Hence we have the maximal M-field when it is perpendicular to z-axis $(\theta = 90°)$, and the minimum when it is parallel to z-axis $(\theta = 0°$ or $\theta = 180°)$. In other words, we have the maximal M-field along the direction perpendicular to the direction of $i(t)$, and the minimum along the direction parallel to the direction of $i(t)$. The results are consistent with Ampere's law, i.e., $\nabla \times \vec{H} = \vec{J}$, which implies the induced M-field is mainly perpendicular to the direction of current density $\vec{J}$.

After deriving the magnetic field $\vec{H}$, we can easily derive the associated electric field $\vec{E}$ at point P from Maxwell's equations. Because $\vec{J} = 0$ in free space, we have

$$\nabla \times \vec{H} = j\omega \,\epsilon_0 \,\vec{E}, \qquad (7.17)$$

where ϵ_0 is the permittivity of the free space. From Eq. (7.17), the electric field at point P is given by

$$\vec{E} = \frac{\nabla \times \vec{H}}{j\omega \,\epsilon_0} = E_R \cdot \hat{a}_R + E_\theta \cdot \hat{a}_\theta, \qquad (7.18)$$

where E_R and E_θ are the components of $\vec{E}$ along $\hat{a}_R$ and $\hat{a}_\theta$, respectively. Because $\vec{E} \perp \vec{H}$, the component of $\vec{E}$ along $\hat{a}_\phi$ is null.

From Eqs. (7.16) and (7.18), we can obtain E_R and E_θ given by

$$E_R = -2M \cos\theta \cdot e^{-j\beta R} \cdot \left[\frac{1}{(j\beta R)^2} + \frac{1}{(j\beta R)^3}\right], \qquad (7.19)$$

$$E_\theta = -M \sin\theta \cdot e^{-j\beta R} \cdot \left[\frac{1}{j\beta R} + \frac{1}{(j\beta R)^2} + \frac{1}{(j\beta R)^3}\right], \qquad (7.20)$$

where

$$M = \frac{Id\ell}{4\pi}\eta_0\beta^2 = 30Id\ell\beta^2. \tag{7.21}$$

Note that η_0 is the wave impedance of free space and $\eta_0 = 120\pi$. From Eqs. (7.19) to (7.21), we find that the magnitude of $\vec{E}$ is proportional to the length $d\ell$, and it decreases when the distance R increases. These features are similar to those of $\vec{H}$.

D   Near Field and Far Field

The **near zone** and the **far zone** stand for two regions of EM field around an antenna. In Fig. 7.7, when the distance between point P and the origin (antenna) is much smaller than a wavelength, i.e., $R \ll \lambda_0$, we say point P locates at the **near zone**, and the associated EM fields are called **near fields**. In this case, we have

$$\beta R = \frac{2\pi}{\lambda_0} \cdot R \ll 1. \tag{7.22}$$

Hence

$$\frac{1}{(\beta R)^3} \gg \frac{1}{(\beta R)^2} \gg \frac{1}{(\beta R)}, \tag{7.23}$$

and

$$e^{-j\beta R} \approx 1. \tag{7.24}$$

From Eqs. (7.23) and (7.24), the M-field and the E-field in Eqs. (7.17), (7.19) and (7.20) can be approximated by

$$H_\phi \approx \frac{Id\ell}{4\pi R^2} \sin\theta, \tag{7.25}$$

$$E_R \approx -j\frac{2M\cos\theta}{(\beta R)^3}, \tag{7.26}$$

$$E_\theta \approx -j\frac{M\sin\theta}{(\beta R)^3}. \tag{7.27}$$

On the other hand, when the distance between point P and the origin (antenna) is much greater than a wavelength, i.e., $R \gg \lambda_0$, we say point P locates at the **far zone** and the associated EM fields are called **far fields**. In this case, we have

$$\frac{1}{\beta R} \gg \frac{1}{(\beta R)^2} \gg \frac{1}{(\beta R)^3}. \tag{7.28}$$

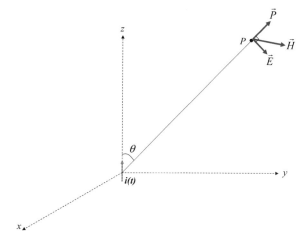

Fig. 7.8 Radiated EM field and the associated Poynting vector

Hence Eqs. (7.16), (7.19), and (7.20) can be approximated by

$$H_\phi \approx j\frac{Idl \cdot \beta \sin\theta}{4\pi R} \cdot e^{-j\beta R}, \tag{7.29}$$

$$E_R \approx \frac{2M\cos\theta}{(\beta R)^2} \cdot e^{-j\beta R}, \tag{7.30}$$

$$E_\theta \approx j\frac{M\sin\theta}{\beta R} \cdot e^{-j\beta R}. \tag{7.31}$$

In most applications such as wireless communications, what we concern is the EM wave behavior at the far zone because users are usually at this region. Hence in the following, we will focus on Eqs. (7.29)–(7.31).

In the far field, because $\beta R \ll (\beta R)^2$, from Eqs. (7.30) and (7.31), we have $|E_\theta| \gg |E_R|$. Hence the E-fled can be further simplified as

$$\vec{E} = E_R \cdot \hat{a}_R + E_\theta \cdot \hat{a}_\theta \approx E_\theta \cdot \hat{a}_\theta. \tag{7.32}$$

Equation (7.32) means that the E-field at the far zone is dominated by E_θ.

In Fig. 7.8, the current $i(t)$ flows along $\hat{z}$. From the above discussions, at the far zone, the E-field mainly has $\hat{a}_\theta$ component and the M-field has $\hat{a}_\phi$ component. Because $\hat{a}_\theta \times \hat{a}_\phi = \hat{a}_R$, the direction of Poynting vector ($\vec{P} = \vec{E} \times \vec{H}$) is along $\hat{a}_R$. It is exactly the direction of EM radiation, which means the antenna deliver EM power directly along the direction of $\hat{a}_R$ at the point P.

In addition, from Eqs. (7.21), (7.29), and (7.31), the ratio of E_θ to H_ϕ is given by

$$\frac{E_\theta}{H_\phi} = 120\pi = \eta_0. \tag{7.33}$$

This result is not surprising because at the far zone, the radiated EM wave is very close to a plane wave. Hence, the ratio of E-field to the M-field is equal to the wave impedance η_0.

Finally, from Eqs. (7.21) and (7.31), we have

$$E_\theta = j30Idl \cdot \beta \sin\theta \cdot \left(\frac{e^{-j\beta R}}{R}\right). \tag{7.34}$$

From Eq. (7.34), the E-field at the far zone depends on θ and is inversely proportional to the distance R. Besides, the associated phase varies with $e^{-j\beta R}$, which is an important property of a radiated EM wave.

Example 7.3
Suppose we have a tiny conducting line at the origin and it points toward $+z$ as shown in Fig. 7.8. Its length is $dl = 1$ cm and the current is given by $i(t) = 2 \cdot \cos 2\pi f t$ ampere, where $f = 100$ MHz. Suppose point P is at (R, θ, ϕ), where $R = 1$ km, $\theta = 30°$ and $\phi = 70°$. Please derive the E-field and M-field at point P.

Solution
First, the corresponding wavelength of the operating frequency $f = 100$ MHz is given by

$$\lambda_0 = \frac{c}{f} = \frac{3 \times 10^8}{100 \times 10^6} = 3\,\text{m}.$$

Because $R \gg \lambda_0$, the point P locates at the far zone. From Eq. (7.34), we have

$$E_\theta = j30Idl \cdot \beta \sin\theta \cdot \left(\frac{e^{-j\beta R}}{R}\right).$$

Since

$$\beta = \frac{2\pi}{\lambda_0} = \frac{2\pi}{3}$$

$$\beta R = \frac{2\pi}{3} \times 1000 = \frac{2000\pi}{3}$$

$$\sin\theta = \sin 30° = \frac{1}{2},$$

we can derive the E-field given by

$$E_\theta = j30 \cdot (2) \cdot (0.01) \cdot \left(\frac{2\pi}{3}\right) \cdot \left(\frac{1}{2}\right) \cdot \frac{e^{-j\frac{2000\pi}{3}}}{1000} = j\frac{\pi}{5000} \cdot e^{-j\frac{2\pi}{3}}$$

$$= \frac{\pi}{5000} \cdot e^{-j\frac{\pi}{6}}.$$

Note that $j = e^{j\frac{\pi}{2}}$. Finally we get

$$\vec{E} = E_\theta \cdot \hat{a}_\theta = \frac{\pi}{5000} \cdot e^{-j\frac{\pi}{6}} \cdot \hat{a}_\theta.$$

From Eq. (7.33), we immediately get

$$H_\phi = \frac{E_\theta}{\eta_0} = \frac{1}{600,000} \cdot e^{-j\frac{\pi}{6}}$$

and the M-field is given by

$$\vec{H} = H_\phi \cdot \hat{a}_\varphi = \frac{1}{600,000} \cdot e^{-j\frac{\pi}{6}} \cdot \hat{a}_\phi.$$

■

7.2 Dipole Antenna

When a current flows in a tiny conducting line, according to Ampere's law ($\nabla \times \bar{H} = \bar{J}$), EM waves radiate. When considering radiated EM waves of an antenna, the scenario is more complicated because an antenna consists of lots of tiny conducting lines. Besides, the length and the distribution of current of each conducting line all affect the resultant radiated EM fields. In this section, we investigate a simple antenna, **dipole antenna**, in order to explore the fundamental properties of antennas.

A   Far Field

As shown in Fig. 7.9, we have a dipole antenna connecting to a TX line. The dipole antenna is composed of two tiny conducting lines and each has the length h. These two conducting lines can be treated as the parts of the TX line but bended toward $+z$ and $-z$, respectively. Because of the bending, an EM wave radiates into the space.

Suppose the center of the antenna in Fig. 7.9 is at $z = 0$ and the terminals of two conducting lines are at $z = h$ and $z = -h$, respectively. Because both terminals are **open circuit**, the associated currents at $z = h$ and $z = -h$ must be zero. If the input signal is a sinusoidal signal, due to the requirement of zero current at $z = h$ and

$z = -h$, the current phasor in two conducting lines is given by

$$I(z) = I_0 \sin \beta(h - |z|), \quad -h \le z \le h, \tag{7.35}$$

where $\beta = 2\pi/\lambda_0$ is the phase constant, and λ_0 denotes the wavelength in free space. From Eq. (7.35), the current distribution is sinusoidal and vanishes at the two terminals, i.e., $z = \pm h$.

Suppose the dipole antenna locates at the origin (O) as shown in Fig. 7.10. Let a point P locate at (R, θ, ϕ) and its distance from the origin is much greater than the wavelength, i.e., $R \gg \lambda_0$. Therefore, P is at the **far zone**. Now, we want to derive the **far field** of the dipole antenna at point P.

First, we focus on the tiny segment of the dipole antenna between z and $z + dz$. Its length is dz and $dz \rightarrow 0$. The current flowing in this segment is $I(z)$. In Fig. 7.10, the distance between the origin and point P is R, and the distance between the tiny segment and point P is R'. Note that R and R' are not equal, although they may be very close. Next, it is reasonable to assume $R \gg h$. By Law of Cosine, we can easily get the relationship between R' and R given by

$$R' = \sqrt{R^2 + z^2 - 2Rz \cos \theta} \approx R - z \cos \theta. \tag{7.36}$$

Because the length of the tiny segment is dz and the current is $I(z)$, from the result in Sect. 7.1, the radiated E-field at point P is $d\vec{E} = dE_\theta \cdot \hat{a}_\theta$ and

$$dE_\theta = j30\beta \sin \theta \cdot I(z)dz \cdot \frac{e^{-j\beta R'}}{R'}. \tag{7.37}$$

As we consider the whole dipole from $z = -h$ to $z = h$, the total radiated E-field at point P is the sum of all the radiation from each tiny segment, given by

$$\vec{E} = E_\theta \cdot \hat{a}_\theta, \tag{7.38}$$

Fig. 7.9 Dipole antenna

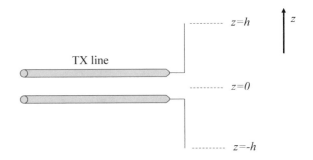

Fig. 7.10 Dipole antenna and a point at far zone

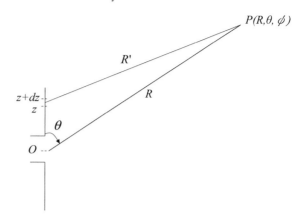

where

$$E_\theta = \int dE_\theta = j30\beta \sin\theta \cdot \int_{-h}^{h} I(z) \cdot \frac{e^{-j\beta R'}}{R'} dz. \tag{7.39}$$

The integral in Eq. (7.39) is not easy to carry out and the precise result shall be derived by a numerical method. However, if we make a reasonable assumption, we can convert it into an integrable function as follows.

First, we assume $^1/_{R'} \approx {}^1/_R$ so that $1/R'$ in Eq. (7.39) can be moved out of the integral. In addition, using Eq. (7.35), we rewrite Eq. (7.39) as

$$E_\theta = \frac{j30 I_0 \beta \sin\theta}{R} \cdot \int_{-h}^{h} \sin\beta(h - |z|) \cdot e^{-j\beta R'} dz. \tag{7.40}$$

Inserting Eq. (7.36) into Eq. (7.40), we have

$$E_\theta = \frac{j30 I_0 \beta \sin\theta}{R} \cdot e^{-j\beta R} \cdot \int_{-h}^{h} \sin\beta(h - |z|) \cdot e^{j\beta z \cos\theta} dz. \tag{7.41}$$

Although Eq. (7.41) seems quite complicated, it is actually integrable. After some derivations, we finally get E_θ, given by

$$E_\theta = j\frac{60 I_0 e^{-j\beta R}}{R} \cdot F(\theta), \tag{7.42}$$

where

$$F(\theta) = \frac{\cos(\beta h \cos\theta) - \cos\beta h}{\sin\theta}. \tag{7.43}$$

Fig. 7.11 EM field of a
dipole antenna

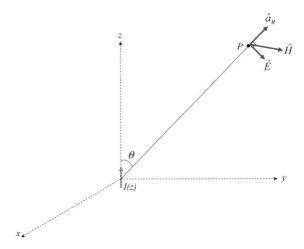

In Eq. (7.42), R is the distance between the point P and the dipole antenna. Hence, the magnitude of E_θ is inversely proportional to the distance R. This is an important property of the far field. Besides, from Eq. (7.43), E_θ depends on θ, where θ is the angle between $\overline{OP}$ and z-axis. Note that z-axis is the pointing direction of the dipole antenna. In addition, E_θ does not depend on ϕ, which means that different ϕ have an identical E-field.

Finally, we consider a special case, $\theta = 0$, in Eq. (7.43). In this case, the numerator and denominator of $F(\theta)$ are both zeroes. From the fundamental theorem of calculus, when $\theta \to 0$, we have

$$
F(\theta) = \lim_{\theta \to 0} \frac{\cos(\beta h \cos \theta) - \cos \beta h}{\sin \theta} = \frac{\frac{d}{d\theta}[\cos(\beta h \cos \theta) - \cos \beta h]}{\frac{d}{d\theta} \sin \theta}\Big|_{\theta=0}
$$

$$
= \frac{\sin(\beta h \cos \theta) \cdot (\beta h \sin \theta)}{\cos \theta}\Big|_{\theta=0}
$$

$$
= 0 \tag{7.44}
$$

Similarly, we can prove that when $\theta = \pi$, $F(\theta)$ also attains zero. Hence, the E-field of a dipole antenna at $\theta = 0$ or $\theta = \pi$ vanishes no matter what value of h is.

In Fig. 7.11, we show the EM field at point P. The direction of E-field is $\hat{a}_\theta$ and the direction of M-field is $\hat{a}_\phi$. The resultant EM wave propagates along $\hat{a}_R$.

B Radiation Pattern

When the length of a dipole antenna changes, we will have a different current distribution in the antenna and different E-field at the far zone. Several cases are discussed in the following.

Fig. 7.12 Current distribution of dipole antenna when $2h = \lambda_0/2$

$$2h = \frac{\lambda_0}{2}$$

(1) Half-wavelength dipole antenna

In Fig. 7.9, the length of the antenna is $2h$. Suppose the antenna length is equal to half a wavelength, i.e., $2h = \lambda_0/2$, we call it a **half-wavelength dipole antenna**. Because $2h = \lambda_0/2$, we have

$$\beta h = \frac{2\pi}{\lambda_0} \cdot \frac{\lambda_0}{4} = \frac{\pi}{2}. \tag{7.45}$$

Since $\beta h = \pi/2$, from Eq. (7.35), we obtain the current distribution given by

$$I(z) = I_0 \sin \beta h \left(1 - \frac{|z|}{h} \right) = I_0 \sin \frac{\pi}{2} \left(1 - \frac{|z|}{h} \right). \tag{7.46}$$

The dashed line in Fig. 7.12 shows the current distribution in Eq. (7.46). Obviously in this case, the current vanishes at the both ends ($z = \pm h$) and achieves the maximum I_0 at the center ($z = 0$).

In addition, from Eq. (7.43) when $\beta h = \pi/2$, we have

$$F(\theta) = \frac{\cos(\beta h \cos \theta) - \cos \beta h}{\sin \theta} = \frac{\cos\left(\frac{\pi}{2} \cos \theta \right)}{\sin \theta}. \tag{7.47}$$

$F(\theta)$ shows how the E-field changes with θ. When $\theta = 0$, it corresponds to E-field along $+z$-axis. When $\theta = \frac{\pi}{2}$, it corresponds to E-field perpendicular to z-axis. When $\theta = \pi$, it corresponds to E-field along $-z$-axis. In the following, we list five cases and reveal how $F(\theta)$ changes with θ.

- When $\theta = 0$, $F(\theta) = 0$.
- When $\theta = \frac{\pi}{4}$, $F(\theta) = \frac{\cos\left[\frac{\pi}{2} \cdot \cos \frac{\pi}{4} \right]}{\sin \frac{\pi}{4}} = 0.63$
- When $\theta = \frac{\pi}{2}$, $F(\theta) = \frac{\cos\left[\frac{\pi}{2} \cdot \cos \frac{\pi}{2} \right]}{\sin \frac{\pi}{2}} = 1$ (maximum)

Fig. 7.13 Radiation pattern of dipole antenna when $2h = \lambda_0/2$

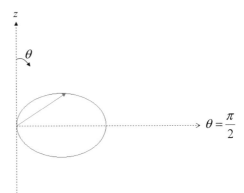

- When $\theta = \frac{3\pi}{4}$, $F(\theta) = \dfrac{\cos\left[\frac{\pi}{2}\cdot\cos\frac{3\pi}{4}\right]}{\sin\frac{3\pi}{4}} = 0.63$
- When $\theta = \pi$, $F(\theta) = 0$

To sum up, Fig. 7.13 shows how $F(\theta)$ changes with θ. The minimum of $F(\theta)$ occurs when $\theta = 0$ and $\theta = \pi$, which means that in the direction parallel to the antenna, the associated E-field vanishes. On the other hand, the maximum occurs when $\theta = \frac{\pi}{2}$, which means that in the direction perpendicular to the antenna, the associated E-field achieves maximum. In fact, we call |F(θ)|, as illustrated in Fig. 7.13, the **radiation pattern** or **antenna pattern** of the dipole antenna. A radiation pattern effectively shows how the E-field of an antenna changes with angle θ.

(2) Dipole antenna of other lengths

For a half-wavelength dipole antenna, the length of the antenna is $2h = \lambda_0/2$ and the radiation pattern is shown in Fig. 7.13. When the length increases, the corresponding radiation pattern will change too.

First, we consider the length $2h = \lambda_0$. In this case, we have

$$\beta h = \frac{2\pi}{\lambda_0} \cdot \frac{\lambda_0}{2} = \pi. \tag{7.48}$$

From Eq. (7.35), we have the current distribution given by

$$I(z) = I_0 \sin \pi \left(1 - \frac{|z|}{h}\right). \tag{7.49}$$

The distribution of $I(z)$ is drawn in Fig. 7.14. Compared with half-wavelength dipole antenna, the maximum of $I(z)$ occurs at $z = \pm h/2$ and the minimum, i.e., $I(z) = 0$, occurs at $z = 0$ and $z = \pm h$. In addition, inserting $\beta h = \pi$ into Eq. (7.43) we have

$$F(\theta) = \frac{\cos(\pi \cos \theta) + 1}{\sin \theta}. \tag{7.50}$$

Fig. 7.14 Current distribution of dipole antenna when $2h = \lambda_0$

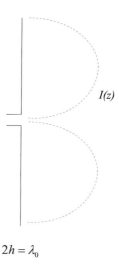

$I(z)$

$2h = \lambda_0$

Fig. 7.15 Radiation pattern of dipole antenna when $2h = \lambda_0$

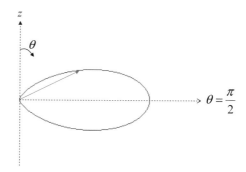

The radiation pattern is determined by Eq. (7.50) and is shown in Fig. 7.15. We find that the pattern is similar to that of a half-wavelength dipole antenna as shown in Fig. 7.13. The radiated energy concentrates on $\theta = \pi/2$, that is the direction perpendicular to the antenna.

Next, let the length keep increasing so that $2h = 3\lambda_0/2$. Then we have

$$\beta h = \frac{2\pi}{\lambda_0} \cdot \frac{3\lambda_0}{4} = \frac{3}{2}\pi. \tag{7.51}$$

In this case, the current distribution is given by

$$I(z) = I_0 \sin \frac{3\pi}{2} \left(1 - \frac{|z|}{h}\right). \tag{7.52}$$

The current distribution is a little bit complicated as shown in Fig. 7.16. In addition, inserting $\beta h = 3\pi/2$ into Eq. (7.43), we have

Fig. 7.16 Current
distribution of dipole
antenna when $2h = 3\lambda_0/2$

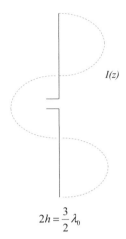

$2h = \dfrac{3}{2}\lambda_0$

Fig. 7.17 Radiation pattern
of dipole antenna when
$2h = 3\lambda_0/2$

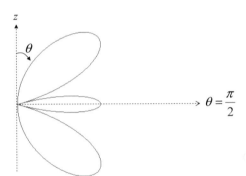

$$F(\theta) = \frac{\cos\left(\frac{3\pi}{2}\cos\theta\right)}{\sin\theta}. \tag{7.53}$$

The radiation pattern ($|F(\theta)|$) is shown in Fig. 7.17. Comparing it with Figs. 7.13 and 7.15, we find a big difference. The maximum of $|F(\theta)|$ occurs at $\theta = {}^{\pi}/_{4}$ and $\theta = {}^{3\pi}/_{4}$, instead of $\theta = {}^{\pi}/_{2}$. It means that the radiated energy mainly concentrates at $\theta = 45°$ and $\theta = 135°$, instead of the direction perpendicular to the dipole antenna. From the above cases, we can adjust the direction of radiated energy by adopting different antenna lengths.

Finally, let the length keep increasing so that $2h = 2\lambda_0$. Then we have

$$\beta h = \frac{2\pi}{\lambda_0} \cdot \lambda_0 = 2\pi. \tag{7.54}$$

In this case, the current distribution is given by

Fig. 7.18 Radiation pattern of dipole antenna when $2h = 2\lambda_0$

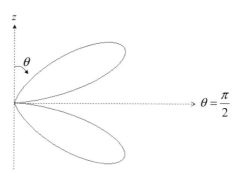

$$I(z) = I_0 \sin 2\pi \left(1 - \frac{|z|}{h} \right). \tag{7.55}$$

Readers are encouraged to take a practice to plot the current distribution $I(z)$ in Eq. (7.55). It shall be interesting and give more insight of the current distribution in an antenna. In addition, inserting $\beta h = 2\pi$ into Eq. (7.43), we have

$$F(\theta) = \frac{\cos(2\pi \cos \theta) - 1}{\sin \theta}. \tag{7.56}$$

The radiation pattern ($|F(\theta)|$) is shown in Fig. 7.18. The radiated energy concentrates around $\theta = 0.32\pi$ and $\theta = 0.68\pi$, and vanishes at $\theta = \pi/2$, which comprises the biggest difference from Fig. 7.17.

In the above, when the length of an antenna changes, the current distribution and the associated radiation pattern change as well. Hence by adjusting antenna length, we can attain a different radiation pattern.

Example 7.4

Suppose the length of a dipole antenna is $2h$ and the current distribution is $I(z) = I_0 \sin \beta(h - |z|)$, where $h = \frac{5}{8}$ m, $\beta = 2\pi/\lambda_0$, and $I_0 = 2$ A. Please derive the magnitude of E-field, i.e., $|\vec{E}|$, in the following cases when $R = 10$ km and $\theta = 60°$: (a) $f = 1$ MHz, (b) $f = 30$ MHz, (c) $f = 60$ MHz, (d) $f = 120$ MHz.

Solution

(a) When $f = 1$ MHz, the wavelength is given by

$$\lambda_0 = \frac{c}{f} = \frac{3 \times 10^8}{10^6} = 300 \, \text{m}.$$

From Eqs. (7.42) and (7.43), the magnitude of E-field is given by

$$|\vec{E}| = \frac{60 I_0}{R} \cdot |F(\theta)|,$$

where

$$F(\theta) = \frac{\cos(\beta h \cos\theta) - \cos\beta h}{\sin\theta}.$$

Since $\lambda_0 = 300$ m, $h = \frac{5}{8}$ m, and $\theta = 60°$, we have

$$\beta h = \frac{2\pi}{\lambda_0} \cdot h = \frac{\pi}{240},$$

$$F(\theta) = \frac{\cos(\beta h \cos\theta) - \cos\beta h}{\sin\theta} = 7.42 \times 10^{-5}.$$

Hence the magnitude of E-field at $R = 10$ km is given by

$$|\vec{E}| = \frac{60 I_0}{R} \cdot |F(\theta)| = \frac{60 \cdot 2}{10 \times 10^3} \cdot (7.42 \times 10^{-5}) = 8.9 \times 10^{-7} \text{ V/m}.$$

(b) When $f = 30$ MHz, the wavelength is given by

$$\lambda_0 = \frac{c}{f} = \frac{3 \times 10^8}{30 \times 10^6} = 10 \text{ m}.$$

In this case, the length $2h$ of antenna is equal to $\frac{\lambda_0}{8}$.
Since $\lambda_0 = 10$ m and $\theta = 60°$, we have

$$\beta h = \frac{2\pi}{\lambda_0} \cdot h = \frac{\pi}{8},$$

$$F(\theta) = \frac{\cos(\beta h \cos\theta) - \cos\beta h}{\sin\theta} = 6.57 \times 10^{-2}.$$

Hence the magnitude of E-field at $R = 10$ km is given by

$$|\vec{E}| = \frac{60 I_0}{R} \cdot |F(\theta)| = \frac{60 \cdot 2}{10 \times 10^3} \cdot (6.57 \times 10^{-2}) = 7.88 \times 10^{-4} \text{ V/m}$$

(c) When $f = 60$ MHz, the wavelength is given by

$$\lambda_0 = \frac{c}{f} = \frac{3 \times 10^8}{60 \times 10^6} = 5 \text{ m}.$$

In this case, the length of antenna $(2h)$ is equal to $\frac{\lambda_0}{4}$, which is a quarter-wavelength dipole antenna. Since $\lambda_0 = 5$ m and $\theta = 60°$, we have

$$\beta h = \frac{2\pi}{\lambda_0} \cdot h = \frac{\pi}{4},$$

$$F(\theta) = \frac{\cos(\beta h \cos\theta) - \cos\beta h}{\sin\theta} = 0.25.$$

Hence the magnitude of E-field at $R = 10$ km is given by

$$|\vec{E}| = \frac{60 I_0}{R} \cdot |F(\theta)| = \frac{60 \cdot 2}{10 \times 10^3} \cdot (0.25) = 3 \times 10^{-3} \text{ V/m.}$$

(d) When $f = 120$ MHz, the wavelength is given by

$$\lambda_0 = \frac{c}{f} = \frac{3 \times 10^8}{120 \times 10^6} = 2.5 \text{ m.}$$

In this case, the length of antenna ($2h$) is equal to $\frac{\lambda_0}{2}$, which is a half-wavelength dipole antenna. Since $\lambda_0 = 2.5$ m and $\theta = 60°$, we have

$$\beta h = \frac{2\pi}{\lambda_0} \cdot h = \frac{\pi}{2},$$

$$F(\theta) = \frac{\cos(\beta h \cos\theta) - \cos\beta h}{\sin\theta} = 0.82.$$

Hence the magnitude of E-field at $R = 10$ km is given by

$$|\vec{E}| = \frac{60 I_0}{R} \cdot |F(\theta)| = \frac{60 \cdot 2}{10 \times 10^3} \cdot (0.82) = 9.84 \times 10^{-3} \text{ V/m.}$$

∎

Although a dipole antenna is simple, it reveals the fundamental principles of antennas. When the length increases, the radiation pattern may change significantly. This is because the current distribution in the antenna changes, and thus the resulted radiation pattern changes as well. Most antennas are much more complicated than a dipole antenna, and it is hard to get their radiation patterns via mathematical analysis. In fact, numerical methods with the aid of computer are widely adopted to attain the radiation patterns of practical antennas. In addition, from the result of Example 7.4, we learn that when the length of an antenna is much smaller than the operating wavelength, the radiation efficiency is very poor. In practice, the length of an antenna usually needs to be at least a quarter-wavelength, i.e., $\lambda_0/4$, in order to achieve an acceptable radiation efficiency. It explains why a handset phone needs to operate at a high frequency (several GHz).

7.3 Radiation Pattern

When investigating the radiation of an antenna, we usually adopt spherical coordinate system because the radial distance R, the elevation angle θ, and the azimuth angle ϕ can effectively describe the radiated fields. In far field, a radiated wave approximates a plane wave and the magnitudes of both E-field and M-field are inversely proportional to the radial distance R. Hence, the relationship between radiated fields and radial distance R is simple. However, the **radiation pattern** depends on angular pair (θ, ϕ) and is much more complicated. It is usually the critical point of antenna design. In the previous section, we have learned the radiation pattern of a dipole antenna. In this section, we extend our learning to general radiation patterns of antennas. The properties and parameters of radiation patterns will be introduced.

A   Plot of Radiation Pattern

A dipole antenna is the simplest antenna and the associated E-field only depends on the elevation angle θ, i.e., $E = E(\theta)$. For most antennas, it is not the case. The E-field usually depends on both θ and ϕ, and can be expressed as a function of both angles, i.e., $E = E(\theta, \phi)$. When we draw $|E(\theta, \phi)|$ with respect to θ and ϕ, it is a 3-D plot, which is not easy to get the insight readily. Therefore, we usually express a radiation pattern in two plots. One displays the relationship between E and θ, and another one displays the relationship between E and ϕ, as given below.

1. For a specific ϕ_0, we draw $|E(\theta)|$, where $0 \leq \theta \leq \pi$.
2. For a specific θ_0, we draw $|E(\phi)|$, where $-\pi \leq \phi \leq \pi$.

In both cases, they are 2-D plots and are easy to extract the properties of an antenna. In the first case, we select $\phi = \phi_0$ and plot E-field as a function of θ. In the second case, we select $\theta = \theta_0$ and plot E-field as a function of ϕ.

An example of $|E(\theta)|$ is shown in Fig. 7.19 for a half-wavelength dipole antenna, where ϕ_0 is arbitrarily selected. From Fig. 7.19, the maximum of E-field occurs when $\theta = \pi/2$ and the minimum occurs when $\theta = 0$. Besides, the associated $|E(\varphi)|$ is illustrated in Fig. 7.20 when we select $\theta = \pi/2$. From the discussions in Sect. 7.2, the E-field of a dipole antenna does not depend on ϕ. Hence $E(\phi)$ is a constant regarding ϕ, where $-\pi \leq \phi \leq \pi$. It can be seen in Fig. 7.20 where $E(\phi)$ forms a circle.

The above examples also show that we can easily understand how a radiated E-field depends on θ and ϕ, respectively.

Fig. 7.19 Example radiation pattern (I)

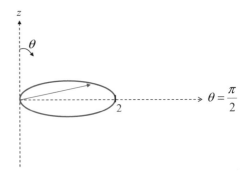

Fig. 7.20 Example radiation
pattern (II)

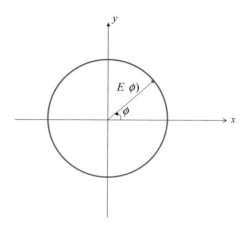

B Spherical Coordinate and Rectangular Coordinate

In Fig. 7.19, we display a radiation pattern in a spherical coordinate system, where
the dipole antenna points along z-axis and θ is the elevation angle. The advantage of
this display is to show the radiation pattern versus θ in space and give us the insight
immediately. For example, from Fig. 7.19, we can easily realize that the minimal
E-field occurs along z-axis and the E-field achieves the maximum when it is perpendicular to z-axis. However, the disadvantage is that we cannot tell the magnitude
of E-field immediately. For example in Fig. 7.19, we cannot tell the magnitude of
E-field when $\theta = \pi/3$. Hence in some applications, a spherical coordinate system is
not a good choice.

In order to tell the magnitude of E-field in different angles immediately, we may
adopt rectangular coordinate system to display a radiation pattern. We let θ be the
x-axis and $|E(\theta)|$ be the y-axis. In Fig. 7.21, we specify radiation pattern $|E(\theta)|$ for
each angle θ in the rectangular coordinate. In this case, we can easily tell the value of
$|E(\theta)|$ for a given θ. For example, we can immediately specify the point of maximum
$|E(\theta)|$ when $\theta = \theta_{\max}$ and the points of half maximum when $\theta = \theta_m$ and $\theta = \theta_n$.

In practice, we usually use both coordinates to display a radiation pattern. They
complement each other and effectively reveal the important features of an antenna.

Fig. 7.21 Displaying
radiation pattern in
rectangular coordinate
system

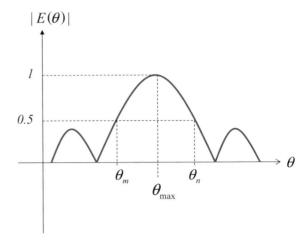

Fig. 7.22 Main beam,
sidelobes and beamwidth of
radiation pattern (I)

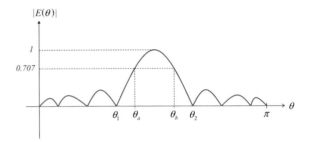

C Beamwidth

A half-wavelength dipole antenna is easy to analyze and the magnitude of E-field
simply depends on the elevation angle θ. For most antennas, the radiation pattern
depends on both θ and the azimuth angle ϕ. Hence, in order to draw the radiation
pattern, we need to fix ϕ and then draw $|E(\theta)|$ in a rectangular coordinate system
as shown in Fig. 7.22, where $0 \leq \theta \leq \pi$. From Fig. 7.22, we find that the energy
radiates mainly between θ_1 and θ_2. We call the beam between θ_1 and θ_2 as the **main
beam**, and the other beams as the **sidelobes**. In general, an antenna has a main beam
and several sidelobes.

Suppose we normalize the radiation pattern (making the maximal magnitude be
unity). Then we can find two elevation angles θ_a and θ_b corresponding to $\frac{1}{\sqrt{2}} = 0.707$.
The angular width between θ_a and θ_b is defined as **beamwidth**, abbreviated as BW,
given by

$$BW = |\theta_a - \theta_b|. \tag{7.57}$$

Fig. 7.23 Main beam,
sidelobes and beamwidth of
radiation pattern (II)

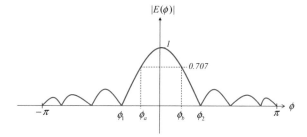

Beamwidth is an important parameter of an antenna because it tells us how concentrating the radiated energy is.

Similar concepts also apply to $|E(\phi)|$ of an antenna in Fig. 7.23 for a given elevation angle θ, where $-\pi \leq \phi \leq \pi$. From Fig. 7.23, we find the radiated energy concentrates between ϕ_1 and ϕ_2, and hence the main beam is defined between them. In addition, if we normalize the radiation pattern, then we find two azimuth angles ϕ_a and ϕ_b corresponding to $\frac{1}{\sqrt{2}}$. The angular width between ϕ_a and ϕ_b is the **beamwidth**, given by

$$BW = |\phi_a - \phi_b|. \tag{7.58}$$

In most applications, a main beam determines the coverage of an antenna and the sidelobes express the excess energy. Hence we usually enhance the main beam and reduce sidelobes when designing an antenna.

7.4 Directivity

If an EM wave were a human being, it should prefer an antenna to a TX line. Because a TX line is like a prison which confines an EM wave in a specified space, and an antenna is like a door of the prison, where an EM wave can escape from the door to the sky.

When we release an EM wave by an antenna and try to make it propagate to a specific direction, we rely on the **directivity** of an antenna. In this section, we will introduce a parameter that can well define the directivity of an antenna.

A Average Radiation Power

The power flow of a radiated EM field is defined by its Poynting vector, given by

$$\vec{P} = \vec{E} \times \vec{H}. \tag{7.59}$$

Fig. 7.24 Schematic plot of radiation power of antenna

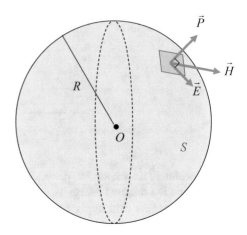

In Eq. (7.59), the direction of $\vec{P}$ is exactly the direction of radiated power flow, and it is also the propagation direction of the EM wave. Note that $\vec{P}$ represents the power density, but not power. Its unit is Watt/m², which means the power per unit area. For a plane wave, because $\vec{E} \perp \vec{H}$, from Eq. (7.59) we have

$$|\vec{P}| = |\vec{E}| \cdot |\vec{H}| = \frac{|\vec{E}|^2}{\eta_0}, \tag{7.60}$$

where $\eta_0 = 120\pi = 377\,\Omega$ is the wave impedance in free space.

In Fig. 7.24, suppose we have a dipole antenna located at the origin (O) and the radiated wave propagates outward omnidirectionally. We consider the wavefront of the same phase at the surface S of a sphere having the radius R. When $R \gg \lambda_0$, an EM field at the surface S shall be treated as far field, and the associated EM wave approximates a plane wave, where $\vec{E} = E_\theta \cdot \hat{a}_\theta$ and $\vec{H} = H_\phi \cdot \hat{a}_\phi$. Obviously, the direction of the associated Poynting vector is $\hat{a}_R$. Because $\hat{a}_R$ is perpendicular to the spherical surface, the Poynting vector $\vec{P}$ is perpendicular to the spherical surface.

In Fig. 7.24, suppose $\vec{E}$ is the phasor of E-field at a point on the surface S. From the result of Sect. 3.3, the **time average power density** is given by

$$P_d = \frac{|\vec{E}|^2}{2\eta_0}. \tag{7.61}$$

When doing the surface integral of P_d over S, we get the **total radiation power** of the antenna given by

$$P_r = \oint_S P_d \cdot ds = \oint_S \frac{|\vec{E}|^2}{2\eta_0} ds. \tag{7.62}$$

In a spherical coordinate system, we have $ds = R^2 \sin\theta \cdot d\theta d\phi$, where $0 \leq \theta \leq \pi$ and $-\pi \leq \phi \leq \pi$. Hence Eq. (7.62) can be rewritten as

$$P_r = \frac{R^2}{2\eta_0} \int_{-\pi}^{\pi} \int_0^{\pi} |\vec{E}|^2 \sin\theta \cdot d\theta d\phi. \tag{7.63}$$

Equation (7.63) applies to all kinds of antennas and implies that the total radiation power P_r can be derived by the E-field over a spherical surface S.

In the following, we discuss the radiation power of two kinds of antennas.

Case 1: Omnidirectional antenna.

First, we consider an **omnidirectional antenna** whose E-field is a constant for all directions, i.e., $|\vec{E}| = E_0$, which does not depend on θ and ϕ. Inserting the constant into Eq. (7.63), we have

$$P_r = \frac{R^2}{2\eta_0} \cdot E_0^2 \cdot \int_{-\pi}^{\pi} \int_0^{\pi} \sin\theta \cdot d\theta d\phi = \left(4\pi R^2\right) \cdot \left(\frac{E_0^2}{2\eta_0}\right). \tag{7.64}$$

Note that $\int_0^{\pi} \sin\theta d\theta = 2$. In Eq. (7.64), $4\pi R^2$ is the area of S and $E_0^2 / 2\eta_0$ is the average power density on S. Their product is the total radiation power P_r.

Equation (7.64) is simple and easy to apply. By using Eq. (7.64), we can derive P_r when E_0 is known. On the other hand, we can derive E_0 when P_r is given. Unfortunately, Eq. (7.64) cannot apply to general cases because most antennas are directional antennas which will be discussed in the following.

Case 2: Directional antenna

Most antennas have the ability to direct radiated power in a given direction and the radiated E-fields depend on (θ, ϕ). In Fig. 7.24, suppose we have a **directional antenna** at the origin and the E-field on the spherical surface S is denoted by $\vec{E} = \vec{E}(\theta, \phi)$. Let $P_d(\theta, \phi)$ be the time average power density of a point on S, and Ω be the average of $P_d(\theta, \phi)$ over S. Then we have

$$\Omega = \frac{\oint_S P_d(\theta, \phi) \cdot ds}{area} = \frac{P_r}{4\pi R^2}. \tag{7.65}$$

Note that in Eq. (7.65), the surface integral of $P_d(\theta, \phi)$ over S is P_r and the area of S is $4\pi R^2$. Because P_r is the total radiation power, Ω is actually the average power density on the surface S. The unit of Ω is Watt/m^2.

Next, we define

$$\Omega = \frac{E_{av}^2}{2\eta_0}, \tag{7.66}$$

where E_{av} is the average magnitude of E-field over S. From Eqs. (7.65) and (7.66), we get the relationship between P_r and E_{av} given by

$$P_r = \left(4\pi R^2\right) \cdot \left(\frac{E_{av}^2}{2\eta_0}\right). \tag{7.67}$$

Comparing Eqs. (7.67) with (7.64), we find that E_{av} of a directional antenna has its counterpart E_0 of an omnidirectional antenna. It implies that we may treat a directional antenna as an omnidirectional antenna having an E-field magnitude E_{av} over S.

Finally, it is worth to mention that $|\vec{E}(\theta, \phi)|$ is the E-field magnitude along the direction defined by θ and ϕ. On the other hand, E_{av} is the average E-field magnitude over S and does not depend on (θ, ϕ). Exploring the relationship between $|\vec{E}(\theta, \phi)|$ and E_{av} can help us define the directivity of an antenna. It will be discussed in the next subsection.

Example 7.5
In Fig. 7.24, suppose an omnidirectional antenna locates at the origin. At the spherical surface S having $R = 5$ km, we have $P_d = 0.02\ \text{W/m}^2$. Please derive the total radiation power P_r and the E-field magnitude E_0.

Solution
Because P_d is the power density over the surface S, we have

$$P_r = 4\pi R^2 \cdot P_d = 4\pi \cdot (5 \times 10^3)^2 \cdot (0.02) = 6.3 \times 10^6\ \text{W}$$

Since the relationship between P_d and E_0 is given by

$$P_d = \frac{E_0^2}{2\eta_0},$$

we get

$$E_0 = \sqrt{2\eta_0 P_d} = \sqrt{2 \cdot (377) \cdot (0.02)} = 3.88\ \text{V/m}.$$

∎

Example 7.6
In Fig. 7.24, suppose we have a directional antenna having the total radiation power $P_r = 20$ W over the spherical surface S having $R = 10$ km. Please derive Ω and E_{av}.

Solution

First, from Eq. (7.65), we have

$$\Omega = \frac{P_r}{4\pi R^2} = \frac{20}{4\pi \cdot (10^4)^2} = 1.6 \times 10^{-8} \, \text{W/m}^2.$$

Next, from Eq. (7.66), we get

$$E_{av} = \sqrt{2\eta_0 \Omega} = \sqrt{2 \cdot (377) \cdot (1.6 \times 10^{-8})} = 3.47 \times 10^{-3} \, \text{V/m}.$$

■

B Directivity

Suppose we have a directional antenna, whose $|\vec{E}(\theta, \phi)|$ at far field is inversely proportional to the distance R and given by

$$|\vec{E}(\theta, \phi)| = \frac{Y(\theta, \phi)}{R}, \tag{7.68}$$

where $0 \leq \theta \leq \pi$ and $0 \leq \phi \leq 2\pi$. In Eq. (7.68), $Y(\theta, \phi)$ is a positive function of θ and ϕ, and it does not depend on R. For example, suppose we have a half-wavelength dipole antenna as discussed in Sect. 7.2. The associated $|\vec{E}(\theta, \phi)|$ at a distance R is given by

$$|\vec{E}(\theta, \phi)| = \frac{60 I_0}{R} \cdot \frac{\cos\left(\frac{\pi}{2} \cos \theta\right)}{\sin \theta}, \tag{7.69}$$

where I_0 is the current in antenna. In this case, the function $Y(\theta, \phi)$ is given by

$$Y(\theta, \phi) = 60 I_0 \cdot \frac{\cos\left(\frac{\pi}{2} \cos \theta\right)}{\sin \theta}. \tag{7.70}$$

Note that Eq. (7.68) applies to all kinds of antennas, but different antennas have different $Y(\theta, \phi)$. When inserting Eq. (7.68) into Eq. (7.63), we get the total radiation power given by

$$P_r = \frac{1}{2\eta_0} \int_0^{2\pi} \int_0^\pi Y^2(\theta, \phi) \sin \theta \cdot d\theta d\phi. \tag{7.71}$$

Equation (7.71) shows that we can obtain total radiation power directly from $Y(\theta, \phi)$ without involving the distance R. From Eqs. (7.67) and (7.71), we have

$$E_{av}^2 = \frac{\eta_0 P_r}{2\pi R^2} = \frac{1}{4\pi R^2} \cdot \int_0^{2\pi} \int_0^{\pi} Y^2(\theta, \phi) \sin\theta \cdot d\theta d\phi. \tag{7.72}$$

In Eq. (7.72), E_{av} depends on R. It is reasonable because the greater the distance R, the greater the surface area of S, and hence the smaller the E_{av}.

Now, we define **directional gain** (or **antenna gain**) as

$$G_D(\theta, \phi) = \frac{\left|\vec{E}(\theta, \phi)\right|^2}{E_{av}^2}, \tag{7.73}$$

where $\left|\vec{E}(\theta, \phi)\right|$ is the E-field magnitude along the direction (θ, ϕ) and E_{av} is the average E-field magnitude. In Eq. (7.73), because the power density is proportional to $\left|\vec{E}(\theta, \phi)\right|^2$, the directional gain G_D is actually a **ratio of power density along the direction (θ, ϕ) to the average power density over S**.

Next, inserting Eqs. (7.68) and (7.72) into Eq. (7.73), we get

$$G_D(\theta, \phi) = \frac{4\pi Y^2(\theta, \phi)}{\int_0^{2\pi} \int_0^{\pi} Y^2(\theta, \phi) \sin\theta \cdot d\theta d\phi}. \tag{7.74}$$

Obviously, G_D depends on (θ, ϕ), but not on R, which is the reason why we introduce $Y(\theta, \phi)$. If $G_D(\theta, \phi) > 1$, it means the power density along the direction (θ, ϕ) is greater than the average power density over S. On the other hand, if $G_D(\theta, \phi) < 1$, it means the power density along the direction (θ, ϕ) is less than the average power density over S. In above, the directional gain G_D provides a simple and effective index for an engineer to realize radiated power distribution in different directions.

For an omnidirectional antenna, we have $Y(\theta, \phi) = Y_0$, which is a constant. In this case, Eq. (7.74) can be simplified as

$$G_D(\theta, \phi) = \frac{4\pi Y_0^2}{\int_0^{2\pi} \int_0^{\pi} Y_0^2 \sin\theta \cdot d\theta d\phi} = \frac{4\pi Y_0^2}{4\pi Y_0^2} = 1. \tag{7.75}$$

It means the directional gain of an omnidirectional antenna is unity for all directions.

For a directional antenna, directional gain G_D might be greater than unity in some directions and smaller than unity in some other directions. If an antenna have a great G_D in a specific direction (θ_0, ϕ_0), it means a large amount of power radiates along this direction. Hence this antenna is very "directive". In order to quantify this feature, we define a parameter called **directivity** given by

$$D = G_{D(max)} = \frac{4\pi Y^2(\theta, \phi)_{max}}{\int_0^{2\pi} \int_0^{\pi} Y^2(\theta, \phi) \sin\theta \cdot d\theta d\phi}, \tag{7.76}$$

where $G_{D(\max)}$ is the maximum of $G_D(\theta, \phi)$. The greater the directivity D, the more directive an antenna is. Obviously, for an omnidirectional antenna, the directivity $D = 1$. For a directional antenna, the directivity is greater than one, i.e., $D > 1$.

Example 7.7
Suppose an antenna has a total radiation power $P_r = 10$ W and along the direction $(\theta = \frac{\pi}{6}, \phi = \frac{\pi}{4})$, the directional gain is $G_D(\theta, \phi) = 3$. Please derive the E-field magnitude along this direction at the surface of a sphere whose radius is $R = 2$ km.

Solution
First, from Eq. (7.67), we have

$$P_r = (4\pi R^2) \cdot \left(\frac{E_{av}^2}{2\eta_0} \right).$$

Hence

$$E_{av}^2 = \frac{\eta_0 P_r}{2\pi R^2} = \frac{(120\pi) \cdot (10)}{2\pi \cdot (2 \times 10^3)^2} = 1.5 \times 10^{-4}.$$

Next, from Eq. (7.73), along the direction $(\theta = \frac{\pi}{6}, \phi = \frac{\pi}{4})$, the directional gain is given by

$$G_D(\theta, \phi) = \frac{\left| \vec{E}(\theta, \phi) \right|^2}{E_{av}^2} = 3.$$

Hence the radiated E-field magnitude along this direction is given by

$$\left| \vec{E}(\theta, \phi) \right| = \sqrt{G_D \cdot E_{av}^2} = 0.021 \text{ V/m}.$$

■

Example 7.8
We use a small segment of conducting line having the length dl as an antenna. The magnitude of radiated E-field at far field is given by

$$|\vec{E}| = \frac{30I \cdot dl \cdot \beta}{R} \cdot \sin\theta,$$

where I is the current in the antenna, R is the distance, and $\beta = 2\pi/\lambda_0$ is the phase constant. Please derive the directional gain of this antenna along the directions $(\theta = \frac{\pi}{6}, \phi = \frac{\pi}{2})$ and $(\theta = \frac{\pi}{2}, \phi = \frac{\pi}{2})$, respectively.

Solution
From Eq. (7.68), we have

$$|\vec{E}(\theta, \phi)| = \frac{30 I dl \beta}{R} \cdot \sin \theta = Y(\theta, \phi) \cdot \frac{1}{R}.$$

Hence

$$Y(\theta, \phi) = K \cdot \sin \theta,$$

where

$$K = 30 I dl \beta.$$

In order to get the directional gain, we first derive the double integral

$$\int_0^{2\pi} \int_0^{\pi} Y^2(\theta, \phi) \sin \theta \cdot d\theta d\phi = \int_0^{2\pi} d\phi \int_0^{\pi} K^2 \sin^3 \theta d\theta = \frac{8\pi}{3} \cdot K^2.$$

Note that $\int_0^{\pi} \sin^3 \theta d\theta = \frac{4}{3}$.
Next, from Eq. (7.74), the directional gain is given by

$$G_D(\theta, \phi) = \frac{4\pi Y^2(\theta, \phi)}{\int_0^{2\pi} \int_0^{\pi} Y^2(\theta, \phi) \sin \theta \cdot d\theta d\phi} = \frac{4\pi \cdot K^2 \sin^2 \theta}{\frac{8\pi}{3} K^2} = \frac{3}{2} \sin^2 \theta.$$

Hence along the direction $(\theta = \frac{\pi}{6}, \phi = \frac{\pi}{2})$, we have the directional gain given by

$$G_D(\theta, \phi) = \frac{3}{2} \sin^2 \left(\frac{\pi}{6}\right) = \frac{3}{8}.$$

And along the direction $(\theta = \frac{\pi}{2}, \phi = \frac{\pi}{2})$, we have the directional gain given by

$$G_D(\theta, \phi) = \frac{3}{2} \sin^2 \left(\frac{\pi}{2}\right) = \frac{3}{2}.$$

∎

Example 7.9
For a half-wavelength dipole antenna, the magnitude of E-field at far field is given by

$$|\vec{E}(\theta, \phi)| = \frac{60 I}{R} \cdot \frac{\cos(\frac{\pi}{2} \cos \theta)}{\sin \theta},$$

where I is the current in the antenna and R is the distance. Please derive the directivity.

Solution

From Eq. (7.68), we have

$$|\vec{E}(\theta, \phi)| = \frac{60I}{R} \cdot \frac{\cos(\frac{\pi}{2}\cos\theta)}{\sin\theta} = Y(\theta, \phi) \cdot \frac{1}{R}.$$

Hence

$$Y(\theta, \phi) = K \cdot \frac{\cos(\frac{\pi}{2}\cos\theta)}{\sin\theta},$$

where

$$K = 60I.$$

In order to get the directivity, we first derive the double integral

$$\int_0^{2\pi} \int_0^{\pi} Y^2(\theta, \phi) \sin\theta \cdot d\theta d\phi = \int_0^{2\pi} \int_0^{\pi} K^2 \frac{\cos^2(\frac{\pi}{2}\cos\theta)}{\sin\theta} d\theta d\phi = (2.44)\pi K^2.$$

Note that $\int_0^{\pi} \frac{\cos^2(\frac{\pi}{2}\cos\theta)}{\sin\theta} d\theta = 1.22$.

Next, because

$$Y(\theta, \phi)\text{max} = K \cdot \left.\frac{\cos(\frac{\pi}{2}\cos\theta)}{\sin\theta}\right|_{\theta=\frac{\pi}{2}} = K,$$

from Eq. (7.74), the directivity is given by

$$D = \frac{4\pi Y^2(\theta, \phi)\text{max}}{\int_0^{2\pi} \int_0^{\pi} Y^2(\theta, \phi) \sin\theta \cdot d\theta d\phi} = \frac{4\pi K^2}{(2.44)\pi K^2} = 1.64.$$

∎

7.5 Radiation Efficiency

The goal of an antenna is to radiate EM waves into the space. Hence in practice, we want to know whether it can achieve the goal efficiently. Because an antenna consists of conducting elements, the arrangement of conducting elements and the distribution of the associated currents all have impact on **radiation efficiency**. In addition, the analysis of radiation efficiency is usually complicated.

Fig. 7.25 Signal transmission from source to antenna with TX line

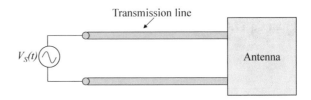

In this section, we first take a circuitry-based approach to investigate the power consumption of an antenna. Then we introduce radiation efficiency and the associated parameters.

A Power Consumption of an Antenna

In Fig. 7.25, we convey a signal $V_S(t)$ by a TX line to an antenna. The antenna radiates EM waves into the space. When we take a circuitry-based approach, the antenna is regarded as a load of the TX line and it has the impedance Z_a given by

$$Z_a = R_a + jX_a, \tag{7.77}$$

where R_a is the real part of Z_a and X_a is the imaginary part. From the principle of circuitry, we know that only the real part R_a consumes power and the imaginary part X_a does not. Hence, all the power consumption of an antenna is determined by R_a.

Because an antenna consists of conducting elements and these conducting elements consume power due to **surface resistance**, the power consumption of R_a consists of loss by radiation and loss by surface resistance. Hence R_a can be decomposed as two parts:

$$R_a = R_{rad} + R_{loss}, \tag{7.78}$$

where R_{rad} is **radiation resistance** representing the radiation loss, and R_{loss} is **loss resistance** representing the power consumption of surface resistance.

B Relationship Between Powers

As shown in Fig. 7.26, we consider an antenna as a load of TX line and its impedance is Z_a. Suppose the characteristic resistance of the TX line is R_0, where $Z_a \neq R_0$. In this case, reflection occurs at the input of the antenna. Therefore, we have both forward wave and backward wave in the TX line. Let V_+ and V_- be the forward voltage and the backward voltage at the antenna input, respectively. From the results of Chap. 6, we have

$$V_- = \Gamma V_+, \tag{7.79}$$

where Γ is a reflection coefficient determined by R_0 and Z_a. It is given by

$$\Gamma = \frac{Z_a - R_0}{Z_a + R_0}. \tag{7.80}$$

Let V_a be the voltage across Z_a. Then we have

$$V_a = V_+ + V_- = V_+ \cdot (1 + \Gamma). \tag{7.81}$$

Suppose P_F is the average power of the forward wave. From the property of a TX line, we have the forward current $I_+ = V_+/R_0$. Hence P_F is given by

$$P_F = \frac{1}{2}|V_+ I_+| = \frac{|V_+|^2}{2R_0}. \tag{7.82}$$

Similarly, the average power of the backward wave is given by

$$P_B = \frac{|V_-|^2}{2R_0}. \tag{7.83}$$

In addition, suppose I_a is the current through Z_a. Then we have

$$I_a = \frac{V_a}{Z_a}. \tag{7.84}$$

Because $Z_a = R_a + jX_a$ and only R_a consumes power, the average power consumption of Z_a is given by

$$P_a = \frac{1}{2}R_a|I_a|^2. \tag{7.85}$$

Fig. 7.26 Forward wave and backward wave coexist when reflection occurs at antenna

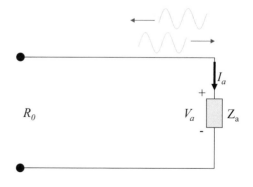

Note that because of X_a, the voltage across R_a is not V_a. Hence we need to derive I_a first, and then P_a.

From Eqs. (7.83) to (7.85), we have

$$P_B + P_a = \frac{|V_-|^2}{2R_0} + \frac{R_a}{2} \cdot \frac{|V_a|^2}{|Z_a|^2} = \frac{1}{2R_0}\left(|V_-|^2 + R_0 R_a \frac{|V_a|^2}{|Z_a|^2}\right). \qquad (7.86)$$

From Eqs. (7.79) to (7.82), we can rewrite Eq. (7.86) as

$$P_B + P_a = \frac{|V_+|^2}{2R_0}\left(|\Gamma|^2 + \frac{R_0 R_a}{|Z_a|^2} \cdot |1 + \Gamma|^2\right) = P_F \cdot \left(|\Gamma|^2 + \frac{R_0 R_a}{|Z_a|^2} \cdot |1 + \Gamma|^2\right). \qquad (7.87)$$

Using Eq. (7.80), it can be proved the following equality holds:

$$|\Gamma|^2 + \frac{R_0 R_a}{|Z_a|^2} \cdot |1 + \Gamma|^2 = 1. \qquad (7.88)$$

Finally, inserting Eq. (7.88) into Eq. (7.87), we have

$$P_F = P_B + P_a, \qquad (7.89)$$

which means

Average power of forward wave = Average power of backward wave + Average power consumed by antenna.

It is a reasonable result because when the forward wave comes to an antenna, some portion is reflected back and the remaining is consumed by the antenna.

From the above result, we realize that when designing an antenna, we shall reduce reflection as much as we can, so that we can convey signal power efficiently to an antenna. Therefore, we usually design an impedance matching circuit between the TX line and the antenna in order to avoid reflection.

Example 7.10
In Fig. 7.26, suppose we have $R_0 = 50\,\Omega$ and $Z_a = 30 + j10\,\Omega$. If the forward voltage is $V_+ = 10e^{j\frac{\pi}{3}}$ at the antenna input. Please derive the average power consumption of the antenna.

Solution

First, from Eq. (7.80), we derive the reflection coefficient given by

$$\Gamma = \frac{Z_a - R_0}{Z_a + R_0} = \frac{-2 + j}{8 + j}.$$

Next, from Eq. (7.81), we can derive the voltage and the current across the antenna, given by

$$V_a = V_+(1 + \Gamma) = (10e^{j\frac{\pi}{3}}) \cdot \frac{6 + 2j}{8 + j},$$

$$I_a = \frac{V_a}{Z_a} = (10e^{j\frac{\pi}{3}}) \cdot \frac{6 + 2j}{8 + j} \cdot \frac{1}{30 + j10},$$

respectively. Finally from Eq. (7.85), we have

$$P_a = \frac{1}{2}R_a|I_a|^2 = \frac{1}{2} \cdot (30) \cdot \left(\frac{4}{65}\right) = \frac{12}{13} \text{ W}.$$

Meanwhile, we can try another approach to derive P_a:

First, from Eq. (7.82), we get the average power of forward wave given by

$$P_F = \frac{|V_+|^2}{2R_0} = \frac{100}{2 \cdot 50} = 1.$$

Next, from Eq. (7.83), we get the average power of backward wave given by

$$P_B = \frac{|V_-|^2}{2R_0} = \frac{|V_+|^2|\Gamma|^2}{2R_0} = \frac{(100) \cdot (\frac{1}{13})}{2 \cdot (50)} = \frac{1}{13}.$$

Because $P_B + P_a = P_F$, we get the average power consumption of the antenna given by

$$P_a = P_F - P_B = \frac{12}{13} \text{ W}.$$

∎

C *Radiation Efficiency*

Suppose P_a is the average power consumption of an antenna. Then it can be decomposed by

$$P_a = P_{rad} + P_{loss}, \tag{7.90}$$

where P_{rad} is the average radiation power and P_{loss} is the average power consumption of surface resistance.

Now, we define **radiation efficiency** of an antenna as the ratio of P_{rad} to P_a, given by

$$\chi = \frac{P_{rad}}{P_a}, \tag{7.91}$$

where $\chi \le 1$. When χ is larger, the antenna is more efficient. For a well-designed antenna, χ approaches one.

In Fig. 7.26, the average radiation power of the antenna is given by

$$P_{rad} = \frac{1}{2} R_{rad} |I_a|^2. \tag{7.92}$$

On the other hand, the average power consumption of surface resistance is given by

$$P_{loss} = \frac{1}{2} R_{loss} |I_a|^2. \tag{7.93}$$

From Eqs. (7.91) to (7.93), we have radiation efficiency given by

$$\chi = \frac{P_{rad}}{P_{rad} + P_{loss}} = \frac{R_{rad}}{R_{rad} + R_{loss}}. \tag{7.94}$$

From Eq. (7.94), when designing an antenna, we try to increase the radiation resistance R_{rad} and reduce the surface resistance R_{loss} so that $R_{rad} \gg R_{loss}$.

Example 7.11

Suppose we have a tiny segment of conducting line as an antenna and its length is dl. From the previous result, the magnitude of E-field at far field is given by

$$|\vec{E}| = \frac{30I \cdot dl \cdot \beta}{R} \cdot \sin\theta,$$

where I is the current on the conducting line, R is the distance, and $\beta = 2\pi/\lambda_0$. Let the radius of this conducting line be a, conductivity σ, and operating frequency f. Please derive the radiation resistance R_{rad} and surface resistance R_{loss}.

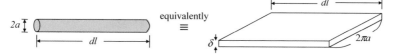

Fig. 7.27 Equivalent thin plate for a conducting line due to skin effect

Solution

First, from Eq. (7.63) of Sect. 7.4, the total radiation power P_r is derived by

$$P_r = \frac{R^2}{2\eta_0} \int_{-\pi}^{\pi} \int_0^{\pi} |\vec{E}|^2 \sin\theta d\theta d\phi$$
$$= (30I \cdot dl \cdot \beta)^2 \cdot \left(\frac{1}{2\eta_0}\right) \cdot (2\pi) \cdot \int_0^{\pi} \sin^3\theta d\theta$$
$$= 10I^2 \cdot dl^2 \cdot \beta^2. \tag{7.95}$$

Note that $\eta_0 = 120\pi$ and $\int_0^{\pi} \sin^3\theta d\theta = \frac{4}{3}$. Because P_r represents the average radiation power of this antenna, from Eqs. (7.92) and (7.95), we have

$$P_r = P_{rad} = \frac{1}{2}I^2 R_{rad} = 10I^2 \cdot dl^2 \cdot \beta^2.$$

Hence the radiation resistance is given by

$$R_{rad} = 20dl^2 \cdot \beta^2 = 80\pi^2 \left(\frac{dl}{\lambda_0}\right)^2.$$

On the other hand, the **skin depth** of the conducting line is given by (referring Sect. 3.4)

$$\delta = \frac{1}{\sqrt{\pi f \mu_0 \sigma}}.$$

As shown in Fig. 7.27, this conducing line having length dl can be equivalently treated as a very thin conducting plate having length dl, width $2\pi a$, and thickness δ. From fundamental electricity, its surface resistance is given by

$$R_{loss} = \frac{1}{\sigma} \cdot \frac{dl}{A},$$

where the cross-sectional area is $A = \delta \cdot 2\pi a$. Hence we have

$$R_{loss} = \frac{1}{\sigma} \cdot \frac{dl}{\delta \cdot 2\pi a} = \frac{dl}{2a} \cdot \sqrt{\frac{\mu_0 f}{\pi \sigma}}$$

∎

From the above example, we learn that the radiation resistance R_{rad} depends on the length of antenna and operating frequency. The surface resistance R_{loss} depends on conductivity and operating frequency. Hence for a specific antenna, its radiation efficiency may change when it operates at different frequencies.

Example 7.12
Suppose we have a tiny conducting segment as an antenna and its length is dl. It has conductivity $\sigma = 5.8 \times 10^7$ S/m, radius $a = 3$ mm, and operating frequency $f = 100$ MHz. When

(a) $dl = 0.01$ m,
(b) $dl = 0.1$ m,

please derive the radiation efficiency χ.

Solution
First, the corresponding wavelength of $f = 100$ MHz is given by

$$\lambda_0 = \frac{c}{f} = \frac{3 \times 10^8}{100 \times 10^6} = 3 \, \text{m}.$$

(a) When $dl = 0.01$ m, from the results in Example 7.11, we have

$$R_{rad} = 80\pi^2 \left(\frac{dl}{\lambda_0}\right)^2 = 80\pi^2 \cdot \left(\frac{0.01}{3}\right)^2 = 0.0088 \, \Omega,$$

$$R_{loss} = \frac{dl}{2a} \cdot \sqrt{\frac{\mu_0 f}{\pi \sigma}} = \frac{0.01}{2 \cdot (0.003)} \cdot \sqrt{\frac{(4\pi \times 10^{-7}) \cdot 10^8}{\pi \cdot (5.8 \times 10^7)}} = 0.0014 \, \Omega.$$

Hence the radiation efficiency is given by

$$\chi = \frac{R_{rad}}{R_{rad} + R_{loss}} = \frac{0.0088}{0.0088 + 0.0014} = 0.86.$$

(b) When $dl = 0.1$ m, from the results in Example 7.11, we have

$$R_{rad} = 80\pi^2 \left(\frac{dl}{\lambda_0}\right)^2 = 80\pi^2 \cdot \left(\frac{0.1}{3}\right)^2 = 0.88 \, \Omega,$$

$$R_{loss} = \frac{dl}{2a} \cdot \sqrt{\frac{\mu_0 f}{\pi \sigma}} = \frac{0.1}{2 \cdot (0.003)} \cdot \sqrt{\frac{(4\pi \times 10^{-7}) \cdot 10^8}{\pi \cdot (5.8 \times 10^7)}} = 0.014 \, \Omega.$$

Hence the radiation efficiency is given by

$$\chi = \frac{R_{rad}}{R_{rad} + R_{loss}} = \frac{0.88}{0.88 + 0.014} = 0.98.$$

■

Example 7.13

Suppose we have a half-wavelength dipole antenna and the magnitude of E-field at far field is given by

$$|\vec{E}| = \frac{60I}{R} \cdot \frac{\cos(\frac{\pi}{2}\cos\theta)}{\sin\theta},$$

where I is the current and R is the distance. Please derive the radiation resistance.

Solution

First, from Eq. (7.63) of Sect. 7.4, the total radiation power P_r is derived by

$$P_r = \int_0^{2\pi} \int_0^{\pi} \left(\frac{1}{2\eta_0}|\vec{E}|^2\right) \cdot R^2 \sin\theta d\theta d\phi$$
$$= (60I)^2 \cdot \left(\frac{1}{2\eta_0}\right) \cdot (2\pi) \cdot \int_0^{\pi} \frac{\cos^2(\frac{\pi}{2}\cos\theta)}{\sin\theta} d\theta.$$
$$= 30I^2 \cdot (1.22)$$

Note that $\int_0^{\pi} \frac{\cos^2(\frac{\pi}{2}\cos\theta)}{\sin\theta} d\theta = 1.22$.
Next, from Eq. (7.92), we have

$$P_r = P_{rad} = \frac{1}{2}I^2 R_{rad} = 30I^2 \cdot (1.22).$$

Hence the radiation resistance is given by

$$R_{rad} = 60 \cdot (1.22) = 73\,\Omega.$$

■

7.6 Receiving Antenna

In the previous sections, we use an antenna to radiate EM waves and learn the associated principles. This antenna is called a **transmitting antenna**. On the other hand, we can use an antenna as a receiving unit so that the received EM wave can be

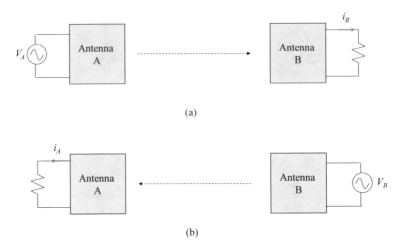

(a)

(b)

Fig. 7.28 Transmitting/receiving antenna pair and the interchange of their roles

converted into current. In this case, it is called a **receiving antenna**. By intuition, the investigation of the latter shall be more complicated than that of the former because the received EM waves need to be converted into current—a complicated topic we have not learned before. Fortunately, scientists have found a useful theorem enabling us to realize the properties of a receiving antenna from those of a corresponding transmitting antenna. It also helps us further understand a wireless communication system composed of both a transmitting antenna and a receiving antenna.

A   Reciprocity Theorem

Before investigating a receiving antenna, we introduce an important theorem called **Reciprocity Theorem**.

First, as shown in Fig. 7.28a, we have two identical antennas at different positions in space. They are called antenna A and antenna B, respectively. Next, suppose we apply a voltage V_A to antenna A and it generates an EM wave. This EM wave propagates from antenna A to antenna B and induces a current i_B on antenna B. Conversely, as shown in Fig. 7.28b, if we apply a voltage V_B to antenna B and it generates an EM wave. This EM wave propagates from antenna B to antenna A and induces a current i_A on antenna A. When $V_A = V_B$, **because antenna A and antenna B are geometrically symmetric in space, the effect of antenna A on antenna B is identical to that of antenna B on antenna A.** In other words, the induced currents should be equal, i.e., $i_B = i_A$.

The above example reveals the key principle of reciprocity theorem: when we use the antenna A as a transmitter having an applied voltage V, the current i is induced in the receiving antenna B. Then conversely, when we use the antenna B as a transmitter having an applied voltage V, the same current i will be induced in the receiving antenna A. In other words, **the relationship between applied voltage and the induced current are unchanged if the roles of transmitting antenna and receiving antenna are interchanged**.

Furthermore, reciprocity theorem can be extended in linear mediums. Suppose we have a transmitting antenna A having an applied voltage V_A, the current i_B is induced in an antenna B. Conversely, when we have antenna B as a transmitter having an applied voltage V_B, the current i_A is induced in antenna A. If $V_A = V_B$, then by reciprocity theorem, $i_B = i_A$. For a linear medium, if $V_A = k V_B$, then $i_B = k i_A$, where k is a constant. Hence we have the following equality.

$$\frac{V_A}{i_B} = \frac{k V_B}{k i_A} = \frac{V_B}{i_A}. \tag{7.96}$$

Equation (7.96) tells us that the ratio of the applied voltage to the induced current is unchanged when the roles of transmitting antenna and receiving antenna are interchanged.

In addition, Eq. (7.96) can be used to explore the relationship between transmitting power and receiving power of antennas. Suppose P_{TA} is the transmitting power of antenna A (applied voltage V_A) and P_{RB} is the receiving power of antenna B (induced current i_B). Conversely, let P_{TB} be the transmitting power of antenna B (applied voltage V_B) and P_{RA} be the receiving power of antenna A (induced current i_A). Then, from Eq. (7.96), we have

$$\left(\frac{V_A}{i_B}\right)^2 = \left(\frac{V_B}{i_A}\right)^2. \tag{7.97}$$

Because $P \propto V^2$ at the transmitting antenna and $P \propto i^2$ at the receiving antenna, we finally get

$$\frac{P_{TA}}{P_{RB}} = \frac{P_{TB}}{P_{RA}}. \tag{7.98}$$

Equation (7.98) tells us an important result: **the ratio of transmitting power to receiving power is unchanged when the roles of transmitting antenna and receiving antenna are interchanged.**

In the above, we assume antenna A and antenna B are two identical antennas. In fact, Eq. (7.98) holds for two different antennas as well. It is a general formula for any transmitting/receiving antenna pairs. In the following, we utilize Eq. (7.98) to explore an important property of antennas.

Fig. 7.29 An antenna
system having ANT1 as
transmitting antenna and
ANT2 as receiving antenna

B Effective Area

In Fig. 7.29, we have two antennas, ANT1 and ANT2, where ANT1 locates at point
A and ANT2 locates at point B. Their distance is R and R is much greater than the
operating wavelength λ_0. In spherical coordinate system, suppose point B locates in
the direction (θ, ϕ) with respect to point A and conversely, point A locates at the
direction (θ', ϕ') with respect to point B, where $(\theta, \phi) \neq (\theta', \phi')$.

First, we let ANT1 be a transmitting antenna and ANT2 be a receiving antenna.
Suppose P_{T1} is the transmitting power of ANT1. Then the receiving power density
at point B, denoted as Ω, is given by

$$\Omega = G_1(\theta, \phi) \cdot \frac{P_{T1}}{4\pi R^2} \ (\text{W/m}^2), \tag{7.99}$$

where $G_1(\theta, \phi)$ is the **directional gain** of ANT1 regarding (θ, ϕ) direction and
$P_{T1}/4\pi R^2$ is the average power density. Equation (7.99) can be interpreted as follows.
Imagine that we have a sphere having radius R and ANT1 locates at the center. Point
B locates on the spherical surface whose area is $4\pi R^2$. Hence, the average power
density on the surface of the sphere is $P_{T1}/4\pi R^2$ when considering omnidirectional
radiation. We multiply $P_{T1}/4\pi R^2$ by the directional gain $G_1(\theta, \phi)$ and obtain the
power density at point B.

Let P_{R2} be the receiving power of ANT2. Then we want to answer the following
question:

How do we derive P_{R2} when the power density Ω is known?

It is not an easy question to deal with because P_{R2} depends not only on Ω, but
also on many properties of ANT2 such as its structure and orientation. In order to
simplify the problem, scientists have an interesting idea:

Imagine that the radiating EM wave of ANT1 is like a group of fishes swimming from ANT1
to ANT2. Then ANT2 is like a fishing net (trap) whose opening is toward ANT1. Hence,
the greater the opening, the more the fishes being caught. The number of caught fishes is
proportional to the area of the opening, that can be regarded as the receiving power of ANT2.

Based on the above idea, the receiving power of ANT2 can be derived by

Fig. 7.30 An antenna system having ANT2 as transmitting antenna and ANT1 as receiving antenna

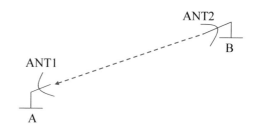

$$P_{R2} = A_2(\theta', \phi') \cdot \Omega, \tag{7.100}$$

where $A_2(\theta', \phi')$ is called the **effective area** of ANT2 regarding (θ', ϕ') direction. Intuitively, $A_2(\theta', \phi')$ is like the opening area of a fishing net and Ω is like the fish density. The product of $A_2(\theta', \phi')$ and Ω is equal to the number of caught fishes, which represents the receiving power of ANT2.

We shall notice that P_{R2} in Eq. (7.100) is defined as the **maximum receiving power** of ANT2 for a given Ω. Hence $A_2(\theta', \phi')$ is the **effective receiving area under optimal conditions**. The unit of $A_2(\theta', \phi')$ is square meters (m^2).

In Fig. 7.30, we interchange the roles of ANT1 and ANT2 so that ANT2 becomes a transmitting antenna and ANT1 becomes a receiving antenna. Let P_{T2} be the transmitting power of ANT2. Following the same reasoning, the power density at point A, denoted as Ω', is given by

$$\Omega' = G_2(\theta', \phi') \cdot \left(\frac{P_{T2}}{4\pi R^2} \right). \tag{7.101}$$

where $G_2(\theta', \phi')$ is the directional gain of ANT2 regarding (θ', ϕ') direction. Also, the receiving power of ANT1 can be derived by

$$P_{R1} = A_1(\theta, \phi) \cdot \Omega', \tag{7.102}$$

where $A_1(\theta, \phi)$ is the effective area of ANT1 regarding (θ, ϕ) direction.

From previous discussions, once the effective area is known, the receiving power can be readily obtained. However, it is usually not easy to derive the effective area of an antenna, as can be seen in the following example.

Fig. 7.31 Plot of Example
7.14

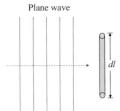

Plane wave

dl

Example 7.14

In Fig. 7.31, we have a tiny segment of conducting line as a receiving antenna
and its length is dl. Suppose the incident EM wave is a uniform plane wave and
the propagation direction is perpendicular to the antenna. In addition, the incident
E-field is parallel to the antenna and the magnitude is $E_i \cos \omega t$. Please derive the
effective area of this antenna.

Solution

First, because the incident E-field is perpendicular to the antenna, the induced voltage
across the antenna is given by

$$V(t) = (E_i \cos \omega t) \cdot dl = V_0 \cdot \cos \omega t,$$

where

$$V_0 = E_i \cdot dl.$$

From the viewpoint of circuitry, the equivalent circuit of the receiving antenna is
provided in Fig. 7.32. It can be regarded as a single voltage source $V(t)$ in series with
an equivalent resistance R_{eq}. From Eq. (7.78) of Sect. 7.5, the equivalent resistance
R_{eq} is given by

Fig. 7.32 Equivalent circuit
of the antenna

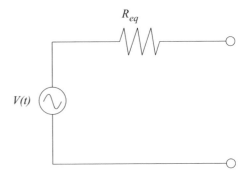

R_{eq}

$V(t)$

Fig. 7.33 Equivalent circuit
when considering load
impedance

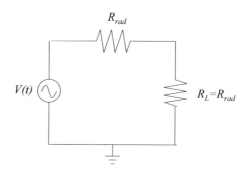

$$R_{eq} = R_{rad} + R_{loss},$$

where R_{rad} is the radiation resistance and R_{loss} is the surface resistance. Assume the
antenna is a perfect conductor and thus $R_{loss} = 0$. Then $R_{eq} = R_{rad}$.

From the results of Example 7.11, we have

$$R_{rad} = 80\pi^2 \left(\frac{dl}{\lambda_0}\right)^2.$$

In practice, we need to consider a load resistance (R_L) in series with the equivalent
circuit (receiving antenna) as shown in Fig. 7.33. According to circuitry, when $R_L = R_{rad}$, we can obtain the maximum power at R_L. In this case, the load current is given
by

$$I_L = \frac{V_0}{2R_L} \cdot \cos \omega t = I_0 \cdot \cos \omega t,$$

where

$$I_0 = \frac{V_0}{2R_L}.$$

Since $R_L = R_{rad}$, the maximum receiving power of the antenna is given by

$$P_R = \frac{1}{2} \cdot R_L I_0^2 = \frac{V_0^2}{8R_{rad}}.$$

Let A be the effective area of the receiving antenna. From Eq. (7.100), we have

$$P_R = A \cdot \Omega,$$

where Ω is the average power density of the incident EM wave. The average power
density Ω is given by

$$\Omega = \frac{E_i^2}{2\eta_0} = \frac{E_i^2}{240\pi}.$$

Hence the effective area is derived by

$$A = \frac{P_R}{\Omega} = \frac{30\pi V_0^2}{R_{rad} E_i^2}.$$

Because $V_0 = E_i \cdot dl$ and $R_{rad} = 80\pi^2 (\frac{dl}{\lambda_0})^2$, we finally get

$$A = \frac{3\lambda_0^2}{8\pi}.$$

■

From Example 7.14, we find that even for a tiny conducting line as an antenna, the derivation of the effective area is complicated. In general, a receiving antenna is much more complex than a tiny conducting line. Hence we need to find an efficient way to derive the effective area of an antenna. It is provided in the next subsection.

C   Relationship Between Directional Gain and Effective Area

From reciprocity theorem, we find that a specific relation exists between antenna directional gain $G(\theta, \phi)$ and effective area $A(\theta, \phi)$. By utilizing this relation, we can avoid complicated computation and obtain the effective area efficiently. We explore the relation as follows.

First, as shown in Fig. 7.29, ANT1 is a transmitting antenna and ANT2 is a receiving antenna. From Eqs. (7.99) and (7.100), we have

$$P_{R2} = A_2(\theta', \phi') \cdot G_1(\theta, \phi) \cdot \frac{P_{T1}}{4\pi R^2}, \tag{7.103}$$

where P_{T1} is the transmitting power of ANT1, and P_{R2} is the maximum receiving power of ANT2.

Conversely, as shown in Fig. 7.30, ANT2 is a transmitting antenna and ANT1 is a receiving antenna. From Eqs. (7.101) and (7.102), we have

$$P_{R1} = A_1(\theta, \phi) \cdot G_2(\theta', \phi') \cdot \frac{P_{T2}}{4\pi R^2}, \tag{7.104}$$

where P_{T2} is the transmitting power of ANT2, and P_{R1} is the maximum receiving power of ANT1.

From Eqs. (7.103) and (7.104), we have the following equality:

$$\frac{P_{R2}}{P_{R1}} = \frac{A_2(\theta', \phi')}{A_1(\theta, \phi)} \cdot \frac{G_1(\theta, \phi)}{G_2(\theta', \phi')} \cdot \frac{P_{T1}}{P_{T2}}. \tag{7.105}$$

Next, from Eq. (7.98), the ratio of transmitting power to receiving power is unchanged when the roles of transmitting antenna and receiving antenna are interchanged. Hence we have

$$\frac{P_{T1}}{P_{R2}} = \frac{P_{T2}}{P_{R1}} \Rightarrow \frac{P_{T1}}{P_{T2}} = \frac{P_{R2}}{P_{R1}}. \tag{7.106}$$

Inserting Eqs. (7.106) into (7.105), we get

$$\frac{A_2(\theta', \phi')}{A_1(\theta, \phi)} \cdot \frac{G_1(\theta, \phi)}{G_2(\theta', \phi')} = 1. \tag{7.107}$$

Equation (7.107) can be rewritten as

$$\frac{A_1(\theta, \phi)}{G_1(\theta, \phi)} = \frac{A_2(\theta', \phi')}{G_2(\theta', \phi')} = \text{constant}. \tag{7.108}$$

Equation (7.108) means that the ratio of effective area to the directional gain of ANT1 is equal to that of ANT2. Notice that Eq. (7.108) applies to all types of antennas because we do not specify the types of ANT1 and ANT2. In addition, Eq. (7.108) applies to arbitrary directions because (θ, ϕ) and (θ', ϕ') are not restricted in the derivation.

To sum up, Eq. (7.108) can be rewritten as

$$\frac{A(\theta, \phi)}{G(\theta, \phi)} = \text{constant} \tag{7.109}$$

which means **the ratio of effective area to directional gain is a universal constant for all types of antennas**. Besides, the constant does not depend on transmitting/receiving directions.

In practice, when we use an antenna as a transmitting antenna, the directional gain $G(\theta, \phi)$ is easy to get. It is an important specification of an antenna. On the other hand, when we use the antenna as a receiving antenna, we need to know the effective area $A(\theta, \phi)$. Suppose the universal constant in Eq. (7.109) is known, we can simply derive $A(\theta, \phi)$ from $G(\theta, \phi)$.

Because Eq. (7.109) holds for all types of antennas, we can consider the simplest antenna, and derive the ratio of $A(\theta, \phi)$ to $G(\theta, \phi)$. As shown in Example 7.14, we use a tiny conducting line as a receiving antenna and the incident EM wave is perpendicular to the antenna, i.e., $\theta = \frac{\pi}{2}$. From the result, the effective area is given by

$$A = \frac{3\lambda_0^2}{8\pi}. \tag{7.110}$$

In addition, from Example 7.8, the directional gain of this antenna depends on θ and given by

$$G(\theta) = \frac{3}{2}\sin^2\theta. \tag{7.111}$$

When $\theta = \frac{\pi}{2}$, we get

$$G = \frac{3}{2}. \tag{7.112}$$

From Eqs. (7.110) and (7.112), we finally obtain

$$\frac{A}{G} = \frac{\lambda_0^2}{4\pi}. \tag{7.113}$$

Because this constant applies to all types of antennas and arbitrary transmitting/receiving directions, we can rewrite Eq. (7.113) as

$$\frac{A(\theta, \phi)}{G(\theta, \phi)} = \frac{\lambda_0^2}{4\pi}. \tag{7.114}$$

Equation (7.114) is the most important identity for antenna applications. It elaborates the relation between directional gain and effective area. Using Eq. (7.114), we can derive $A(\theta, \phi)$ from $G(\theta, \phi)$, and then attain the maximum receiving power.

From Eq. (7.114), the effective area $A(\theta, \phi)$ is proportional to transmitting directional gain $G(\theta, \phi)$. That means, if an antenna can transmit EM wave effectively in some direction, it can receive EM wave effectively in this direction as well. This is an important result because in wireless communications or radar applications, we usually use one antenna as a transmitter in some scenarios and as a receiver in other scenarios. Hence, Eq. (7.114) can actually simplify the design of an antenna system and reduce cost. For example, in satellite communications, when a ground segment successfully tracks a satellite, the associated antenna can not only transmit, but also receive signal effectively.

Example 7.15

We have a half-wavelength dipole antenna whose directional gain depends on θ and is given by

$$G(\theta) = 1.64 \cdot \left(\frac{\cos(\frac{\pi}{2}\cos\theta)}{\sin\theta}\right)^2.$$

When operating frequency $f = 100$ MHz and

(a) $\theta = 0$
(b) $\theta = \frac{\pi}{3}$
(c) $\theta = \frac{\pi}{2}$

Please derive the effective area in each case.

Solution

First, the wavelength can be derived by

$$\lambda_0 = \frac{c}{f} = \frac{3 \times 10^8}{10^8} = 3\,\text{m}.$$

Next, from Eq. (7.114), we have

$$A(\theta) = \frac{\lambda_0^2}{4\pi} \cdot G(\theta) = (0.72) \cdot G(\theta).$$

(a) When $\theta = 0$, we get the directional gain given by

$$G(\theta) = \lim_{\theta \to 0} \left[1.64 \cdot \left(\frac{\cos(\frac{\pi}{2}\cos\theta)}{\sin\theta} \right)^2 \right] = 0.$$

Hence the effective area is given by

$$A(\theta) = 0$$

(b) When $\theta = \frac{\pi}{3}$, we get the directional gain given by

$$G(\theta) = 1.64 \cdot \left(\frac{\cos(\frac{\pi}{2}\cos\frac{\pi}{3})}{\sin\frac{\pi}{3}} \right)^2 = 1.09.$$

Hence the effective area is given by

$$A(\theta) = (0.72) \cdot (1.09) = 0.78\,\text{m}^2$$

(c) When $\theta = \frac{\pi}{2}$, we get the directional gain given by

$$G(\theta) = 1.64 \cdot \left(\frac{\cos(\frac{\pi}{2}\cos\frac{\pi}{2})}{\sin\frac{\pi}{2}} \right)^2 = 1.64.$$

Hence the effective area is given by

$$A(\theta) = (0.72) \cdot (1.64) = 1.18\,\text{m}^2$$

■

D   Friis Transmission Formula

In wireless communications, we usually need to derive the maximum received power P_R when given a transmitted power P_T. Suppose the distance between transmitting antenna and receiving antenna is R. The receiving antenna locates in the direction (θ, ϕ) with respect to the transmitting antenna and conversely, the transmitting antenna locates in the direction (θ', ϕ') with respect to the receiving antenna. Then from Eq. (7.103), we have

$$P_R = G_T(\theta, \phi) \cdot \frac{P_T}{4\pi R^2} \cdot A_R(\theta', \phi'), \qquad (7.115)$$

where $G_T(\theta, \phi)$ is the directional gain of transmitting antenna along (θ, ϕ) and $A_R(\theta', \phi')$ is the effective area of receiving antenna along (θ', ϕ'). Note that $P_T/4\pi R^2$ denotes the average power density on the surface of a sphere having the radius R.

Next, utilizing Eq. (7.114), we can rewrite Eq. (7.115) as

$$P_R = P_T \cdot G_T(\theta, \phi) \cdot G_R(\theta', \phi') \cdot \left(\frac{\lambda_0}{4\pi R} \right)^2, \qquad (7.116)$$

where $G_R(\theta', \phi')$ is the directional gain of receiving antenna along (θ', ϕ'). Equation (7.116) is called **Friis transmission formula**. It effectively reveals the relationship between transmitted power P_T and received power P_R in a simple form. Utilizing Friis transmission formula, when both directional gains are known, we can derive P_R from P_T, and vice versa. Hence it is a very useful formula in practical communication systems.

Finally, it shall be noted that since the increase of operating frequency f means the decrease of wavelength λ_0, Eq. (7.116) often induces a misunderstanding: the received power reduces when the operating frequency increases because $P_R \propto \lambda_0^2$. This misunderstanding can be explained with the following example. Assume the operating frequency is $f = 100$ MHz ($\lambda = 3$ m), and the transmitting antenna as well as the receiving antenna are both quarter-wavelength dipole antennas ($\lambda/4 = 75$ cm). Suppose the maximum received power is P_{R1}. When the operating frequency increases so that $f = 200$ MHz, the corresponding wavelength reduces as half ($\lambda = 1.5$ m). In this case, the antenna becomes a half-wavelength dipole antenna. Hence, the directional gains $G_T(\theta, \phi)$ and $G_R(\theta', \phi')$ in Eq. (7.116) have been changed. Therefore, although $P_R \propto \lambda_0^2$ in Eq. (7.116), the received power P_{R2} (when $f = 200$ MHz) might not be smaller than P_{R1} (when $f = 100$ MHz) because the characteristics of the antennas has changed at different operating frequencies.

Example 7.16

Suppose we have a geostationary communication satellite with the altitude 36,000 km. The directional gain of the transmitting antenna at the satellite is 20 dB and that of the receiving antenna at the ground station is 50 dB. Let the transmitted power be $P_T = 20$ W and the operating frequency be $f = 8.3$ GHz. Please derive the maximum received power P_R.

Solution

First, we have

$$G_T = 20\,\text{dB} = 10^2, \qquad \left[10 \cdot \log 10^2 = 20\,\text{dB}\right]$$

and $G_R = 50\,\text{dB} = 10^5$.

$$\lambda_0 = \frac{c}{f} = \frac{3 \times 10^8}{8.3 \times 10^9} = 0.036\,\text{m},$$

$$\left(\frac{\lambda_0}{4\pi R}\right)^2 = \left[\frac{0.036}{4\pi \cdot (3.6 \times 10^7)}\right]^2 = 6.3 \times 10^{-21}.$$

From Friis transmission formula, we get the maximum received power given by

$$P_R = P_T \cdot G_T \cdot G_R \cdot \left(\frac{\lambda_0}{4\pi R}\right)^2 = (20) \cdot (10^2) \cdot (10^5) \cdot (6.3 \times 10^{-21}) = 1.26 \times 10^{-12}\,\text{W}.$$

■

Summary

7.1: We learn spherical coordinate system, and the near field and the far field of a tiny conducting line are studied.

7.2: Dipole antenna
We learn EM field and properties of a dipole antenna.

7.3: Radiation pattern
We learn how to express radiation pattern of an antenna using spherical coordinate system and rectangular coordinate system.

7.4: Directivity
We learn average radiation power and directivity of an antenna.

7.5: Radiation efficiency
We learn the power relationship between a transmission line and an antenna, and the associated impedances. Then radiation efficiency of an antenna follows.

7.6: Receiving antenna
We learn reciprocity theorem and the application for a transmitting antenna and a receiving antenna. We further get a critical relationship between directional gain and effective area of an antenna.

Exercises

1. Suppose a point A has the Cartesian coordinate as follows. Please derive its corresponding spherical coordinate:

    (a) $A = (1, 0, \sqrt{3})$.
    (b) $A = (1, -1, 2)$.
    (c) $A = (-1, \sqrt{2}, 0)$.

    (Hint: Example 7.1).

2. Suppose a point A has the spherical coordinate as follows. Please derive its corresponding Cartesian coordinate:

    (a) $A = (5, 45°, 60°)$.
    (b) $A = (6, 120°, -30°)$.
    (c) $A = (8, 30°, 90°)$.

    (Hint: Example 7.2).

3. Let $(\hat{a}_R, \hat{a}_\theta, \hat{a}_\phi)$ be three unit vectors at a point A in spherical coordinate, and the coordinate of A is (R, θ, ϕ). It can be proved that using Cartesian coordinate, $(\hat{a}_R, \hat{a}_\theta, \hat{a}_\phi)$ can be represented by

    $$\hat{a}_R = \sin\theta\cos\phi \cdot \hat{x} + \sin\theta\sin\phi \cdot \hat{y} + \cos\theta \cdot \hat{z},$$
    $$\hat{a}_\theta = \cos\theta\cos\phi \cdot \hat{x} + \cos\theta\sin\phi \cdot \hat{y} - \sin\theta \cdot \hat{z},$$
    $$\hat{a}_\phi = -\sin\phi \cdot \hat{x} + \cos\phi \cdot \hat{y}.$$

    Using the above formulas, please prove

    (a) $\hat{a}_R \perp \hat{a}_\theta \perp \hat{a}_\phi$,
    (b) $|\hat{a}_R| = |\hat{a}_\theta| = |\hat{a}_\phi| = 1$,
    (c) $\hat{a}_R = \frac{\overrightarrow{OA}}{|\overrightarrow{OA}|}$, where O is the origin.

4. Using the formulas of Exercise 3, please derive $(\hat{a}_R, \hat{a}_\theta, \hat{a}_\phi)$ for the following cases:

    (a) $A = (5, 45°, 60°)$.
    (b) $A = (6, 120°, -30°)$.
    (c) $A = (8, 30°, 90°)$.

5. Suppose a short conducting line is located at the origin and directs along z-direction, whose length $dl = 2$ cm, current $i(t) = 3 \cdot \cos 2\pi ft, f = 100$ MHz. If the spherical coordinate of a point P is (R, θ, ϕ), please derive E-field and M-field at the point P for the following cases:

    (a) $R = 1$ cm, $\theta = 45°$, $\phi = 90°$
    (b) $R = 10$ km, $\theta = 30°$, $\phi = 120°$

    (Hint: Example 7.3).

6. Prove that $\int_0^h [\sin\beta(h - z) \cdot \cos(\beta z \cos\theta)]dz = \frac{\cos(\beta h\cos\theta) - \cos\beta h}{\beta\sin^2\theta}$.

7. A dipole antenna has the center at the origin and points along z-direction. Its length $2h = \frac{3}{4}\lambda_0$ and the current $I(z) = I_0 \sin \beta(h - |z|)$.

   (a) Please derive locations where $I(z)$ reaches its maximum and minimum.
   (b) Please derive radiation pattern $F(\theta)$.
   (c) Please calculate $F(\theta)$ at $\theta = 0, \frac{\pi}{6}, \frac{\pi}{3}, \frac{\pi}{2}, \frac{2\pi}{3}, \frac{5\pi}{6}, \pi$.

   (Hint: Refer to Sect. 7.2).

8. Repeat Exercise 7 with $2h = \frac{5}{3}\lambda_0$.

9. A dipole antenna has the center at the origin and directs along z-direction. Its length $2h = 3$ m and the current $i(t) = 5 \cdot \cos 2\pi f t$. Suppose a point P has the spherical coordinate $(10 \text{ km}, \frac{\pi}{6}, \frac{\pi}{4})$. Please derive E_θ (phasor) at the point P for the following frequencies: (a) $f = 1$ MHz, (b) $f = 100$ MHz, (c) $f = 1$ GHz. (Hint: Example 7.4).

10. A dipole antenna has the center at the origin and directs along z-direction. Its length $2h = 6$ m and the frequency $f = 300$ MHz. If the magnetic field of a point in the far field is $\vec{H} = 2\cos(2\pi f t + 35°) \cdot \hat{a}_\phi$, please derive the followings at this point:

    (a) E-field.
    (b) Poynting vector.
    (c) average power density.

11. The radiation pattern of a dipole antenna is given by

$$F(\theta) = \frac{\cos(\beta h \cos \theta) - \cos \beta h}{\sin \theta},$$

   where β is the propagation constant and $2h$ is the antenna length. Suppose $2h = \frac{\lambda_0}{4}$. Please use polar coordinate to plot the following:

   (a) its E-plane pattern at $\phi = \frac{\pi}{2}$,
   (b) its H-plane pattern at $\theta = \frac{\pi}{4}$.

   (Hint: Refer to Sect. 7.3).

12. Repeat Exercise 11 for $2h = \frac{3}{4}\lambda_0$.

13. An antenna has the radiation pattern as $|E(\theta, \phi)| = |2\sin\theta \sin\phi|$.

    (a) Please use polar coordinate to plot its E-plane pattern at $\phi = \frac{\pi}{2}$.
    (b) Please use rectangular coordinate to plot its E-plane pattern at $\phi = \frac{\pi}{2}$.
    (c) Please derive its beamwidth of E-plane pattern at $\phi = \frac{\pi}{2}$.
    d). If $|E(\theta, \phi)| = |2(1 + \sin\theta)\sin\phi|$, please derive its beamwidth of E-plane pattern at $\phi = \frac{\pi}{2}$.

14. An antenna has the radiation pattern given by $|E(\theta, \phi)| = \left|\sin^2\theta \cdot \frac{\cos\phi}{2+\cos\phi}\right|$.

    (a) Please use polar coordinate to plot its H-plane pattern at $\theta = \frac{\pi}{2}$.
    (b) Please use rectangular coordinate to plot its H-plane pattern at $\theta = \frac{\pi}{2}$.
    (c) Please derive its beamwidth of H-plane pattern at $\theta = \frac{\pi}{2}$.

15. A point in the far field of an antenna in free space has $\vec{E} = E_A \cdot \cos(\omega t + 30°) \cdot \hat{a}_\theta$. Please derive the followings at this point:

   (a) magnetic field,
   (b) Poynting vector,
   (c) maximum power density,
   (d) average power density.

   (Hint: Refer to Sect. 7.4).

16. An antenna has the radiation pattern of E-field given by $|E(\theta, \phi)| = \left|\frac{K}{R} \sin \theta (1 + \cos \phi)\right|$, where K is a constant and R is the distance from this antenna. Please derive the average radiated power of this antenna.

17. An omnidirectional antenna has the total radiated power $P_r = 2\,\mathrm{W}$. If a point located at a distance $R = 5\,\mathrm{km}$ from the antenna, please derive the following parameters at this point:

   (a) average power density.
   (b) magnitude of E-field ($|\vec{E}|$).
   (c) magnitude of M-field ($|\vec{H}|$).

   (Hint: Example 7.5).

18. A directional antenna is located at the origin and the total radiation power $P_r = 10\,\mathrm{W}$. Please derive the average power density (U) and average E-field intensity (E_{av}) at $R = 5\,\mathrm{km}$.
   (Hint: Example 7.6).

19. A directional antenna is located at the origin and the associated E-field is given by

$$E(\theta, \phi) = \frac{K \cdot \sin \theta \cdot \sin^2 \phi}{R},$$

   where K is a constant and R is the distance from the antenna. Please derive the directional gains at $(\theta = \frac{\pi}{6}, \phi = \frac{\pi}{4})$ and $(\theta = \frac{\pi}{2}, \phi = \frac{\pi}{3})$, respectively.
   (Hint: Example 7.8).

20. In Exercise 19, please derive the directivity of the antenna.

21. A directional antenna is located at the origin and the associated E-field is given by

$$E(\theta, \phi) = K \cdot \frac{(1 - \cos \theta) \cdot \sin \phi}{R},$$

   where K is a constant and R is the distance from the antenna. Please derive the directivity of this antenna.
   (Hint: Example 7.9).

22. A transmission line having characteristic resistance $R_0 = 50\,\Omega$ is used to deliver a signal to an antenna. Suppose the impedance of the antenna is $Z_a = 60\,\Omega$ and

the forward voltage at the antenna input is $V_+ = 6$. Please derive the power dissipation of this antenna.
(Hint: Example 7.10).

23. Repeat Exercise 22 for $Z_L = 50 + j25\,\Omega$.

24. A transmission line having the characteristic resistance $R_0 = 50\,\Omega$ is used to deliver a signal to an antenna. Suppose the impedance of the antenna is $Z_a = 70 + j20\,\Omega$ and the measured voltage at the antenna input is $V_a = 3e^{j\frac{\pi}{4}}$. Please derive the power dissipation of this antenna.

25. Suppose the impedance of an antenna is $Z_a = 50 + j50\,\Omega$ and the measured voltage at the antenna input is $V_a = 8 \cdot e^{j\frac{\pi}{3}}$. If the radiation efficiency of this antenna is $\chi = 0.9$, please derive

    (a) radiation resistance and loss resistance.
    (b) average radiated power.
    (c) average magnitude of E-field at a distance of $R = 10$ km away from this antenna.

    (Hint: Refer to Sect. 7.5).

26. If a short conducting line with length dl is used as an antenna. Its radiation resistance and loss resistance are given by

$$R_r = 80\pi^2 \left(\frac{dl}{\lambda_0}\right)^2,$$

$$R_{loss} = \frac{dl}{2a} \cdot \sqrt{\frac{\mu_0 f}{\pi \sigma}},$$

where λ_0 is the wavelength and σ is the conductivity. Suppose $f = 100\,\text{MHz}$, $a = 2\,\text{mm}$, and $\sigma = 5.8 \times 10^7\,\text{S/m}$. Please derive the radiation efficiency in the following cases: (a) $dl = 1\,\text{cm}$, (b) $dl = 5\,\text{cm}$, (c) $dl = 20\,\text{cm}$.
(Hint: Example 7.12).

27. In Exercise 26, if $dl = 10\,\text{cm}$, please derive the radiation efficiency in the following cases: (a) $f = 1\,\text{MHz}$, (b) $f = 10\,\text{MHz}$, (c) $f = 100\,\text{MHz}$.

28. Consider two points A and B in free space. If we take A as the origin, the spherical coordinate of B is (R, θ, ϕ). If we take B as the origin, the spherical coordinate of A is (R', θ', ϕ'). Please derive the relationship between (R, θ, ϕ) and (R', θ', ϕ').

29. Suppose antenna A is located at the origin and the spherical coordinate of antenna B is $(10\,\text{km}, \frac{\pi}{6}, \frac{\pi}{3})$. Let the average transmitted power of antenna A be $P_T = 50\,\text{W}$.

    (a) If antenna A is an omnidirectional antenna, please derive the average power density at antenna B.
    (b) If antenna A is a half-wavelength dipole antenna, please derive the average power density at antenna B.

    (Hint: Refer to Sect. 7.6).

30. Suppose antenna A is located at the origin and the spherical coordinate of antenna B is $(10\,\text{km}, \frac{\pi}{6}, \frac{\pi}{3})$. Suppose the average transmitted power of antenna A is $P_T = 50\,\text{W}$.

    (a) If antenna A is an omnidirectional antenna, please derive the average power density at antenna B.

    (b) If antenna A is a half-wavelength dipole antenna, please derive the average power density at antenna B.

31. We have a dipole antenna whose directional gain is given by

$$G(\theta) = 2.4 \cdot \left[\frac{\cos(\pi \cos \theta)+1}{\sin \theta} \right]^2 .$$

When operating frequency $f = 200\,\text{MHz}$ and

    (a) $\theta = 0$

    (b) $\theta = \frac{\pi}{6}$

    (c) $\theta = \frac{\pi}{2}$

Please derive the corresponding effective area in each case.
(Hint: Example 7.15).

32. We have a directional antenna at the origin and the electric field intensity on the spherical surface with radius R is given by

$$|E(\theta, \phi)| = K \cdot \frac{(1 + \cos \theta) \cdot (1 - 2sin\phi)}{R}$$

where K is a constant. Suppose operating frequency $f = 200\,\text{MHz}$. Please derive the effective area in the following cases.

    (a) $\left(\theta = \frac{\pi}{6}, \phi = \frac{\pi}{4} \right)$

    (b) $\left(\theta = 0, \phi = \frac{3\pi}{2} \right)$

33. Suppose the distance between transmitting antenna and receiving antenna is $R = 2000\,\text{km}$. The directional gains of the transmitting antenna and the receiving antenna are 33 dB and 46 dB, respectively. If the maximum receiving power is $P_R = 10^{-5}(\text{watt})$ when the operating frequency is $f = 500\,\text{MHz}$. Please derive the transmitting power P_T .
(Hint: Friis transmission formula.)

34. Suppose we have a low earth orbit (LEO) satellite with the altitude 600 km. The directional gain of the transmitting antenna on the satellite is 18 dB and that of the receiving antenna of ground station is 45 dB. Let the transmitted power be $P_T = 5\,\text{W}$, and the operating frequency be $f = 8.3\,\text{GHz}$. Please derive the maximum received power P_R.
(Hint: Example 7.16).

Appendix A
Vector Operations

We provide fundamental vector operations in this appendix in order to lay the mathematical background of Maxwell's equations.

Addition and Subtraction of Vectors

Suppose $\vec{A}$ is a vector in 3-D space and is denoted by

$$\vec{A} = A_x \cdot \hat{x} + A_y \cdot \hat{y} + A_z \cdot \hat{z}, \tag{A.1}$$

where $\hat{x}$, $\hat{y}$, and $\hat{z}$ are the three perpendicular unit vectors in space, and A_x, A_y, and A_z are the associated components of each unit vector. The length of a vector $\vec{A}$ is denoted by $\left| \vec{A} \right|$ and from Eq. (A.1), it is given by

$$\left| \vec{A} \right| = \sqrt{A_x^2 + A_y^2 + A_z^2}. \tag{A.2}$$

Next, suppose we have another vector $\vec{B}$ given by

$$\vec{B} = B_x \cdot \hat{x} + B_y \cdot \hat{y} + B_z \cdot \hat{z}. \tag{A.3}$$

Then the addition and subtraction of $\vec{A}$ and $\vec{B}$ are respectively given by

$$\vec{A} + \vec{B} = (A_x + B_x) \cdot \hat{x} + (A_y + B_y) \cdot \hat{y} + (A_z + B_z) \cdot \hat{z}, \tag{A.4}$$

$$\vec{A} - \vec{B} = (A_x - B_x) \cdot \hat{x} + (A_y - B_y) \cdot \hat{y} + (A_z - B_z) \cdot \hat{z}. \tag{A.5}$$

© Springer Nature Switzerland AG 2020
M.-S. Kao and C.-F. Chang, *Understanding Electromagnetic Waves*,
https://doi.org/10.1007/978-3-030-45708-2

Fig. A.1 Illustrating dot
product of two vectors

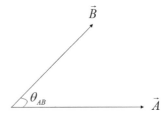

Scalar Multiplication and Dot Product

Let k be a real number. The scalar multiplication of a vector by k is given by

$$k\vec{A} = k(A_x \cdot \hat{x} + A_y \cdot \hat{y} + A_z \cdot \hat{z}) = kA_x \cdot \hat{x} + kA_y \cdot \hat{y} + kA_z \cdot \hat{z}. \qquad (A.6)$$

Next, suppose we have two vectors given by $\vec{A} = A_x \cdot \hat{x} + A_y \cdot \hat{y} + A_z \cdot \hat{z}$ and $\vec{B} = B_x \cdot \hat{x} + B_y \cdot \hat{y} + B_z \cdot \hat{z}$ as shown in Fig. A.1. The **dot product** of $\vec{A}$ and $\vec{B}$ is defined as

$$\vec{A} \cdot \vec{B} = \left|\vec{A}\right| \cdot \left|\vec{B}\right| \cos\theta_{AB}, \qquad (A.7)$$

where θ_{AB} is the angle between $\vec{A}$ and $\vec{B}$, and $0 \leq \theta_{AB} \leq \pi$. From the definition in Eq. (A.7), it can be proved that the dot product is equivalent to the following.

$$\vec{A} \cdot \vec{B} = A_x B_x + A_y B_y + A_z B_z. \qquad (A.8)$$

Hence, the dot product $\vec{A} \cdot \vec{B}$ is a scalar and can be easily calculated by Eq. (A.8).

Differentiation of Vector Fields

Suppose we have a 3-D vector field $\vec{A} = A_x \cdot \hat{x} + A_y \cdot \hat{y} + A_z \cdot \hat{z}$ and its three axial components are functions of the location, i.e., $A_x = A_x(x, y, z)$, $A_y = A_y(x, y, z)$, $A_z = A_z(x, y, z)$. If we take the first partial derivative of $\vec{A}$ with respect to x, we get

$$\frac{\partial \vec{A}}{\partial x} = \frac{\partial}{\partial x}(A_x \cdot \hat{x} + A_y \cdot \hat{y} + A_z \cdot \hat{z}) = \frac{\partial A_x}{\partial x} \cdot \hat{x} + \frac{\partial A_y}{\partial x} \cdot \hat{y} + \frac{\partial A_z}{\partial x} \cdot \hat{z}. \qquad (A.9)$$

Similarly, we get

$$\frac{\partial \vec{A}}{\partial y} = \frac{\partial A_x}{\partial y} \cdot \hat{x} + \frac{\partial A_y}{\partial y} \cdot \hat{y} + \frac{\partial A_z}{\partial y} \cdot \hat{z}, \tag{A.10}$$

$$\frac{\partial \vec{A}}{\partial z} = \frac{\partial A_x}{\partial z} \cdot \hat{x} + \frac{\partial A_y}{\partial z} \cdot \hat{y} + \frac{\partial A_z}{\partial z} \cdot \hat{z}. \tag{A.11}$$

Furthermore, if we take the second partial derivative of $\vec{A}$ with respect to x, we obtain

$$\frac{\partial^2 \vec{A}}{\partial x^2} = \frac{\partial}{\partial x}\left(\frac{\partial \vec{A}}{\partial x}\right) = \frac{\partial^2 A_x}{\partial x^2} \cdot \hat{x} + \frac{\partial^2 A_y}{\partial x^2} \cdot \hat{y} + \frac{\partial^2 A_z}{\partial x^2} \cdot \hat{z}. \tag{A.12}$$

Similarly, we obtain

$$\frac{\partial^2 \vec{A}}{\partial y^2} = \frac{\partial^2 A_x}{\partial y^2} \cdot \hat{x} + \frac{\partial^2 A_y}{\partial y^2} \cdot \hat{y} + \frac{\partial^2 A_z}{\partial y^2} \cdot \hat{z}, \tag{A.13}$$

$$\frac{\partial^2 \vec{A}}{\partial z^2} = \frac{\partial^2 A_x}{\partial z^2} \cdot \hat{x} + \frac{\partial^2 A_y}{\partial z^2} \cdot \hat{y} + \frac{\partial^2 A_z}{\partial z^2} \cdot \hat{z}. \tag{A.14}$$

Example 1

Suppose we have a vector field given by $\vec{A} = (3xy^2) \cdot \hat{x} + (yz+5x) \cdot \hat{y} + (2z+7y^3) \cdot \hat{z}$.
Please calculate $\frac{\partial \vec{A}}{\partial x}$ and $\frac{\partial^2 \vec{A}}{\partial y^2}$.

Answer

First, we have three axial components given by

$$A_x = 3xy^2$$
$$A_y = yz + 5x$$
$$A_z = 2z + 7y^3$$

When we take the first partial derivative of $\vec{A}$ with respect to x, we have

$$\begin{aligned}\frac{\partial A_x}{\partial x} &= 3y^2 \\ \frac{\partial A_y}{\partial x} &= 5 \\ \frac{\partial A_z}{\partial x} &= 0\end{aligned} \quad .$$

Hence we get

$$\frac{\partial \vec{A}}{\partial x} = \frac{\partial A_x}{\partial x} \cdot \hat{x} + \frac{\partial A_y}{\partial x} \cdot \hat{y} + \frac{\partial A_z}{\partial x} \cdot \hat{z} = (3y^2) \cdot \hat{x} + 5\hat{y}.$$

In addition, when we take the first partial derivative with respect to y, we have

$$\begin{array}{l} \frac{\partial^2 A_x}{\partial y^2} = 6x \\ \frac{\partial^2 A_y}{\partial y^2} = 0 \\ \frac{\partial^2 A_z}{\partial y^2} = 42y \end{array} \quad .$$

Hence we get

$$\frac{\partial^2 \vec{A}}{\partial y^2} = \frac{\partial^2 A_x}{\partial y^2} \cdot \hat{x} + \frac{\partial^2 A_y}{\partial y^2} \cdot \hat{y} + \frac{\partial^2 A_z}{\partial y^2} \cdot \hat{z} = (6x) \cdot \hat{x} + (42y) \cdot \hat{z}.$$

Integral of Vector Fields

In this subsection, we introduce three types of integral of vector fields: **line integral, surface integral** and **volume integral**. In Fig. A.2, let $\vec{A} = A_x \cdot \hat{x} + A_y \cdot \hat{y} + A_z \cdot \hat{z}$ be a vector field and C be a curve. The line integral of $\vec{A}$ along C is given by

$$\int_C \vec{A} \cdot \vec{dl}, \tag{A.15}$$

where $\vec{dl}$ denotes a tangential vector along C and its length approaches zero. It is then reasonable to express $\vec{dl}$ by

$$\vec{dl} = dx \cdot \hat{x} + dy \cdot \hat{y} + dz \cdot \hat{z}, \tag{A.16}$$

Fig. A.2 Illustration of line integral

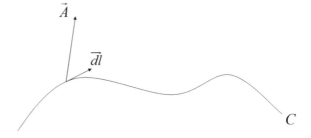

where dx, dy, and dz approach zero. Next, in Eq. (A.15), because $\vec{A} \cdot \vec{dl}$ is the inner product of $\vec{A}$ and $\vec{dl}$, we have

$$\int_C \vec{A} \cdot \vec{dl} = \int_C (A_x dx + A_y dy + A_z dz). \tag{A.17}$$

From Eq. (A.17), we find that a line integral $\int_C \vec{A} \cdot \vec{dl}$ is a scalar, and the physical meaning is the sum of inner products of $\vec{A}$ with each tiny segment $\vec{dl}$ along the curve C.

Furthermore, when C is a **closed curve** as shown in Fig. A.3, the line integral of $\vec{A}$ along C is denoted by

$$\oint_C \vec{A} \cdot \vec{dl}. \tag{A.18}$$

Notice that the difference between Eqs. (A.15) and (A.18) is the circle on the symbol of integral.

Next, a surface integral is illustrated in Fig. A.4. Suppose S is a **surface** and the surface integral of a vector field $\vec{A}$ over S is given by

$$\int_S \vec{A} \cdot \vec{ds}, \tag{A.19}$$

where $\vec{A} \cdot \vec{ds}$ is the inner product of $\vec{A}$ and $\vec{ds}$, and $\vec{ds}$ is a vector representing a surface element of S and given by

$$\vec{ds} = ds \cdot \hat{a}_n. \tag{A.20}$$

Fig. A.3 Illustrating line integral of a closed curve

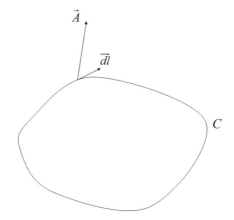

Fig. A.4 Illustration of
surface integral

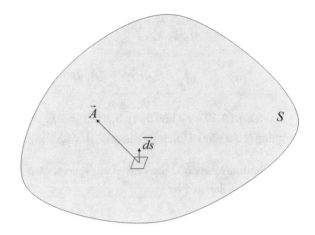

In Eq. (A.20), ds is a tiny surface element whose area approaches zero, and $\hat{a}_n$ denotes a unit vector perpendicular to the surface element. Inserting Eqs. (A.20) into (A.19), we have

$$\int_S \vec{A} \cdot \vec{ds} = \int_S (\vec{A} \cdot \hat{a}_n) ds. \qquad (A.21)$$

In Eq. (A.21), each inner product of $\vec{A}$ and $\hat{a}_n$ is a scalar, and hence the overall result is a scalar as well.

Furthermore, if S is a **closed surface**, the surface integral of $\vec{A}$ over S is denoted by

$$\oint_S \vec{A} \cdot \vec{ds}. \qquad (A.22)$$

Readers may compare the difference between Eqs. (A.22) and (A.19).

Finally, a volume integral is illustrated in Fig. A.5. Suppose V is a volume and the volume integral of a vector field $\vec{A}$ over V is given by

$$\int_V \vec{A} \cdot dv, \qquad (A.23)$$

where dv is a tiny volume element given by

$$dv = dx\,dy\,dz. \qquad (A.24)$$

Hence Eq. (A.23) can be written by

Fig. A.5 Illustration of
volume integral

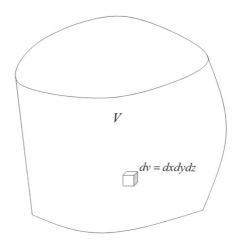

V

$dv = dxdydz$

$$\int_V \overrightarrow{A} \cdot dv = \int_V \overrightarrow{A} \cdot dxdydz. \tag{A.25}$$

Because $dxdydz$ is a scalar, the integral $\int_V \overrightarrow{A} \cdot dv$ is a vector equal to the sum of product of $\overrightarrow{A}$ and each volume element dv of V. Furthermore, suppose $F(x, y, z)$ is a **scalar function**. Then, the volume integral of F over V is given by

$$\int_V F \cdot dv = \int_V F(x, y, z) \cdot dxdydz. \tag{A.26}$$

In this case, the result $\int_V F(x, y, z) \cdot dv$ is a scalar equal to the sum of product of $F(x, y, z)$ and each volume element dv of V.

Appendix B
Electric Field and Magnetic Field

An electromagnetic field (EM field) can be viewed as a combination of an electric field (E-field) and a magnetic field (M-field). It is produced by stationary charges or moving ones (electric currents) and extends unboundedly throughout the space. The major feature of an EM field is its interaction on a charged particle: *when we place a charged particle in an EM field, the particle experiences the associated electromagnetic force.* In order to characterize an EM field, scientists define four physical entities as follows.

$\vec{E}$: **electric field intensity,**

$\vec{D}$: **electric flux density,**

$\vec{H}$: **magnetic field intensity,**

$\vec{B}$: **magnetic flux density**

The first two $(\vec{E}, \vec{D})$ characterize an E-field and the other two $(\vec{H}, \vec{B})$ characterize a magnetic field. We are going to explain the physical meanings of these four entities below.

1. E-field ($\vec{E}$ and $\vec{D}$)

First, we introduce electric field intensity $\vec{E}$. Suppose we place a particle with charge q in the electric field. Then the particle experiences an electric force given by

$$\vec{F} = q\vec{E}.\tag{B.1}$$

From Eq. (B.1), when the charge q is positive, the direction of the force $\vec{F}$ is identical to that of $\vec{E}$. On the other hand, when the charge q is negative, the force $\vec{F}$ has opposite direction of $\vec{E}$. In addition, the greater the $\vec{E}$, the greater the force $\vec{F}$.

© Springer Nature Switzerland AG 2020
M.-S. Kao and C.-F. Chang, *Understanding Electromagnetic Waves*,
https://doi.org/10.1007/978-3-030-45708-2

Fig. B.1 The electric field produced by a positive charge

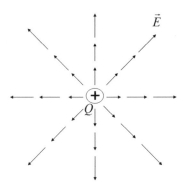

In Eq. (B.1), the unit of q is *Coulomb*, the unit of $\overrightarrow{E}$ is V/m, and the unit of $\overrightarrow{F}$ is *Newton*. Suppose we have a charged particle with $q = 1$ (*Coulomb*) in an E-field having the intensity $\overrightarrow{E} = 1(V/m)$. Then the particle experiences an electric force with $\overrightarrow{F} = 1$ N. When the mass of the particle is m, from Newton's second law of motion, i.e., $\overrightarrow{F} = m\vec{a}$, the acceleration $\vec{a}$ of the particle is given by

$$\vec{a} = \frac{\overrightarrow{F}}{m} = \frac{q\overrightarrow{E}}{m}. \tag{B.2}$$

Hence, the acceleration is proportional to the electric field intensity $\overrightarrow{E}$, in the same direction of $\overrightarrow{E}$.

From above, the electric field intensity $\overrightarrow{E}$ tells us the strength and the direction of the E-field. At a point in space, $\overrightarrow{E}$ reveals the strength of the electric force a charged particle may experience, and how the particle behaves under the force.

Next, we discuss how $\overrightarrow{E}$ is produced. Suppose we have an electric charge Q at point A. The charge will produce an electric field indefinitely extending throughout the space and exert interaction on surrounding charged particles. For example, let B be another point and the distance to point A is r. Then the electric field intensity produced by charge Q at point B is given by

$$\overrightarrow{E} = \left(\frac{Q}{4\pi \in r^2} \right) \cdot \hat{n}_{AB}, \tag{B.3}$$

where $\in$ is **permittivity** of the medium and $\hat{n}_{AB}$ denotes the unit vector from point A to point B. In Fig. B.1, when Q is positive, $\overrightarrow{E}$ radiates outward from point A. On the other hand, when Q is negative as shown in Fig. B.2, $\overrightarrow{E}$ radiates inward toward point A. Besides, the magnitude of $\overrightarrow{E}$ is inversely proportional to r^2 and the permittivity $\in$ of the medium.

Finally, we introduce the physical meaning of electric flux density. From Eq. (B.3), $\overrightarrow{E}$ depends on $\in$, which is determined by the medium. If we define a physical

Fig. B.2 The electric field
produced by a negative
charge

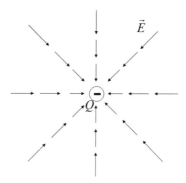

entity $\overrightarrow{D}$ at point B so that

$$\overrightarrow{D} = \epsilon \, \overrightarrow{E},$$ (B.4)

Then from Eq. (B.3), we have

$$\overrightarrow{D} = \left(\frac{Q}{4\pi r^2}\right) \cdot \hat{n}_{AB}.$$ (B.5)

Obviously, $\overrightarrow{D}$ only depends on charge Q and the distance r, but has nothing to do with ϵ. Hence $\overrightarrow{D}$ does not depend on the property of the medium. We call $\overrightarrow{D}$ the **electric flux density**.

Because $\overrightarrow{D}$ does not depend on the medium, it can characterize E-field in a more effective way than $\overrightarrow{E}$ in some cases. For example, when we formulate **Gauss's law** by using $\overrightarrow{E}$, we have

$$\nabla \cdot \overrightarrow{E} = \frac{\rho}{\epsilon},$$ (B.6)

where ρ is the **charge density**. Instead, when we formulate Gauss's law by using $\overrightarrow{D}$, we simply have

$$\nabla \cdot \overrightarrow{D} = \rho.$$ (B.7)

Obviously, Eq. (B.7) is simpler than Eq. (B.6) and it shows that $\nabla \cdot \overrightarrow{D}$ only depends on the charge density ρ and has nothing to do with the medium.

Fig. B.3 Illustration of
magnetic force on a moving
charged particle

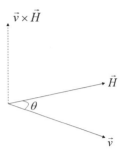

In above, $\vec{E}$ and $\vec{D}$ are two physical entities characterizing an E-field. They have the same direction and $\vec{D}$ is proportional to $\vec{E}$ with a ratio $\in$. Once the permittivity $\in$ is given, we can derive $\vec{D}$ from $\vec{E}$, and vice versa.

2. M-field ($\vec{H}$ and $\vec{B}$)

Unlike an electric field, a magnetic filed exerts force only on moving charged particles. No magnetic force is exerted on stationary charged particles. In Electromagnetism, **magnetic field intensity** $\vec{H}$ and **magnetic flux density** $\vec{B}$ are two physical entities characterizing a magnetic field. First, suppose we place a particle with charge q in a magnetic field having the intensity $\vec{H}$. When the particle is stationary, $\vec{H}$ has no interaction with it. However, when the particle moves with a velocity $\vec{v}$, $\vec{H}$ will exert a magnetic force $\vec{F}$ on the particle given by

$$\vec{F} = \mu \cdot q \cdot (\vec{v} \times \vec{H}), \tag{B.8}$$

where μ is the **permeability** of the medium and $\vec{v} \times \vec{H}$ denotes the **cross product** of $\vec{v}$ and $\vec{H}$. As shown in Fig. B.3, the force $\vec{F}$ is determined by the right-hand rule and perpendicular to $\vec{v}$ and $\vec{H}$.

From Eq. (B.8), when q is positive, the direction of $\vec{F}$ is identical to that of $\vec{v} \times \vec{H}$; when q is negative, the force $\vec{F}$ has opposite direction of $\vec{v} \times \vec{H}$. Besides, $\vec{F}$ is always perpendicular to $\vec{v}$, i.e., the exerted magnetic force is perpendicular to the instant motion. Therefore, $\vec{F}$ keeps changing the moving direction of the charged particle.

Also from Eq. (B.8), we get the magnitude of $\vec{F}$ given by

$$\left|\vec{F}\right| = \left|\mu q \cdot \vec{v} \times \vec{H}\right| = \mu\left(|q| \cdot |\vec{v}| \cdot \left|\vec{H}\right| \cdot \sin\theta\right), \tag{B.9}$$

where θ is the angle between $\vec{H}$ and $\vec{v}$. In Eq. (B.9), when the moving direction is identical to that of the magnetic field, that is, $\vec{v}$ has the same direction as that of $\vec{H}$,

Fig. B.4 Magnitude of magnetic force with respect to the angle between magnetic field and moving direction of a charged particle

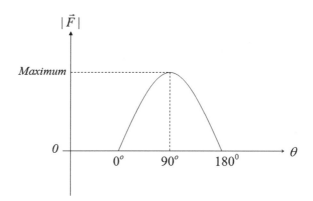

then $\theta = 0$ and thus $\vec{F} = 0$. In this case, the magnetic field does not exert any force on the particle. In addition, when the moving direction is perpendicular to that of the magnetic field, that is, $\vec{v}$ is perpendicular to $\vec{H}$, then $\theta = 90°$. In this case, the particle experiences the maximal magnetic force. The above description is illustrated in Fig. B.4.

Finally, we discuss the physical meaning of magnetic flux density $\vec{B}$. From Eq. (B.8), the exerted force of a charged particle in a magnetic field depends on permeability of a medium. If we define a magnetic flux density $\vec{B}$ by

$$\vec{B} = \mu \vec{H}. \tag{B.10}$$

Then Eq. (B.8) becomes

$$\vec{F} = q \cdot (\vec{v} \times \vec{B}). \tag{B.11}$$

Unlike Eq. (B.8), the exerted magnetic force of a charged particle does not depend on permeability in Eq. (B.11). Hence, by defining $\vec{B}$ as in Eq. (B.10), the exerted force will depend simply on $\vec{B}$, instead of property of a medium.

From Eqs. (B.10) and (B.11), $\vec{B}$ is simply another entity to characterize a magnetic field. In some cases, it is more effective to characterize a magnetic field by using $\vec{B}$. For example, when we formulate Faraday's law by using $\vec{H}$, we have

$$\nabla \times \vec{E} = -\mu \frac{\partial \vec{H}}{\partial t}. \tag{B.12}$$

When we formulate it by using $\vec{B}$, we have

$$\nabla \times \vec{E} = -\frac{\partial \vec{B}}{\partial t}. \tag{B.13}$$

Obviously, Eq. (B.13) is simpler than Eq. (B.12). Eq. (B.13) means that a time-varying $\vec{B}$ will generate an electric field, and this phenomenon does not depend on the property of the medium regarding $\vec{B}$.

Finally, it is worth to mention that when investigating EM waves, we mainly use $\vec{E}$ and $\vec{H}$ to characterize electric fields and magnetic fields, respectively, with $\vec{D}$ and $\vec{B}$ as auxiliary entities. Moreover, although we introduce E-field and M-field separately with ($\vec{E}$, $\vec{D}$) and ($\vec{H}$, $\vec{B}$), they actually mutually interact via Maxwell's Equations. Their mutual interaction is the core of EM wave that is going to be studied in this book.

Appendix C
Physical Meaning of Refractive Index

The refractive index n can be treated as a constant as we usually did in most applications. However, it actually depends on the frequency and hence is a function of frequency. In order to understand why the refractive index is a function of frequency, we need to understand the physical meaning of the refractive index n. In the following, we will start from the material property of dielectric medium to understand its response to an applied E-field. Then we describe the physical meaning of refractive index n.

First, we know that every ordinary element is composed of atoms, which consists of a centering nucleus and surrounding electrons. A nucleus consists of neutrons and protons and is electrically positive. The surrounding electrons are electrically negative, and most of them are bound to the nucleus and called **bound electrons**. Only few of them are free to move and called **free electrons**. Because the number of bound electrons are much more than that of free electrons, the EM property of a dielectric medium or insulator is mainly determined by bound electrons.

In Fig. C.1, we have a dielectric medium without any external electric field. In this case, the internal nuclei are arranged regularly and the associated bound electrons are randomly distributed around each nucleus. Hence, we do not have any net electric field inside the medium.

Now, we apply an external electric field on the medium as shown in Fig. C.2. The internal bound electrons inside the dielectric medium will be attracted by the

Fig. C.1 A dielectric medium without external electric field

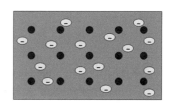

● : nucleus

◌ : electron

© Springer Nature Switzerland AG 2020
M.-S. Kao and C.-F. Chang, *Understanding Electromagnetic Waves*,
https://doi.org/10.1007/978-3-030-45708-2

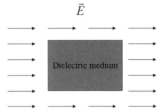

Fig. C.2 A dielectric medium with external electric field

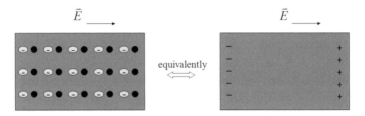

Fig. C.3 Dielectric polarization caused by external electric field

Fig. C.4 Illustration of the induced electric dipole inside a dielectric medium

external electric field $\vec{E}$. These bound electrons do not flow but only slightly shift from their average equilibrium positions and cause dielectric polarization as shown in Fig. C.3. In this case, the dielectric medium can be regarded as an **electric dipole** with positive charge at the right-hand side and negative charges at the left-hand side as shown in Fig. C.4. This "dipole effect" depends on the EM properties of the medium and becomes more significant when the applied field $\vec{E}$ grows. Hence, the induced $\vec{E}_{dipole}$ is proportional to the applied electric field $\vec{E}$, given by

$$\vec{E}_{dipole} = -\beta \vec{E}, \qquad (C.1)$$

where β is a real number between 0 and 1. The negative sign means that the induced $\vec{E}_{dipole}$ and the applied $\vec{E}$ have opposite directions.

Let $\vec{E}_m$ represents the net electric field in the dielectric medium. According to **superposition theorem**, we have

$$\vec{E}_m = \vec{E} + \vec{E}_{dipole} = (1 - \beta)\vec{E}. \tag{C.2}$$

Because $0 \leq \beta < 1$, the magnitude of internal net electric field $\vec{E}_m$ is smaller than that of the applied electric field $\vec{E}$, i.e., $\left|\vec{E}_m\right| < \left|\vec{E}\right|$. In other words, the induced $\vec{E}_{dipole}$ actually "resists" the affect of the applied electric field in a dielectric medium. This can be realized as $\vec{E}_{dipole}$ and $\vec{E}$ have opposite signs in Eq. (C.1).

To simplify the dipole effect inside a dielectric medium, we use the permittivity $\in$ to describe the effect. For a fixed applied electric field, if the medium has more significant induced dipole effect, it has a greater permittivity, and vice versa. In other words, the $\in$ reflects how significant the induced dipole effect may be.

Because $\in = \in_r \in_0$, and according to the definition of refractive index in Sect. 2.4, we have

$$n = \sqrt{\in_r}. \tag{C.3}$$

Therefore, the greater the refractive index, the more significant dipole effect is induced in the medium, i.e., the greater the $\vec{E}_{dipole}$. This is the physical meaning of n.

In the following, we explain why the refractive index n is a function of frequency. In Fig. C.4, suppose the applied electric field varies with time given by

$$\vec{E}(t) = \vec{E}_0 \cos \omega t, \tag{C.4}$$

where ω is the frequency.

When the frequency ω is low, the bound electrons can respond promptly with $\vec{E}(t)$ and hence

$$\vec{E}_{dipole} = -\beta \vec{E}(t). \tag{C.5}$$

However, when the frequency ω keeps growing, the bound electrons gradually cannot respond as fast as the variation of $\vec{E}(t)$. Hence, the induced $\vec{E}_{dipole}$ depends on the frequency ω and can be represented by

$$\vec{E}_{dipole} = -\beta(\omega)\vec{E}(t), \tag{C.6}$$

where $\beta(\omega)$ is a function of frequency. Because n depends on β, it is also a function of frequency, i.e., $n = n(\omega)$. It establishes our claim.

It is worth to mention that although n is a function of frequency, its variation with frequency is quite small in most applications. Hence we usually treat it as a constant. In specific situations such as discussing phase velocity v_p and group velocity v_g, we must treat n as a function of frequency. Otherwise, the phase velocity v_p must be equal to the group velocity v_g and thus we cannot tell their difference, which is necessary when learning the propagation of EM wave in dispersive mediums. In the remaining context, we treat n as a constant unless otherwise specified in order to simplify the analysis.

Appendix D
Propagation Constant

According to the electrical conductivity, we usually categorize materials into three types: conductors, semiconductors, and insulators (dielectric mediums). The conductivity of a conductor is very large and the conductivity of an insulator is very small. The conductivity of a semiconductor is between them. Generally, the propagation constant is given by

$$\gamma = j\omega \sqrt{\mu \in \left(1 + \frac{\sigma}{j\omega \in}\right)}, \tag{D.1}$$

where σ is the conductivity, $\in$ is the permittivity, μ is the permeability, and ω is the frequency. Eq. (D.1) holds for conductors, semiconductors, and insulators. For a metal, the conductivity σ is large and in most frequency spectrum, $\sigma \gg \omega \in$. Hence Eq. (D.1) can be simplified as

$$\gamma \approx j\omega \sqrt{\mu \in \left(\frac{\sigma}{j\omega \in}\right)} = \alpha + j\beta, \tag{D.2}$$

where

$$\alpha = \beta = \sqrt{\pi f \mu \sigma}. \tag{D.3}$$

Because the real part of γ **is not zero, the associated electric field will decay quickly and result in skin effect**.

For an insulator, because σ is usually very small and in most frequency spectrum, $\omega \in \gg \sigma$, we can simplify γ as

$$\gamma \approx j\omega \sqrt{\mu \in} = jk, \tag{D.4}$$

© Springer Nature Switzerland AG 2020
M.-S. Kao and C.-F. Chang, *Understanding Electromagnetic Waves*,
https://doi.org/10.1007/978-3-030-45708-2

where $k = \omega\sqrt{\mu \in}$. Because the real part of γ **is approximately zero, little energy loss occurs when an EM wave propagates in an insulator**.

In practice, we do not have a "perfect conductor" ($\sigma \to \infty$) or a "perfect insulator" ($\sigma = 0$). When dealing with EM wave, if $\sigma >> \omega \in$, we call the medium a **good conductor**. On the other hand, if $\sigma << \omega \in$, we call the medium a **good insulator**. In most frequency spectrum, metals such as gold, silver, copper, and iron are good conductors. Their propagation constant is approximated by Eq. (D.2). On the other hand, air, plastics, and woods are good insulators. Their propagation constant is approximated by Eq. (D.4). Utilizing these approximations, we can simplify our analysis when dealing with EM waves in conductors or insulators. Besides, some mediums have conductivity between conductors and insulators, for instance, semiconductors. In this case, we cannot simplify the associated propagation constant by neither Eq. (D.2) nor Eq. (D.4). The propagation constant shall be expressed by Eq. (D.1).

Next, we can rewrite Eq. (D.1) by

$$\gamma = j\omega\sqrt{\mu \in} \cdot \sqrt{1 + \frac{\sigma}{j\omega \in}} = jk \cdot \sqrt{1 - j\frac{\sigma}{\omega \in}}. \tag{D.5}$$

Then we define a parameter called **loss tangent** given by

$$\tan\theta = \frac{\sigma}{\omega \in} \tag{D.6}$$

In Eq. (D.6), we can use an angle θ to express the relationship between σ and $\omega \in$ as follows. If $\omega \in >> \sigma$, then

$$\tan\theta = \frac{\sigma}{\omega \in} \approx 0 \quad \Rightarrow \quad \theta \approx 0. \tag{D.7}$$

In this case, the corresponding medium is a good insulator. On the other hand, if $\sigma >> \omega \in$, then

$$\tan\theta = \frac{\sigma}{\omega \in} >> 1 \quad \Rightarrow \quad \theta \approx 90°. \tag{D.8}$$

In this case, the corresponding medium is a good conductor. Therefore, when θ is larger, the medium is more like a good conductor, and the decay of EM waves increases as it propagates. Hence we call θ the **loss angle.**.

Finally, we summarize the propagation constant of a conductor, an insulator, and a semiconductor as follows.

1. For a conductor, the propagation constant is approximated by

$$\gamma \approx j\alpha + j\beta,$$

where $\alpha = \beta = \sqrt{\pi f \mu \sigma}$.

2. For an insulator, the propagation constant is approximated by

$$\gamma \approx j\omega\sqrt{\mu \in} = jk,$$

where $k = \omega\sqrt{\mu \in}$.

3. For a semiconductor, the propagation constant is given by

$$\gamma = j\omega\sqrt{\mu \in \left(1 + \frac{\sigma}{j\omega \in}\right)}.$$

Index

Printed in the United States
by Baker & Taylor Publisher Services